Graduate Texts in Mathematics   263

T0207291

Graduate Texts in Mathematics

Graduate Texts in Mathematics bridge the gap between passive study and creative understanding, offering graduate-level introductions to advanced topics in mathematics. The volumes are carefully written as teaching aids and highlight characteristic features of the theory. Although these books are frequently used as textbooks in graduate courses, they are also suitable for individual study.

For further volumes:
http://www.springer.com/series/136

Kehe Zhu

Analysis on Fock Spaces

 Springer

Kehe Zhu
Department of Mathematics and Statistics
State University of New York
Albany, NY
USA

ISSN 0072-5285
ISBN 978-1-4899-7340-5   ISBN 978-1-4419-8801-0 (eBook)
DOI 10.1007/978-1-4419-8801-0
Springer New York Heidelberg Dordrecht London

Mathematics Subject Classification (2010): 30H20

Printed on acid-free paper

Springer is part of Springer Science+Business Media (www.springer.com)

Preface

Several natural L^p spaces of analytic functions have been widely studied in the past few decades, including Hardy spaces, Bergman spaces, and Fock spaces. The terms "Hardy spaces" and "Bergman spaces" are by now standard and well established. But the term "Fock spaces" is a different story. I am aware of at least two other terms that refer to the same class of spaces: Bargmann spaces and Segal–Bargmann spaces. There is no particular reason, other than personal tradition, why I use "Fock spaces" instead of the other variants. I have not done and do not intend to do any research in order to justify one choice over the others.

Numerous excellent books now exist on the subject of Hardy spaces. Several books about Bergman spaces, including some of my own, have also appeared in the past few decades. But there has been no book on the market concerning the Fock spaces. The purpose of this book is to fill that vacuum. There seems to be an honest need for such a book, especially when many results are by now complete. It is at least desirable to have the most important results and techniques summarized in one book, so that newcomers, especially graduate students, have a convenient reference to the subject.

There are certainly common themes to the study of the three classes of spaces mentioned above. For example, the notions of zero sets, interpolating sets, Hankel operators, and Toeplitz operators all make perfect sense in each of the three cases. But needless to say, the resulting theories and results as well as the techniques devised often depend on the underlying spaces. I will not say anything about the various differences between the Hardy and Bergman theories; experts in these fields are well aware of them.

What makes Fock spaces a genuinely different subject is mainly the flatness of the domain on which these spaces are defined: the complex plane with the Euclidean metric in our setup. Hardy and Bergman spaces are usually defined on curved spaces, for example, bounded domains or half-spaces with a non-Euclidean metric. Another major difference between the Fock theory and the Hardy/Bergman theory is the behavior of the reproducing kernel in the L^2 case: the Fock L^2 space possesses an exponential kernel, while the Hardy and Bergman L^2 spaces both have a polynomial kernel.

Let me mention a few particular phenomena that are unique to the analysis on Fock spaces, as opposed to the more well-known Hardy and Bergman space settings.

First, the Fock kernel $e^{\alpha z\overline{w}}$ is neither bounded above nor bounded below, even when one of the two variables is fixed. In the Hardy and Bergman theories, the kernel function $(1 - z\overline{w})^{\alpha}$ is both bounded above and bounded below when one of the two variables is fixed. This makes many estimates in the Fock space setting much more difficult. On the other hand, the exponential decay of $e^{-\alpha|z|^2}$ makes it much easier to prove the convergence of certain integrals and infinite series in the Fock space setting than their Hardy and Bergman space counterparts.

Second, in the Fock space setting, there are no bounded analytic or harmonic functions other than the trivial ones (constants). Therefore, many techniques in the Hardy and Bergman space theories that are based on approximation by bounded functions are no longer valid.

Third, and more technically, in the theory of Hankel and Toeplitz operators on the Fock space, there is no "cutoff" point when characterizing membership in the Schatten classes, while "cutoff" exists in both the Hardy and Bergman settings. Also, for a bounded symbol function φ, the Hankel operator H_φ on the Fock space is compact if and only if $H_{\overline{\varphi}}$ is compact. This is something unique for the Fock spaces.

Fourth, because analysis on Fock spaces takes place on the whole complex plane, certain techniques and methods from Fourier analysis become available. One such example is the relationship between Toeplitz operators on the Fock space and pseudodifferential operators on $L^2(\mathbb{R})$.

And finally, I want to mention the role that Fock spaces play in quantum physics, harmonic analysis on the Heisenberg group, and partial differential equations. In particular, the normalized reproducing kernels in the Fock space are exactly the so-called coherent states in quantum physics, the parametrized Berezin transform on the Fock space provides a solution to the initial value problem on the complex plane for the heat equation, and weighted translation operators give rise to a unitary representation of the Heisenberg group on the Fock space.

I chose to develop the whole theory in the context of one complex variable, although pretty much everything we do in the book can be generalized to the case of finitely many complex variables. The case of Fock spaces of infinitely many variables is a subject of its own and will not be discussed at all in the book.

I have tried to keep the prerequisites to a minimum. A standard graduate course in each of real analysis, complex analysis, and functional analysis should prepare the reader for most of the book. There are, however, several exceptions. One is Lindelöf's theorem which determines when a certain entire function is of finite type, and the other is the Calderón–Vaillancourt theorem concerning the boundedness of certain pseudodifferential operators. These two results are included in Chap. 1 without proof. Used without proof are also a couple of theorems from abstract algebra when we characterize finite-rank Hankel and Toeplitz operators in Chaps. 6 and 7, and a couple of theorems from the general theory of interpolation when we describe the complex interpolation spaces for Fock spaces in Chap. 2.

I have included some exercises at the end of each chapter. Some of these are extensions or supplements to the main text, some are routine estimates omitted in the main proofs, some are "lemmas" taken out of research papers, while others are estimates or lemmas that I came up during the writing of the book that were eventually abandoned because of better approaches found later. I have tried my best to give a reference whenever a nontrivial result appears in the exercises.

I have tried to include as many relevant references as possible. But I am sure that the Bibliography is not even nearly complete. I apologize in advance if your favorite paper or reference is missing here. I did not omit it on purpose. I either overlooked it or was not aware of it. The same is true with the brief comments I make at the end of each chapter. I have tried my best to point the reader to sources that I consider to be original or useful, but these comments are by no means authoritative and are more likely biased because of my limitations in history and knowledge.

As usual, my family has been very supportive during the writing of this book. I am very grateful to them—my wife Peijia and our sons Peter and Michael—for their encouragement, understanding, patience, and tolerance. During the writing of the book, I also received help from Lewis Coburn, Josh Isralowitz, Haiying Li, Alex Schuster, Kristian Seip, Dan Stevenson, and Chunjie Wang. Thank you all!

Albany, NY, USA   Kehe Zhu

Contents

1 Preliminaries 1
  1.1 Entire Functions 3
  1.2 Lattices in the Complex Plane 9
  1.3 Weierstrass σ-Functions 13
  1.4 Pseudodifferential Operators 19
  1.5 The Heisenberg Group 25
  1.6 Notes 27
  1.7 Exercises 29

2 Fock Spaces 31
  2.1 Basic Properties 33
  2.2 Some Integral Operators 43
  2.3 Duality of Fock Spaces 53
  2.4 Complex Interpolation 59
  2.5 Atomic Decomposition 63
  2.6 Translation Invariance 75
  2.7 A Maximum Principle 81
  2.8 Notes 87
  2.9 Exercises 89

3 The Berezin Transform and BMO 93
  3.1 The Berezin Transform of Operators 95
  3.2 The Berezin Transform of Functions 101
  3.3 Fixed Points of the Berezin Transform 113
  3.4 Fock–Carleson Measures 117
  3.5 Functions of Bounded Mean Oscillation 123
  3.6 Notes 133
  3.7 Exercises 135

4 Interpolating and Sampling Sequences 137
  4.1 A Notion of Density 139
  4.2 Separated Sequences 143

4.3   Stability Under Weak Convergence 151
4.4   A Modified Weierstrass σ-Function 159
4.5   Sampling Sequences 165
4.6   Interpolating Sequences 177
4.7   Notes 187
4.8   Exercises 189

5   **Zero Sets for Fock Spaces** 193
5.1   A Necessary Condition 195
5.2   A Sufficient Condition 197
5.3   Pathological Properties 199
5.4   Notes 209
5.5   Exercises 211

6   **Toeplitz Operators** 213
6.1   Trace Formulas 215
6.2   The Bargmann Transform 221
6.3   Boundedness 229
6.4   Compactness 237
6.5   Toeplitz Operators in Schatten Classes 245
6.6   Finite Rank Toeplitz Operators 255
6.7   Notes 263
6.8   Exercises 265

7   **Small Hankel Operators** 267
7.1   Small Hankel Operators 269
7.2   Boundedness and Compactness 271
7.3   Membership in Schatten Classes 275
7.4   Finite Rank Small Hankel Operators 281
7.5   Notes 283
7.6   Exercises 285

8   **Hankel Operators** 287
8.1   Boundedness and Compactness 289
8.2   Compact Hankel Operators with Bounded Symbols 293
8.3   Membership in Schatten Classes 301
8.4   Notes 327
8.5   Exercises 329

References 331

Index 341

Chapter 1
Preliminaries

In this chapter, we collect several preliminary results about entire functions, lattices in the complex plane, pseudodifferential operators, and the Heisenberg group. The purpose is to fix notation and to facilitate references later on. All the results concerning entire functions, except Lindelöf's theorem, are well known and elementary. The section about Weierstrass σ-functions is self-contained, while the section on pseudodifferential operators is very sketchy.

K. Zhu, *Analysis on Fock Spaces*, Graduate Texts in Mathematics 263,
DOI 10.1007/978-1-4419-8801-0_1,
© Springer Science+Business Media New York 2012

1.1 Entire Functions

This book is about certain spaces of entire functions and certain operators defined on these spaces. So we begin by recalling some elementary results about entire functions. The first few of these results can be found in any graduate-level complex analysis text, and no proof is included here.

Let $\mathbb{C}$ denote the complex plane. If a function f is analytic on the entire complex plane $\mathbb{C}$, we say that f is an entire function. One of the fundamental results in complex analysis is the following *identity theorem*.

Theorem 1.1. *If f is entire and the zero set of f,*

$$Z(f) = \{z \in \mathbb{C} : f(z) = 0\},$$

has a limit point in $\mathbb{C}$, then $f \equiv 0$ on $\mathbb{C}$.

Another version of the identity theorem is the following:

Theorem 1.2. *Suppose f is an entire function. If there is a point $a \in \mathbb{C}$ such that $f^{(n)}(a) = 0$ for all $n \geq 0$, then $f \equiv 0$ on $\mathbb{C}$.*

When we say that $\{z_n\}$ is the zero sequence of an entire function f, we always assume that any zero of multiplicity k is repeated k times in $\{z_n\}$. As a consequence of the identity theorem, we see that the zero set of an entire function that is not identically zero cannot have any finite limit point and no value occurs infinitely many times in the sequence. Consequently, the zero sequence $\{z_n\}$ of an entire function is either finite or satisfies the condition that $|z_n| \to \infty$ as $n \to \infty$. In particular, we can always arrange the zeros so that $|z_1| \leq |z_2| \leq \cdots \leq |z_n| \leq \cdots$.

The following result is called the *mean value theorem*, which follows from the subharmonicity of the function $|f(z)|^p$ in $|z - a| < R$.

Theorem 1.3. *Suppose f is entire and $0 < p < \infty$. Then*

$$|f(a)|^p \leq \frac{1}{2\pi} \int_0^{2\pi} |f(a + re^{i\theta})|^p \, d\theta \tag{1.1}$$

for all $a \in \mathbb{C}$ and all $r \in [0, \infty)$.

Because r above is arbitrary, we often multiply both sides of (1.1) by some function of r and then integrate with respect to r. For example, if we multiply both sides of (1.1) by r and then integrate from 0 to R, the result is

$$|f(a)|^p \leq \frac{1}{\pi R^2} \int_{|z-a|<R} |f(z)|^p \, dA(z), \tag{1.2}$$

where $z = x + iy$ and $dA(z) = dxdy$ is the Lebesgue area measure. The inequality in (1.2) is the area version of the mean value theorem.

The next result is called *Liouville's theorem*.

Theorem 1.4. *A bounded entire function is necessarily constant. More generally, if a complex-valued harmonic function defined on the entire complex plane is bounded, then it must be constant.*

The lack of bounded entire functions is one of the key differences between the theory of Fock spaces and the more classical theories of Hardy and Bergman spaces.

A central problem in complex analysis is the study of zeros of analytic functions in specific function spaces. An important tool in any such study is the classical *Jensen's formula* below:

Theorem 1.5. *Suppose that*

(a) *f is analytic on the closed disk $|z| \leq r$,*
(b) *f does not vanish on $|z| = r$,*
(c) *$f(0) = 1$, and*
(d) *the zeros of f in $|z| < r$ are $\{z_1, \cdots, z_N\}$, with multiple zeros repeated according to multiplicity,*

Then

$$\sum_{k=1}^{N} \log \frac{r}{|z_k|} = \frac{1}{2\pi} \int_0^{2\pi} \log|f(re^{i\theta})|\, d\theta. \tag{1.3}$$

If $f(0)$ is nonzero but not necessarily 1, Jensen's formula takes the form

$$\log|f(0)| = -\sum_{k=1}^{N} \log \frac{r}{|z_k|} + \frac{1}{2\pi} \int_0^{2\pi} \log|f(re^{i\theta})|\, d\theta, \tag{1.4}$$

where $\{z_1, \cdots, z_N\}$ are zeros of f in $0 < |z| < r$. More generally, if f has a zero of order k at the origin, then Jensen's formula takes the following form:

$$\log \frac{|f^{(k)}(0)|}{k!} + k\log r = -\sum_{k=1}^{N} \log \frac{r}{|z_k|} + \frac{1}{2\pi} \int_0^{2\pi} \log|f(re^{i\theta})|\, d\theta,$$

where $\{z_1, \cdots, z_N\}$ are zeros of f in $0 < |z| < r$.

Let f be an entire function. We can factor out the zeros of f in a canonical way, a process that is usually referred to as *Weierstrass factorization*. The basis for the Weierstrass factorization theorem is a collection of simple entire functions called elementary factors. More specifically, we define

$$E_0(z) = 1 - z,$$

and for any positive integer n,

$$E_n(z) = (1 - z)\exp\left(z + \frac{z^2}{2} + \cdots + \frac{z^n}{n}\right).$$

If a is any nonzero complex number, it is clear that $E(z/a)$ has a unique, simple zero at $z = a$.

Theorem 1.6. *Let $\{z_n\}$ be a sequence of nonzero complex numbers such that the sequence $\{|z_n|\}$ is nondecreasing and tends to ∞. Then it is possible to choose a sequence $\{p_n\}$ of nonnegative integers such that*

$$\sum_{n=1}^{\infty}\left(\frac{r}{|z_n|}\right)^{p_n+1} < \infty \qquad (1.5)$$

for all $r > 0$. Furthermore, the infinite product

$$P(z) = \prod_{n=1}^{\infty} E_{p_n}\left(\frac{z}{z_n}\right) \qquad (1.6)$$

converges uniformly on every compact subset of $\mathbb{C}$, the function P is entire, and the zeros of P are exactly $\{z_n\}$, counting multiplicity.

Note that the choice $p_n = n - 1$ will always satisfy (1.5). In many cases, however, there are "better" choices. In particular, if $\{z_n\}$ is the zero sequence of an entire function f and if there exists an integer p such that

$$\sum_{n=1}^{\infty}\frac{1}{|z_n|^{p+1}} < \infty, \qquad (1.7)$$

we say that f is of *finite rank*. If p is the smallest integer such that (1.7) is satisfied, then f is said to be of rank p. A function with only a finite number of zeros has rank 0. A function is of *infinite rank* if it is not of finite rank.

If f is of finite rank p and $\{z_n\}$ is the zero sequence of f, then (1.7) is satisfied with $p_n = p$. The product $P(z)$ associated with this canonical choice of $\{p_n\}$ will be called the standard form.

Theorem 1.7. *Let f be an entire function of finite rank p. If P is the standard product associated with the zeros of f, then there exist a nonzero integer m and an entire function g such that*

$$f(z) = z^m P(z) e^{g(z)}. \qquad (1.8)$$

The integer m is unique, and the entire function g is unique up to an additive constant of the form $2k\pi i$.

For an entire function of finite rank, we say that (1.8) is the standard factorization of f, or the *Weierstrass factorization of f*.

Let f be an entire function of finite rank p. If the entire function g in the standard factorization (1.8) of f is a polynomial of degree q, then we say that f has *finite genus*. In this case, the number $\mu = \max(p,q)$ is called the genus of f.

Let f be an entire function. For any $r > 0$, we write

$$M(r) = M_f(r) = \sup\{|f(z)| : |z| = r\}.$$

We say that f is of *order* ρ if

$$\rho = \limsup_{r \to \infty} \frac{\log \log M(r)}{\log r}.$$

It is clear that $0 \le \rho \le \infty$. When $\rho < \infty$, f is said to be of *finite order*; otherwise, f is of *infinite order*.

There are two useful characterizations for entire functions to be of finite order, the first of which is the following:

Theorem 1.8. *An entire function f is of finite order if and only if there exist positive constants a and r such that*

$$|f(z)| < \exp(|z|^a), \qquad |z| > r.$$

In this case, the order of f is the infimum of the set of all such numbers a.

The following characterization of entire functions of finite order is traditionally referred to as the *Hadamard factorization theorem*.

Theorem 1.9. *An entire function f is of finite order ρ if and only if it is of finite genus μ. Moreover, the order and genus of f satisfy the following relations: $\mu \le \rho \le \mu + 1$.*

When $0 < \rho < \infty$, we define

$$\sigma = \limsup_{r \to \infty} \frac{\log M(r)}{r^\rho}.$$

If $\sigma < \infty$, we say that f is of *finite type*. More specifically, we say that f is of order ρ and type σ. If $\sigma = \infty$, we say that f is of *maximum type* or *infinite type*.

Let $\{z_n\}$ denote the zero sequence, excluding 0, of an entire function f. The infimum of all positive numbers s such that

$$\sum_{n=1}^{\infty} \frac{1}{|z_n|^s} < \infty$$

will be denoted by $\rho_1 = \rho_1(f)$. The smallest positive integer s satisfying the convergence condition above will be denoted by $m + 1$.

Theorem 1.10. *For any entire function f that is not identically zero, we have the following relations among the constants defined above:*

(a) $\rho_1 - 1 \le m \le \rho$.
(b) If ρ is not an integer, then $\rho = \rho_1$.
(c) $m = [\rho_1]$ *if ρ_1 is not an integer.*

Here, $[x]$ denotes the greatest integer less than or equal to x.

The following result is sometimes called *Lindelöf's theorem*. This result is not so standard in the sense that it does not appear in most elementary complex analysis texts. See [38] for a proof.

Theorem 1.11. *Suppose that* ρ *is a positive integer, f is an entire function of order* ρ, $f(0) \neq 0$, *and* $\{z_n\}$ *is the zero sequence of f. Then f is of finite type if and only if the following two conditions hold:*

(a) $n(r) = O(r^\rho)$ *as* $r \to \infty$, *where* $n(r)$ *is the number (counting multiplicity) of zeros of f in* $|z| \leq r$.
(b) *The partial sums*

$$S(r) = \sum_{|z_n| \leq r} \frac{1}{z_n^\rho}$$

are bounded in r.

Lindelöf's theorem will be useful for us in Chap. 5 when we study zero sequences for functions in Fock spaces. The reader should be mindful of the fact that there are several results in complex analysis that are called Lindelöf's theorem. In most cases, these results are certain generalizations of the classical maximum modulus principle.

1.2 Lattices in the Complex Plane

The complex plane is flat, and lattices in it are easy to describe. We will need to use rectangular lattices on several occasions later on. In this section, we fix notation and collect basic facts about lattices in the complex plane.

The simplest lattice in $\mathbb{C}$ is the standard integer lattice

$$\mathbb{Z}^2 = \{m + in : m \in \mathbb{Z}, n \in \mathbb{Z}\},$$

where $\mathbb{Z} = \{0, \pm 1, \pm 2, \cdots, \}$ is the integer group. All lattices we use in the book are isomorphic to $\mathbb{Z}^2$.

Let ω be any complex number, and let ω_1 and ω_2 be any two nonzero complex numbers such that their ratio is not real. For any integers m and n, let $\omega_{mn} = \omega + m\omega_1 + n\omega_2$. The set

$$\Lambda = \Lambda(\omega, \omega_1, \omega_2) = \{\omega_{mn} : m \in \mathbb{Z}, n \in \mathbb{Z}\}$$

is then called the lattice generated by ω, ω_1, and ω_2.

The initial parallelogram at ω spanned by ω_1 and ω_2 has vertices

$$\omega, \quad \omega + \omega_1, \quad \omega + \omega_2, \quad \omega + \omega_1 + \omega_2,$$

and is centered at

$$c = \omega + \frac{1}{2}(\omega_1 + \omega_2).$$

We shift this parallelogram so that the center becomes ω and the vertices become

$$\omega - \frac{1}{2}(\omega_1 + \omega_2), \ \omega + \frac{1}{2}(\omega_1 - \omega_2), \ \omega + \frac{1}{2}(\omega_2 - \omega_1), \ \omega + \frac{1}{2}(\omega_1 + \omega_2).$$

We denote this new parallelogram by R_{00} and call it the *fundamental region* of $\Lambda(\omega, \omega_1, \omega_2)$. For any integers m and n, let $R_{mn} = R_{00} + \omega_{mn}$, with ω_{mn} being the center of R_{mn}.

Lemma 1.12. *Let $\Lambda = \Lambda(\omega, \omega_1, \omega_2)$ be any lattice in $\mathbb{C}$. For any positive number δ, there exists a positive constant C such that*

$$\sum_{z \in \Lambda} e^{-\delta |z-w|^2} \leq C$$

for all $w \in \mathbb{C}$.

Proof. By translation invariance, it suffices for us to prove the desired inequality for w in the fundamental region R_{00} of Λ. If w is in the relatively compact set R_{00}, then $|w/z| < 1/2$ for all but a finite number of points $z \in \Lambda$. For all such points z, we have

$$|z - w|^2 = |z|^2|1 - (w/z)|^2 \ge \frac{1}{4}|z|^2.$$

Since $\sum_{z \in \Lambda} e^{-\frac{\delta}{4}|z|^2}$ is obviously convergent, we obtain the desired result.   □

Lemma 1.13. *With notation from above, we have*

$$\mathbb{C} = \bigcup \{R_{mn} : m \in \mathbb{Z}, n \in \mathbb{Z}\},$$

and

$$\int_{\mathbb{C}} f(z)\, dA(z) = \sum_{m,n \in \mathbb{Z}} \int_{R_{mn}} f(z)\, dA(z)$$

for every $f \in L^1(\mathbb{C}, dA)$.

Proof. The decomposition of $\mathbb{C}$ into the union of congruent parallelograms is obvious. Since any two different R_{mn} only overlap on a set of zero area, the desired integral decomposition follows immediately.   □

In several situations later, we will need to decompose a given lattice into several sparse sublattices. The following lemma tells us how to do it.

Lemma 1.14. *Let $\Lambda = \Lambda(\omega, \omega_1, \omega_2)$ be a lattice in $\mathbb{C}$. For any positive number R, there exists a positive integer N such that we can decompose Λ into the disjoint union of N sublattices,*

$$\Lambda = \Lambda_1 \cup \cdots \cup \Lambda_N,$$

such that the distance between any two points in each of the sublattices is at least R.

Proof. Fix a positive integer k such that $k|\omega_1| > R$ and $k|\omega_2| > R$. For each $j = (j_1, j_2)$ with $0 \le j_1 \le k$ and $0 \le j_2 \le k$, let

$$\Lambda_j = \Lambda(\omega + j_1\omega_1 + j_2\omega_2, k\omega_1, k\omega_2)$$
$$= \{(\omega + j_1\omega_1 + j_2\omega_2) + (mk\omega_1 + nk\omega_2) : m \in \mathbb{Z}, n \in \mathbb{Z}\}.$$

Then each Λ_j is a sublattice of Λ; the distance between any two points in Λ_j is at least R, and $\Lambda = \cup \Lambda_j$. There are a few duplicates among Λ_j caused by points from the boundary of the parallelogram at ω spanned by $k\omega_1$ and $k\omega_2$. After these duplicates are deleted, we arrive at the desired decomposition for Λ.   □

Most lattices we use in the book are square ones. More specifically, for any given positive parameter r, we consider the case when $\omega = 0$, $\omega_1 = r$, and $\omega_2 = ir$. The resulting lattice is

$$r\mathbb{Z}^2 = \{rm + irn : m \in \mathbb{Z}, n \in \mathbb{Z}\}.$$

We mention two particular cases. First, for $r = \sqrt{\pi/\alpha}$, where α is a positive parameter, the resulting lattices are used in the next section when we introduce the Weierstrass σ-functions. Second, for $r = 1/N$, where N is a positive integer, the resulting lattices will be employed in Chaps. 6–8 when we characterize Hankel and Toeplitz operators in Schatten classes.

For any two points $z = x + iy$ and $w = u + iv$ in $r\mathbb{Z}^2$, we let $\gamma(z, w)$ denote the following path in $r\mathbb{Z}^2$: we first move horizontally from z to $u + iy$ and then vertically from $u + iy$ to $u + iv$. When $z = 0$, we write $\gamma(w)$ in place of $\gamma(0, w)$. The path $\gamma(z, w)$ is of course discrete. We use $|\gamma(z, w)|$ to denote the number of points in $\gamma(z, w)$ and call it the length of $\gamma(z, w)$.

The following technical lemma will play a critical role in Chap. 8.

Lemma 1.15. *For any positive r and σ, there exists a positive constant $C = C_{r,\sigma}$ such that*

$$\sum_{z \in r\mathbb{Z}^2} \sum_{w \in r\mathbb{Z}^2} e^{-\sigma|z-w|^2} \chi_{\gamma(z,w)}(u) \leq C$$

for all $u \in r\mathbb{Z}^2$, where $\chi_{\gamma(z,w)}$ is the characteristic function of $\gamma(z, w)$.

Proof. Without loss of generality, we may assume that $r = 1$. Adjusting the constant σ will then produce the general case.

Also, it is obvious that

$$u + \gamma(z, w) = \gamma(u + z, u + w),$$

which implies that the sum

$$S = \sum_{z \in \mathbb{Z}^2} \sum_{w \in \mathbb{Z}^2} e^{-\sigma|z-w|^2} \chi_{\gamma(z,w)}(u)$$

is actually independent of u. For convenience, we will assume that $u = 0$.

For any z and w, the path $\gamma(z, w)$ consists of a horizontal segment and a vertical segment (one or both are allowed to degenerate). From the definition of $\gamma(z, w)$, we see that the origin 0 lies on the horizontal segment of $\gamma(z, w)$ if and only if one of the following is true:

(1) z is on the negative x-axis and w is in the first or fourth quadrant: $z = -n$, $w = m + ki$, where n and m are nonnegative integers and k is an integer.
(2) z is on the positive x-axis and w is in the second or third quadrant: $z = n$, $w = -m + ki$, where n and m are nonnegative integers and k is an integer.

Similarly, 0 lies on the vertical segment of $\gamma(z,w)$ if and only if one of the following is true:

(3) w is on the positive y-axis and z is in the third or fourth quadrant: $w = ni$, $z = k - mi$, where n and m are nonnegative integers and k is an integer.
(4) w is on the negative y-axis and z is in the first or second quadrant: $w = -ni$, $z = k + mi$, where n and m are nonnegative integers and k is an integer.

In each of the cases above, we have

$$|z - w|^2 = (n+m)^2 + k^2 \geq n^2 + m^2 + k^2.$$

Therefore,

$$S \leq 4 \sum_{n=0}^{\infty} \sum_{m=0}^{\infty} \sum_{k=-\infty}^{\infty} e^{-\sigma(n^2+m^2+k^2)}$$

$$= 4 \sum_{n=0}^{\infty} e^{-\sigma n^2} \sum_{m=0}^{\infty} e^{-\sigma m^2} \sum_{k=-\infty}^{\infty} e^{-\sigma k^2} < \infty.$$

This proves the lemma.   □

1.3 Weierstrass σ-Functions

In this section we introduce several Weierstrass functions on the complex plane and prove their periodicity or quasiperiodicity. In particular, the Weierstrass σ-function will serve as a prototype for functions in Fock spaces and will play an important role in our characterization of interpolating and sampling sequences for Fock spaces.

Lattices in this section are all based at the origin:

$$\Lambda = \Lambda(0, \omega_1, \omega_2) = \{\omega_{mn}\}, \qquad \omega_{mn} = m\omega_1 + n\omega_2.$$

To every such lattice, we associate a function $\mathcal{P}(z) = \mathcal{P}_\Lambda(z)$ as follows:

$$\mathcal{P}(z) = \frac{1}{z^2} + \sum_{m,n}{}' \left[\frac{1}{(z - \omega_{mn})^2} - \frac{1}{\omega_{mn}^2} \right], \tag{1.9}$$

where the summation (with a prime) extends over all integers m and n with $(m, n) \neq (0, 0)$.

Proposition 1.16. *The function $\mathcal{P}$ is an even meromorphic function in the complex plane whose poles are exactly the points in the lattice Λ. Furthermore, $\mathcal{P}$ is doubly periodic with periods ω_1 and ω_2:*

$$\mathcal{P}(z + \omega_1) = \mathcal{P}(z), \qquad \mathcal{P}(z + \omega_2) = \mathcal{P}(z), \tag{1.10}$$

for all $z \in \mathbb{C} - \Lambda$.

Proof. For any small $\delta > 0$, let

$$U_\delta = \{z \in \mathbb{C} : d(z, \Lambda) > \delta, |z| < 1/\delta\}.$$

It is clear that for $z \in U_\delta$ we have

$$\frac{1}{(z - \omega_{mn})^2} - \frac{1}{\omega_{mn}^2} = O\left(\frac{1}{|\omega_{mn}|^3} \right)$$

when $|\omega_{mn}|$ is large. Since

$$\sum_{(m,n) \neq (0,0)} \frac{1}{|\omega_{mn}|^3} < \infty,$$

the series in (1.9) converges uniformly and absolutely to an analytic function in U_δ. Since δ is arbitrary, the series in (1.9) converges to an analytic function $\mathcal{P}$ on $\mathbb{C} - \Lambda$. At each point ω_{mn}, it is clear that $\mathcal{P}$ has a double pole. So $\mathcal{P}$ is meromorphic with double poles at precisely the points of Λ.

To see that $\mathcal{P}$ is doubly periodic with periods ω_1 and ω_2, we differentiate the defining equation (1.9) term by term, which is permissible because the series converges uniformly on compact subsets of $\mathbb{C} - \Lambda$. Thus,

$$\mathcal{P}'(z) = -2 \sum_{m,n} \frac{1}{(z - \omega_{mn})^3}.$$

Since $\{-\omega_{mn} : m \in \mathbb{Z}, n \in \mathbb{Z}\}$ represents the same lattice Λ and the series above converges absolutely (so its terms can be rearranged in any way we like), we see that $\mathcal{P}'$ is an odd function, and so the original function $\mathcal{P}$ is even.

On the other hand, for each $k = 1, 2$, we have

$$\mathcal{P}'(z + \omega_k) = -2 \sum_{m,n} \frac{1}{(z - \omega_{mn} + \omega_k)^3}.$$

Since $\{\omega_{mn} - \omega_k : m \in \mathbb{Z}, n \in \mathbb{Z}\}$ represents the same lattice Λ and the above series converges absolutely for any $z \in \mathbb{C} - \Lambda$, we see that $\mathcal{P}'(z + \omega_k) = \mathcal{P}'(z)$, so $\mathcal{P}'$ is doubly periodic with periods ω_1 and ω_2.

If we integrate the equation $\mathcal{P}'(z + \omega_k) = \mathcal{P}'(z)$ on the connected region $\mathbb{C} - \Lambda$, we will find a constant C_k such that $\mathcal{P}(z + \omega_k) = \mathcal{P}(z) + C_k$ for $k = 1, 2$ and all $z \in \mathbb{C} - \Lambda$. Setting $z = -\omega_k/2$ and using the fact that $\mathcal{P}$ is even, we obtain $C_k = 0$ for $k = 1, 2$. This shows that $\mathcal{P}$ is doubly periodic with periods ω_1 and ω_2. $\qquad\square$

To every lattice $\Lambda = \Lambda(0, \omega_1, \omega_2) = \{\omega_{mn}\}$, we associate another function $\zeta(z) = \zeta_\Lambda(z)$ as follows:

$$\zeta(z) = \frac{1}{z} + \sum_{m,n}' \left[\frac{1}{z - \omega_{mn}} + \frac{1}{\omega_{mn}} + \frac{z}{\omega_{mn}^2} \right]. \tag{1.11}$$

The following proposition lists some of the basic properties of this function, which should not be confused with the famous Riemann ζ-function.

Proposition 1.17. *Each ζ is an odd meromorphic function with simple poles at precisely the points of Λ. Furthermore, for $k = 1, 2$, we have*

$$\zeta(z + \omega_k) = \zeta(z) + \eta_k, \qquad z \in \mathbb{C} - \Lambda, \tag{1.12}$$

where $\eta_k = 2\zeta(\omega_k/2)$.

Proof. Again we fix any small positive number δ and consider the region U_δ defined in the proof of the previous proposition. It is clear that

$$\frac{1}{z - \omega_{mn}} + \frac{1}{\omega_{mn}} + \frac{z}{\omega_{mn}^2} = O\left(\frac{1}{|\omega_{mn}|^3} \right), \qquad z \in U_\delta,$$

as $|\omega_{mn}| \to \infty$. It follows that the series in (1.11) converges to an analytic function in $\mathbb{C} - \Lambda$, and the convergence is uniform and absolute on the relatively compact set U_δ. It is clear that the resulting function ζ has a simple pole at (and only at) each point of Λ.

A rearrangement of terms in the series (1.11) easily shows that ζ is an odd function on $\mathbb{C} - \Lambda$. Differentiating the series (1.11) term by term shows that the two Weierstrass functions $\mathcal{P}$ and ζ are related by the differential equation $\zeta'(z) = -\mathcal{P}(z)$ coupled with the condition

$$\lim_{z \to 0} \left(\zeta(z) - \frac{1}{z} \right) = 0.$$

If we integrate the equation $\mathcal{P}(z + \omega_k) = \mathcal{P}(z)$ on the connected region $\mathbb{C} - \Lambda$, we obtain a constant η_k such that $\zeta(z + \omega_k) = \zeta(z) + \eta_k$ for $k = 1, 2$ and all $z \in \mathbb{C} - \Lambda$. Setting $z = -\omega_k/2$ and using the fact that ζ is odd, we obtain $\eta_k = 2\zeta(\omega_k/2)$. This completes the proof of the proposition. □

Because of the relations in (1.12), we say that the Weierstrass function ζ is quasiperiodic.

Lemma 1.18. *The periods ω_k and the constants η_k are related by the following equation:*

$$\eta_1 \omega_2 - \eta_2 \omega_1 = 2\pi i. \tag{1.13}$$

Proof. If we pull the center $c = (\omega_1 + \omega_2)/2$ of the parallelogram spanned by ω_1 and ω_2 to the origin, the result is another parallelogram $R = R_\Lambda$ with the following vertices:

$$-\frac{1}{2}(\omega_1 + \omega_2), \quad \frac{1}{2}(\omega_2 - \omega_1), \quad \frac{1}{2}(\omega_1 - \omega_2), \quad \frac{1}{2}(\omega_1 + \omega_2).$$

Recall that $R = R_\Lambda$ is the fundamental region of the lattice Λ.

It is clear that ζ is analytic on R, up to the boundary, except a simple pole at the center of R (which is the origin) with residue 1. Therefore,

$$\int_{\partial R} \zeta(z) \, dz = 2\pi i.$$

Break this into integration over the four sides of R and use the quasiperiodicity of ζ. We obtain the desired result. □

To every lattice $\Lambda = \Lambda(0, \omega_1, \omega_2) = \{\omega_{mn}\}$, we associate yet another function $\sigma(z) = \sigma_\Lambda(z)$ as follows:

$$\sigma(z) = z \prod_{m,n}' \left[\left(1 - \frac{z}{\omega_{mn}} \right) \exp \left(\frac{z}{\omega_{mn}} + \frac{z^2}{2\omega_{mn}^2} \right) \right]. \tag{1.14}$$

The following proposition lists some of the basic properties of the Weierstrass σ-functions.

Proposition 1.19. *Each σ is an entire function whose zero set is exactly the lattice $\Lambda = \{\omega_{mn}\}$. Furthermore, σ is odd and quasiperiodic in the following sense:*

$$\sigma(z+\omega_k) = -e^{\eta_k(z+(\omega_k/2))}\sigma(z), \tag{1.15}$$

where $k = 1,2$ and η_k are the constants from the previous proposition.

Proof. It follows from a standard argument involving the Weierstrass product (see Sect. 1.1) that the infinite product in (1.14) converges to an entire function σ and the convergence is uniform and absolute on any compact subset of the complex plane. It is also clear that the zero set of σ is exactly the lattice $\Lambda = \{\omega_{mn}\}$.

Replace z by $-z$ in (1.14) and observe that $\{-\omega_{mn} : m \in \mathbb{Z}, n \in \mathbb{Z}\}$ is exactly the same lattice Λ (arranged differently). We see that the function σ is odd.

To prove the quasiperiodicity of σ, we note that the Weierstrass functions σ and ζ are related by the differential equation

$$\frac{d}{dz}\log\sigma(z) = \zeta(z),$$

coupled with the condition

$$\lim_{z\to 0}\frac{\sigma(z)}{z} = 1.$$

If we integrate the equation

$$\zeta(z+\omega_k) = \zeta(z)+\eta_k$$

in the connected region $\mathbb{C}-\Lambda$ and then exponentiate the result, we obtain a constant c_k such that

$$\sigma(z+\omega_k) = c_k e^{\eta_k z}\sigma(z), \qquad z \in \mathbb{C}.$$

Let $z = -\omega_k/2$ and use the fact that σ is odd. We get $c_k = -e^{\eta_k\omega_k/2}$. □

Finally, in this section, we consider the special case of square lattices. For any positive parameter α, we consider the lattice $\Lambda = \Lambda_\alpha$ given by $\omega_1 = \sqrt{\pi/\alpha}$ and $\omega_2 = \sqrt{\pi/\alpha}i$. Thus,

$$\Lambda_\alpha = \{\sqrt{\pi/\alpha}(m+in) : m \in \mathbb{Z}, n \in \mathbb{Z}\}.$$

In this particular case, we will compute the constants η_k and relate the quasiperiodicity of σ to a certain isometry on Fock spaces.

Proposition 1.20. *Suppose σ is the Weierstrass σ-function associated to the square lattice $\Lambda_\alpha = \{\omega_{mn}\}$, where $\omega_{mn} = \sqrt{\pi/\alpha}(m+in)$, so that $\omega_1 = \sqrt{\pi/\alpha}$ and $\omega_2 = \sqrt{\pi/\alpha}i$. Then $\eta_1 = \sqrt{\pi\alpha}$ and $\eta_2 = -\sqrt{\pi\alpha}i$. Furthermore,*

$$e^{\alpha\overline{\omega}_{mn}z-\frac{\alpha}{2}|\omega_{mn}|^2}\sigma(z-\omega_{mn}) = (-1)^{m+n+mn}\sigma(z) \tag{1.16}$$

for all $z \in \mathbb{C}$ and $\omega_{mn} \in \Lambda_\alpha$.

Proof. In this particular case, we have

$$\omega_{mn} = \sqrt{\pi/\alpha}(m+in) = i\sqrt{\pi/\alpha}(n-im) = i\omega_{nm'},$$

where $m' = -m$. It follows that

$$\zeta(iz) = \frac{1}{iz} + \sum_{m,n}' \left(\frac{1}{iz-\omega_{mn}} + \frac{1}{\omega_{mn}} + \frac{iz}{\omega_{mn}^2} \right)$$

$$= \frac{1}{iz} + \sum_{m,n}' \left(\frac{1}{iz-i\omega_{nm'}} + \frac{1}{i\omega_{nm'}} + \frac{iz}{(i\omega_{nm'})^2} \right)$$

$$= \frac{1}{i} \left[\frac{1}{z} + \sum_{m,n}' \left(\frac{1}{z-\omega_{nm'}} + \frac{1}{\omega_{nm'}} + \frac{z}{\omega_{nm'}^2} \right) \right]$$

$$= \frac{1}{i} \zeta(z).$$

Therefore,

$$\eta_2 = 2\zeta(\omega_2/2) = 2\zeta(i\omega_1/2) = \frac{2}{i}\zeta(\omega_1/2) = \frac{\eta_1}{i}.$$

This, along with (1.13), gives $\eta_1 = \sqrt{\pi\alpha}$ and $\eta_2 = -\sqrt{\pi\alpha}\,i$.

To prove the translation relation in (1.16), observe that

$$\omega_{mn} = m\omega_1 + n\omega_2.$$

It follows from (1.15) and induction that

$$\sigma(z+m\omega_1) = (-1)^m \sigma(z)e^{m\eta_1 z + \frac{1}{2}m^2\eta_1\omega_1}$$

for all positive integers m. Since σ is an odd function, it is then easy to see that the above equation also holds for negative integers m. Similarly,

$$\sigma(z+n\omega_2) = (-1)^n \sigma(z)e^{n\eta_2 z + \frac{1}{2}n^2\eta_2\omega_2}$$

for all integers n. Therefore,

$$\sigma(z+\omega_{mn}) = (-1)^n e^{n\eta_2(z+m\omega_1)+\frac{1}{2}n^2\eta_2\omega_2}\,\sigma(z+m\omega_1)$$

$$= (-1)^{m+n} e^{n\eta_2(z+m\omega_1)+\frac{1}{2}n^2\eta_2\omega_2}\, e^{m\eta_1 z+\frac{1}{2}m^2\eta_1\omega_1}\,\sigma(z)$$

$$= (-1)^{m+n} e^{(n\eta_2+m\eta_1)z+nm\eta_2\omega_1+\frac{1}{2}(n^2\eta_2\omega_2+m^2\eta_1\omega_1)}\,\sigma(z).$$

Plug in

$$\omega_1 = \sqrt{\pi/\alpha}, \quad \omega_2 = \sqrt{\pi/\alpha}\,\mathrm{i}, \quad \eta_1 = \sqrt{\pi\alpha}, \quad \eta_2 = -\sqrt{\pi\alpha}\,\mathrm{i}.$$

We obtain

$$\sigma(z + \omega_{mn}) = (-1)^{m+n+mn} e^{\alpha\overline{\omega}_{mn}z + \frac{\alpha}{2}|\omega_{mn}|^2} \sigma(z)$$

for all $z \in \mathbb{C}$ and all $\omega_{mn} \in \Lambda_\alpha$. Replacing ω_{mn} by $-\omega_{mn}$, we obtain

$$e^{\alpha\overline{\lambda}_{mn}z - \frac{\alpha}{2}|\omega_{mn}|^2} \sigma(z - \omega_{mn}) = (-1)^{m+n+mn} \sigma(z)$$

for all $z \in \mathbb{C}$ and all $\omega_{mn} \in \Lambda_\alpha$.   □

Corollary 1.21. *For any $\alpha > 0$, the Weierstrass function σ associated to Λ_α has the following properties:*

(a) The function $|\sigma(z)|e^{-\frac{\alpha}{2}|z|^2}$ is doubly periodic with periods $\sqrt{\pi/\alpha}$ and $\mathrm{i}\sqrt{\pi/\alpha}$.
(b) $|\sigma(z)|e^{-\frac{\alpha}{2}|z|^2} \sim d(z, \Lambda_\alpha)$, where $d(z, \Lambda_\alpha)$ denotes the Euclidean distance from z to the lattice Λ_α.

Proof. Property (a) follows from the quasiperiodicity of σ; see (1.15) and (1.16). Property (b) then follows from (a) and the fact that each point in Λ_α is a simple zero of σ.   □

As a consequence of condition (b) above, we see that the Weierstrass σ-function associated to Λ_α is of order 2 and of type $\alpha/2$.

1.4 Pseudodifferential Operators

One of the tools we will employ in Chap. 6 when we study Toeplitz operators is the notion of pseudodifferential operators. More specifically, Toeplitz operators on the Fock space are unitarily equivalent to a class of pseudodifferential operators on $L^2(\mathbb{R})$. In this section, we introduce the concept of pseudodifferential operators on the real line and collect several results in this area that will be needed later. The references for this section are Folland's books [92] and [93].

We begin with two well-known operators D and X defined on the space of smooth functions on $\mathbb{R}$ by

$$Xf(x) = xf(x), \qquad Df(x) = \frac{1}{2\alpha i}f'(x), \tag{1.17}$$

where α is any fixed positive constant. The introduction of a parameter α at this point will facilitate and simplify our computations later in association with the Fock spaces. The number $h = \pi/\alpha$ plays the role of Planck's constant in quantum physics.

It is easy to verify that, as densely defined unbounded operators on $L^2(\mathbb{R}, dx)$, both D and X are self-adjoint. This is an easy consequence of integration by parts. The operators

$$Z = X + iD, \qquad Z^* = X - iD, \tag{1.18}$$

will also be useful in our discussions.

If f is a sufficiently good function on $\mathbb{R}$, it is clear how to define $f(D)$ and $f(X)$, respectively. For example, if $f(x) = \sum a_k x^k$ is a polynomial, then

$$f(D) = \sum a_k D^k, \quad f(X) = \sum a_k X^k$$

are perfectly and naturally defined. This easily extends to a large class of symbol functions f. What results in are symbol calculi for the self-adjoint operators D and X.

The notion of pseudodifferential operators arises when we try to establish a symbol calculus for the pair of operators D and X. In other words, if we are given a good function $f(\zeta, x)$ on $\mathbb{R} \times \mathbb{R}$, we wish to define an operator $f(D, X)$ in a natural way. If $f = a\zeta + bx$ is linear, obviously we should just define $f(D, X) = aD + bX$. But we already run into problems when f is just a second-degree polynomial, say

$$f(\zeta, x) = \zeta x = x\zeta,$$

because now we have two natural choices,

$$f(D, X) = DX \quad \text{or} \quad f(D, X) = XD.$$

The operators D and X do not commute, so the two products above are not equal. In fact, it is easy to verify the following commutation relation:

$$[D,X] = DX - XD = \frac{1}{2\alpha i}I, \qquad (1.19)$$

where I is the identity operator.

If $f(\zeta,x)$ is a polynomial in ζ and x, then there are several canonical ways to define $f(D,X)$. For example, if we want the differentiations to come before any multiplication, then we write

$$f(\zeta,x) = \sum a_{mn}x^m\zeta^n$$

and define

$$f(D,X) = \sum a_{mn}X^mD^n.$$

Similarly, if we want to perform multiplications before differentiations, then we write

$$f(\zeta,x) = \sum a_{mn}\zeta^nx^m$$

and define

$$f(D,X) = \sum a_{mn}D^nX^m.$$

Again, the resulting operators are generally different.

It is also possible to carry out the above constructions using the operators Z and Z^* from (1.18) and think of a function of two real variables as depending on z and $\bar{z}$. More specifically, if

$$\sigma(z,\bar{z}) = \sum a_{mn}z^n\bar{z}^m = \sum a_{mn}\bar{z}^mz^n$$

is a polynomial in z and $\bar{z}$, then we can define

$$\sigma_w(Z^*,Z) = \sum a_{mn}Z^{*m}Z^n, \quad \sigma_{aw}(Z,Z^*) = \sum a_{mn}Z^nZ^{*m}. \qquad (1.20)$$

The functional calculi defined this way are called Wick and anti-Wick correspondences. They have been studied extensively in analysis and mathematical physics. There is another important functional calculus for D and X, the John–Nirenberg correspondence, which is especially important in partial differential equations.

We will not pursue any of the above correspondences. Instead, we focus on the so-called *Weyl pseudodifferential operators*. This approach depends on a particular, but natural, choice for the definition of $\sigma(D,X)$ when $\sigma(\zeta,x) = e^{2\pi i(p\zeta+qx)}$, where p and q are real constants. Once this is done, the definition of $\sigma(D,X)$ for more general symbol functions $\sigma(\zeta,x)$ can be given with the help of Fourier and inverse Fourier transforms.

Definition 1.22. For any real coefficients p and q, we define

$$e^{2\alpha i(pD+qX)}f(x) = e^{2\alpha iqx+\alpha ipq}f(x+p),\tag{1.21}$$

or, equivalently,

$$e^{2\pi i(pD+qX)}f(x) = e^{2\pi iqx+\frac{\pi^2}{\alpha}ipq}f\left(x+\frac{\pi p}{\alpha}\right).\tag{1.22}$$

To see the rationale behind the definition above, let

$$g(x,t) = \left[e^{2\alpha it(pD+qX)}f\right](x)$$

denote the formal solution to the differential equation

$$\frac{\partial g}{\partial t} = 2\alpha i(pD+qX)g,\tag{1.23}$$

subject to the initial condition $g(x,0) = f(x)$. Rewrite the equation in (1.23) as

$$\frac{\partial g}{\partial t} - p\frac{\partial g}{\partial x} = 2\alpha iqxg,\tag{1.24}$$

and let $G(t) = g(x(t),t)$ with $x(t) = x - pt$. Then by the chain rule,

$$G'(t) = \frac{\partial g}{\partial t} - p\frac{\partial g}{\partial x},$$

so $G(t)$ satisfies the following equations:

$$G'(t) = 2\alpha iq(x-pt)G(t), \quad G(0) = f(x).$$

It is elementary to solve the above equation and obtain

$$G(t) = f(x)e^{2\alpha qixt-\alpha it^2 pq}.$$

Let $t = 1$. We have

$$g(x-p,1) = f(x)e^{2\alpha iqx-\alpha ipq}.$$

Replace x by $x+p$. We arrive at

$$e^{2\alpha i(pD+qX)}f(x) = g(x,1) = e^{2\alpha iqx+\alpha ipq}f(x+p).$$

This gives a justification for the definition in (1.21).

More generally, if $\sigma(\zeta,x)$ is regular enough so that we can perform the Fourier and inverse Fourier transforms on it, then

$$\sigma(\zeta,x) = \int_{\mathbb{R}} \int_{\mathbb{R}} \widehat{\sigma}(p,q) e^{2\pi i(p\zeta+qx)} \, dp \, dq, \tag{1.25}$$

and we define

$$\sigma(D,X) = \int_{\mathbb{R}} \int_{\mathbb{R}} \widehat{\sigma}(p,q) e^{2\pi i(pD+qX)} \, dp \, dq. \tag{1.26}$$

Here, the integral is an ordinary Bochner integral whenever $\widehat{\sigma}$, the Fourier transform of σ, is in $L^1(\mathbb{R} \times \mathbb{R})$.

Theorem 1.23. *If $\sigma(\zeta,x)$ and $f(x)$ are regular enough, then we have*

$$\sigma(D,X)f(x) = \frac{\alpha}{\pi} \int_{\mathbb{R}} \int_{\mathbb{R}} \sigma\left(\zeta, \frac{x+y}{2}\right) e^{2\alpha i(x-y)\zeta} f(y) \, dy \, d\zeta. \tag{1.27}$$

Proof. The Fourier inversion formula

$$\int_{\mathbb{R}} \int_{\mathbb{R}} e^{2\pi i(u-v)\zeta} f(v) \, dv \, d\zeta = f(u)$$

can be expressed in the language of distributions as

$$\int_{\mathbb{R}} e^{2\pi i x\zeta} \, d\zeta = \delta(x), \tag{1.28}$$

where $\delta(x)$ is classical δ-function. Therefore,

$$\begin{aligned}
\sigma(D,X) & f(x) \\
&= \int_{\mathbb{R}} \int_{\mathbb{R}} \widehat{\sigma}(p,q) e^{2\pi i(pD+qX)} f(x) \, dp \, dq \\
&= \int_{\mathbb{R}} \int_{\mathbb{R}} \widehat{\sigma}(p,q) f\left(x + \frac{\pi p}{\alpha}\right) e^{2\pi i qx + \frac{\pi^2 pqi}{\alpha}} \, dp \, dq \\
&= \int_{\mathbb{R}} \int_{\mathbb{R}} \int_{\mathbb{R}} \int_{\mathbb{R}} \sigma(\zeta,w) e^{-2\pi i(p\zeta+qw)} e^{2\pi i qx + \frac{\pi^2 pq}{\alpha}} f\left(x + \frac{\pi p}{\alpha}\right) \, dp \, dq \, d\zeta \, dw \\
&= \int_{\mathbb{R}} \int_{\mathbb{R}} \int_{\mathbb{R}} \sigma(\zeta,w) \delta\left(x - w + \frac{\pi p}{2\alpha}\right) e^{-2\pi i p\zeta} f\left(x + \frac{\pi p}{\alpha}\right) \, dp \, d\zeta \, dw \\
&= \int_{\mathbb{R}} \int_{\mathbb{R}} \sigma\left(\zeta, x + \frac{\pi p}{2\alpha}\right) e^{-2\pi i p\zeta} f\left(x + \frac{\pi p}{\alpha}\right) \, dp \, d\zeta \\
&= \frac{\alpha}{\pi} \int_{\mathbb{R}} \int_{\mathbb{R}} \sigma\left(\zeta, \frac{1}{2}(x+y)\right) e^{-2\alpha i(y-x)\zeta} f(y) \, dy \, d\zeta,
\end{aligned}$$

which is the desired formula. $\square$

It is thus also natural to simply take (1.27) as the definition of the Weyl pseudodifferential operator $\sigma(D,X)$. We remind the reader that there is a positive parameter α built into our definition of pseudodifferential operators. To see the precise relationship between our rescaled $\sigma(D,X)$ and the classical pseudodifferential operators (as defined in Folland's book [92], for example), we change variables and rewrite (1.27) as follows:

$$\sigma(D,X)f_{1/r}(rx) = \int_{\mathbb{R}} \int_{\mathbb{R}} \sigma_r\left(\zeta, \frac{x+y}{2}\right) e^{2\pi i(x-y)\zeta} f(y)\, dy\, d\zeta, \qquad (1.29)$$

where $r = \sqrt{\pi/\alpha}$. Here, $f_r(x) = f(rx)$ denotes the dilation of f by a positive number r. The integral on the right-hand side of (1.29) is the classical definition of the Weyl pseudodifferential operator with symbol σ_r.

The results in the three theorems below are all invariant under dilation. Therefore, our rescaling does not alter the validity of these classical results.

The pseudodifferential operator $\sigma(D,X)$ is so far only loosely defined. If σ is sufficiently regular and f is compactly supported on $\mathbb{R}$, then the integral in (1.27) converges. For general σ, the integral in (1.27) may or may not converge, and the definition of $\sigma(D,X)f$ may only be defined for f in a certain class. Our main concern here is the following problem: for which functions σ can the pseudodifferential operator $\sigma(D,X)$ be extended to a bounded or compact operator on $L^2(\mathbb{R}, dx)$?

Theorem 1.24. *Suppose $\sigma(\zeta,x)$ is a function on $\mathbb{R} \times \mathbb{R}$ of class C^3 and there exists a positive constant C such that*

$$\sum_{n+m\leq 3} \left| \frac{\partial^{n+m}\sigma}{\partial \zeta^n \partial x^m}(\zeta,x) \right| \leq C$$

for all ζ and x in $\mathbb{R}$. Then the pseudodifferential operator $\sigma(D,X)$ is bounded on $L^2(\mathbb{R}, dx)$.

The above result is usually referred to as the *Calderón–Vaillancourt theorem.* Let $C_0(\mathbb{C}) = C_0(\mathbb{R} \times \mathbb{R})$ be the space continuous functions f on $\mathbb{C} = \mathbb{R} \times \mathbb{R}$ such that $f(z) \to 0$ as $z \to \infty$. The following is the compactness version of the Calderón–Vaillancourt theorem.

Theorem 1.25. *Suppose $\sigma(\zeta,x)$ is a function on $\mathbb{R} \times \mathbb{R}$ of class C^3 and*

$$\frac{\partial^{n+m}\sigma}{\partial \zeta^n \partial x^m} \in C_0(\mathbb{R} \times \mathbb{R})$$

for every pair of nonnegative integers m and n with $n+m \leq 3$. Then the pseudodifferential operator $\sigma(D,X)$ is compact on $L^2(\mathbb{R}, dx)$.

There is also a result concerning membership of the pseudodifferential operators $\sigma(D,X)$ in Schatten classes. We refer the reader to [250] for a brief discussion of Schatten class operators on a Hilbert space.

Theorem 1.26. *Suppose* $1 \leq p < \infty$ *and there exists a positive constant* $k = k(p)$ *such that*

$$\frac{\partial^{n+m}\sigma}{\partial\zeta^n\partial x^m} \in L^p(\mathbb{R} \times \mathbb{R}, dxd\zeta)$$

for all nonnegative integers m *and* n *with* $n + m \leq k$. *Then the pseudodifferential operator* $\sigma(D,X)$ *belongs to the Schatten class* S_p.

1.5 The Heisenberg Group

Although we will not use the Heisenberg group in a critical way anywhere in the book, it is interesting to show how it fits nicely in the theory of Fock spaces. In this brief section, we give its definition and produce a unitary representation based on pseudodifferential operators.

The Heisenberg group $\mathbb{H}$ is the set $\mathbb{C} \times \mathbb{R}$ (or $\mathbb{R}^2 \times \mathbb{R}$) with the following group operation:

$$(z,s) \oplus (w,t) = (z+w, s+t - \text{Im}\,(z\overline{w})),$$

where z and w are complex and s and t are real.

More generally, if n is any positive integer, the Heisenberg group $\mathbb{H}_n$ is the set $\mathbb{C}^n \times \mathbb{R}$ with the group operation

$$(z,s) \oplus (w,t) = (z+w, s+t - \text{Im}\,(\langle z, w \rangle)),$$

where $z = (z_1, \cdots, z_n)$, $w = (w_1, \cdots, w_n)$, and

$$\langle z, w \rangle = z_1 \overline{w}_1 + \cdots + z_n \overline{w}_n.$$

There is a natural representation of the Heisenberg group as unitary operators on the Hilbert space $L^2(\mathbb{R}, dx)$. To simplify notation, let us write

$$\rho(p,q) = e^{2\alpha i(pD + qX)}$$

for real p and q.

Lemma 1.27. *We have*

$$\rho(p_1, q_1)\rho(p_2, q_2) = e^{\alpha i(p_1 q_2 - p_2 q_1)}\rho(p_1 + p_2, q_1 + q_2)$$

for all real numbers p_1, q_1, p_2, and q_2.

Proof. This follows directly from the definition of $\rho(p,q)$ in (1.21). Details are left to the reader. □

Lemma 1.28. *We have*

$$\rho(p_1, q_1)\rho(p_2, q_2) = e^{2\alpha i(p_1 q_2 - p_2 q_1)}\rho(p_2, q_2)\rho(p_1, q_1)$$

for all real numbers p_1, q_1, p_2, and q_2.

Proof. This is a direct consequence of Lemma 1.27. □

Theorem 1.29. *Suppose α is any positive parameter and pseudodifferential operators are defined as in the previous section. For any real p and q, the pseudodif-*

ferential operator $e^{2\alpha i(pD+qX)}$ *is a unitary operator on* $L^2(\mathbb{R}, dx)$. *Furthermore, the mapping*

$$(p+iq,t) \mapsto u(p+iq,t) =: e^{\alpha it}e^{2\alpha i(pD+qX)}$$

is a unitary representation of the Heisenberg group $\mathbb{H}$ *on* $L^2(\mathbb{R}, dx)$.

Proof. By (1.21), the action of each $u(z,t)$ on $L^2(\mathbb{R}, dx)$, where $z \in \mathbb{C}$ and $t \in \mathbb{R}$, is a unimodular constant times a certain translation of $\mathbb{R}$. Since any translation of $\mathbb{R}$ is a unitary operator on $L^2(\mathbb{R}, dx)$, we see that each $u(p+iq,t)$ is a unitary operator on $L^2(\mathbb{R}, dx)$.

Let $z_1 = p_1 + iq_1$ and $z_2 = p_2 + iq_2$. It follows from Lemma 1.27 that

$$u(z_1,t_1)u(z_2,t_2) = e^{\alpha i(t_1+t_2)}\rho(p_1,q_1)\rho(p_2,q_2)$$
$$= e^{\alpha i(t_1+t_2+p_1q_2-p_2q_1)}\rho(p_1+p_2,q_1+q_2)$$
$$= u(z_1+z_2,t_1+t_2-\operatorname{Im}(z_1\bar{z}_2)).$$

This shows that $u(z,t)$ preserves the group operation in the Heisenberg group $\mathbb{H}$.   □

The mapping $u(z,t)$ is called the Schrödinger representation of the Heisenberg group $\mathbb{H}$ on $L^2(\mathbb{R})$. In the next chapter, we will obtain another representation of $\mathbb{H}$, a unitary representation on the Fock space based on weighted translations.

1.6 Notes

The results in the first section, except Lindelöf's theorem, are all well known and can be found in any elementary complex analysis book. In particular, these results can all be found in Conway's book [67].

Lindelöf's theorem will be needed in Chap. 3 when we study zero sequences for Fock spaces. This is probably not a result that can be found in elementary texts. See 2.10.1 of Boas' book [38] for a detailed proof of this result.

The section about lattices in the complex plane is completely elementary. Whenever we really use lattices later on, we restrict our attention to square lattices, although many arguments can easily be adapted to arbitrary lattices, even to sequences that behave like lattices. Perhaps Lemma 1.15 looks peculiar to the reader, but it is critical for the study of Hankel operators in Chap. 8.

Pseudodifferential operators constitute an important subject by itself, and there is extensive literature about them. Of course, we have only touched the surface of this vast area of modern analysis. The connection between pseudodifferential operators and Toeplitz operators on the Fock space is both fascinating and useful. Because of this connection, the study of Toeplitz operators on the Fock space becomes especially interesting and fruitful. In particular, this provides us with extra and unique tools to study Toeplitz operators on the Fock space as opposed to Toeplitz operators on the Hardy and Bergman spaces.

Our presentation in Sect. 1.4 follows Folland's books [92, 93] very closely. A slight modification is made in the definition of pseudodifferential operators here in order to incorporate the weight parameter α into everything. Note that the proof of Theorem 1.23 depends on certain elementary facts from Fourier analysis that we are taking for granted. It should be easy for the interested reader to make the arguments completely rigorous.

The Heisenberg group appears very naturally in many different areas, including Fourier analysis, harmonic analysis, and mathematical physics. The Heisenberg group shows up in this book when we study the action of translations on Fock spaces. Although it is possible for us to avoid the Heisenberg group, we thought it is nice to put things in the right context.

The Weierstrass σ-functions provide a family of examples that will be very useful to us later on when we study zero sets, interpolating sets, and sampling sets. The book [241] contains much more information about the Weierstrass σ-functions as well as several other important classes of entire and meromorphic functions.

1.7 Exercises

1. Suppose f is entire, $f(z) \neq 0$ for some $z \in \mathbb{C}$, and $|z| < r$. Then $\log|f(z)|$ is equal to

$$-\sum_{k=1}^{N} \log \left| \frac{r^2 - \bar{z}_k z}{r(z - z_k)} \right| + \frac{1}{2\pi} \int_0^{2\pi} \mathrm{Re} \left(\frac{re^{i\theta} + z}{re^{i\theta} - z} \right) \log|f(re^{i\theta})| \, d\theta,$$

where $\{z_1, \cdots, z_N\}$ are the zeros of f in $0 < |w| < r$.

2. If u is a bounded (complex-valued) harmonic function on the entire complex plane, then u must be constant.

3. Show that

$$\sum \left\{ \frac{1}{(n^2 + m^2)^p} : n \in \mathbb{Z}, m \in \mathbb{Z} \right\} < \infty$$

if and only if $p > 1$.

4. Suppose $r\mathbb{Z}^2 = \{\omega_{mn}\}$ is any square lattice and R is any other positive radius. Show that there exists a positive constant $C = C(r, R)$ such that

$$\sum_{m,n} \int_{|z - \omega_{mn}| < R} f(z) \, dA(z) \leq C \int_{\mathbb{C}} f(z) \, dA(z)$$

for all nonnegative functions f on $\mathbb{C}$. Here dA is area measure.

5. Verify that $\mathbb{H}$ with the operation defined in Sect. 1.5 is indeed a group.

6. Show that the Heisenberg group is nonabelian.

7. Show that $\rho_1 \leq \rho$. See Sect. 1.1 for definitions of these numbers.

8. Suppose f is entire, $0 < p < \infty$, and $0 < R < \infty$. Show that

$$\int_{|z| > R} |f(z)|^p \, dA(z) < \infty$$

if and only if f is identically zero.

9. Show that both X and D are self-adjoint operators on $L^2(\mathbb{R}, dx)$.

10. Justify every interchange of the order of integration in the proof of Theorem 1.23.

11. Discuss the continuity of the Schrödinger representation, namely, the unitary representation of the Heisenberg group given in Theorem 1.29.

12. Show that for any lattice $\Lambda = \{\omega_{mn}\}$, we have

$$\sum_{m,n} \frac{1}{|\omega_{mn}|^p} < \infty$$

if and only if $p > 2$, where the summation is to exclude the possible occurrence of 0 in the denominator.

13. Prove the commutation relation (1.19).

14. Convince yourself that the formal identity (1.28) is equivalent to the Fourier inversion formula.

Chapter 2
Fock Spaces

In this chapter, we define Fock spaces and prove basic properties about them. The following topics are covered in this chapter: reproducing kernel, integral representation, duality, complex interpolation, atomic decomposition, translation invariance, and a version of the maximum modulus principle.

K. Zhu, *Analysis on Fock Spaces*, Graduate Texts in Mathematics 263,
DOI 10.1007/978-1-4419-8801-0_2,
© Springer Science+Business Media New York 2012

2.1 Basic Properties

For any positive parameter α, we consider the Gaussian measure

$$d\lambda_\alpha(z) = \frac{\alpha}{\pi} e^{-\alpha|z|^2} \, dA(z),$$

where dA is the Euclidean area measure on the complex plane. A calculation with polar coordinates shows that $d\lambda_\alpha$ is a probability measure.

The Fock space F_α^2 consists of all entire functions f in $L^2(\mathbb{C}, d\lambda_\alpha)$. It is easy to show that F_α^2 is a closed subspace of $L^2(\mathbb{C}, d\lambda_\alpha)$. Consequently, F_α^2 is a Hilbert space with the following inner product inherited from $L^2(\mathbb{C}, d\lambda_\alpha)$:

$$\langle f, g \rangle_\alpha = \int_{\mathbb{C}} f(z) \overline{g(z)} \, d\lambda_\alpha(z).$$

Proposition 2.1. *For any nonnegative integer n, let*

$$e_n(z) = \sqrt{\frac{\alpha^n}{n!}} \, z^n.$$

Then the set $\{e_n\}$ is an orthonormal basis for F_α^2.

Proof. A calculation with polar coordinates shows that $\{e_n\}$ is an orthonormal set. Given $f \in F_\alpha^2$ and $n \geq 0$, we have

$$\langle f, e_n \rangle_\alpha = \lim_{R \to \infty} \int_{|z|<R} f(z) \overline{e_n(z)} \, d\lambda_\alpha(z).$$

Since the Taylor series

$$f(z) = \sum_{k=0}^{\infty} a_k z^k$$

converges uniformly on $|z| < R$, we have

$$\int_{|z|<R} f(z) \overline{e_n(z)} \, d\lambda_\alpha(z) = \sum_{k=0}^{\infty} a_k \int_{|z|<R} z^k \overline{e_n(z)} \, d\lambda_\alpha(z).$$

Using polar coordinates again, we obtain

$$\langle f, e_n \rangle_\alpha = \lim_{R \to \infty} a_n \int_{|z|<R} z^n \overline{e_n(z)} \, d\lambda_\alpha(z) = a_n \int_{\mathbb{C}} z^n \overline{e_n(z)} \, d\lambda_\alpha(z).$$

Therefore, the condition that $\langle f, e_n \rangle = 0$ for all $n \geq 0$ implies that $a_n = 0$ for all $n \geq 0$ which in turn implies that $f = 0$. This shows that the system $\{e_n\}$ is complete in F_α^2. $\qquad\square$

As a consequence of the above proposition, the Taylor series of every function f in F_α^2 converges to f in the norm topology of F_α^2.

For any fixed $w \in \mathbb{C}$, the mapping $f \mapsto f(w)$ is a bounded linear functional on F_α^2. This follows easily from the mean value theorem. By the Riesz representation theorem in functional analysis, there exists a unique function K_w in F_α^2 such that $f(w) = \langle f, K_w \rangle_\alpha$ for all $f \in F_\alpha^2$. The function $K_\alpha(z, w) = K_w(z)$ is called the reproducing kernel of F_α^2.

Proposition 2.2. *The reproducing kernel of F_α^2 is given by*

$$K_\alpha(z, w) = e^{\alpha z \overline{w}}, \qquad z, w \in \mathbb{C}.$$

Proof. For any $f \in F_\alpha^2$, we have

$$f(0) = \langle f, e_0 \rangle_\alpha = \int_{\mathbb{C}} f(z) \, d\lambda_\alpha(z).$$

Fix any $w \in \mathbb{C}$ and replace $f(z)$ by $f(w - z)$. We obtain

$$
\begin{aligned}
f(w) &= \frac{\alpha}{\pi} \int_{\mathbb{C}} f(w - z) e^{-\alpha|z|^2} \, dA(z) \\
&= \frac{\alpha}{\pi} \int_{\mathbb{C}} f(z) e^{-\alpha|z - w|^2} \, dA(z) \\
&= e^{-\alpha|w|^2} \int_{\mathbb{C}} f(z) e^{\alpha z \overline{w} + \alpha \overline{z} w} \, d\lambda_\alpha(z).
\end{aligned}
$$

Replace $f(z)$ by $f(z) e^{-\alpha z \overline{w}}$. The result is

$$f(w) = \int_{\mathbb{C}} f(z) e^{\alpha \overline{z} w} \, d\lambda_\alpha(z).$$

The desired result then follows from the uniqueness in Riesz representation.   □

Recall that every closed subspace X of a Hilbert space H uniquely determines an orthogonal projection $P : H \to X$.

Corollary 2.3. *The orthogonal projection*

$$P_\alpha : L^2(\mathbb{C}, d\lambda_\alpha) \to F_\alpha^2$$

is an integral operator. More specifically,

$$P_\alpha f(z) = \int_{\mathbb{C}} K_\alpha(z, w) f(w) \, d\lambda_\alpha(w)$$

for all $f \in L^2(\mathbb{C}, d\lambda_\alpha)$ and all $z \in \mathbb{C}$.

Proof. Fix $f \in L^2(\mathbb{C}, d\lambda_\alpha)$ and $z \in \mathbb{C}$. We have

$$P_\alpha f(z) = \langle P_\alpha f, K_z \rangle_\alpha = \langle f, P_\alpha K_z \rangle_\alpha = \langle f, K_z \rangle_\alpha$$

$$= \int_{\mathbb{C}} f(w) K_\alpha(z, w) \, d\lambda_\alpha(z).$$

This proves the integral representation for P_α.   □

For any $z \in \mathbb{C}$, we let

$$k_z(w) = \frac{K_\alpha(w, z)}{\sqrt{K_\alpha(z, z)}} = e^{\alpha \bar{z} w - \frac{\alpha}{2}|z|^2}$$

denote the normalized reproducing kernel at z. Each k_z is a unit vector in F_α^2. The following change of variables formula will be used many times later in the book.

Corollary 2.4. *Suppose $f \geq 0$ or $f \in L^1(\mathbb{C}, d\lambda_\alpha)$. Then for any $z \in \mathbb{C}$, we have*

$$\int_{\mathbb{C}} f(z \pm w) \, d\lambda_\alpha(w) = \int_{\mathbb{C}} f(w) |k_z(w)|^2 \, d\lambda_\alpha(w),$$

and

$$\int_{\mathbb{C}} f[\pm(z - w)] |k_z(w)|^2 \, d\lambda_\alpha(w) = \int_{\mathbb{C}} f(w) \, d\lambda_\alpha(w).$$

Proof. It is clear that

$$\int_{\mathbb{C}} f(z \pm w) \, d\lambda_\alpha(w) = \frac{\alpha}{\pi} \int_{\mathbb{C}} f(z \pm w) e^{-\alpha|w|^2} \, dA(w)$$

$$= \frac{\alpha}{\pi} \int_{\mathbb{C}} f(w) e^{-\alpha|z - w|^2} \, dA(w)$$

$$= \int_{\mathbb{C}} f(w) e^{-\alpha|z|^2 + \alpha \bar{z} w + \alpha z \bar{w}} \, d\lambda_\alpha(w)$$

$$= \int_{\mathbb{C}} f(w) |k_z(w)|^2 \, d\lambda_\alpha(w).$$

The assumption that $f \geq 0$ or $f \in L^1(\mathbb{C}, d\lambda_\alpha)$ ensures that all integrals above make sense. The proof of the other identity is similar.   □

Corollary 2.5. *Suppose $\alpha > 0$ and β is real. Then*

$$\int_{\mathbb{C}} \left| e^{\beta z \bar{a}} \right| \, d\lambda_\alpha(z) = e^{\beta^2 |a|^2 / 4\alpha}$$

for all $a \in \mathbb{C}$.

Proof. It follows from the definition of the reproducing kernel that

$$K_\alpha(a,a) = \int_{\mathbb{C}} |K_\alpha(a,z)|^2 \, d\lambda_\alpha(z), \qquad a \in \mathbb{C}.$$

Replacing a by $\beta a/(2\alpha)$, we obtain the desired result. $\qquad\square$

For $\alpha > 0$ and $p > 0$, we use the notation L_α^p to denote the space of Lebesgue measurable functions f on $\mathbb{C}$ such that the function $f(z)e^{-\alpha|z|^2/2}$ is in $L^p(\mathbb{C}, dA)$. For $f \in L_\alpha^p$, we write

$$\|f\|_{p,\alpha}^p = \frac{p\alpha}{2\pi} \int_{\mathbb{C}} \left| f(z) e^{-\frac{\alpha}{2}|z|^2} \right|^p \, dA(z). \tag{2.1}$$

Similarly, for $\alpha > 0$ and $p = \infty$, we use the notation L_α^∞ to denote the space of Lebesgue measurable functions f on $\mathbb{C}$ such that

$$\|f\|_{\infty,\alpha} = \operatorname{esssup}\left\{ |f(z)| e^{-\alpha|z|^2/2} : z \in \mathbb{C} \right\} < \infty. \tag{2.2}$$

Obviously, we have $L_\alpha^p = L^p(\mathbb{C}, d\lambda_{p\alpha/2})$ for $0 < p < \infty$. But $L_\alpha^\infty \neq L^\infty(\mathbb{C}, dA)$. When $1 \le p \le \infty$, L_α^p is a Banach space with the norm $\|f\|_{p,\alpha}$. When $0 < p < 1$, L_α^p is a complete metric space with the distance $d(f,g) = \|f-g\|_{p,\alpha}^p$.

For $\alpha > 0$ and $0 < p \le \infty$ we let F_α^p denote the space of entire functions in L_α^p. We will call F_α^p Fock spaces. It is elementary to show that F_α^p is closed in L_α^p. Therefore, F_α^p is a Banach space when $1 \le p \le \infty$, and it is a complete metric space when $0 < p < 1$.

Note that the measure associated with the Fock space F_α^p, $d\lambda_{p\alpha/2}$, depends on both α and p. This is a bit unusual and unnatural at first glance, but there are underlying reasons why Fock spaces should be defined this way, and plenty of past experience suggests that this way of defining the Fock spaces will make the statement of many results a lot easier and a lot more natural.

Lemma 2.6. *Suppose $\alpha > 0$, $\zeta \in \mathbb{C} - \{0\}$, and $0 < p \le \infty$. Then the dilation operator $f(z) \mapsto f(\zeta z)$ is an isometry from L_α^p onto $L_{|\zeta|^2\alpha}^p$, and it is an isometry from F_α^p onto $F_{|\zeta|^2\alpha}^p$.*

Proof. This follows from a simple change of variables. $\qquad\square$

The following result gives the optimal rate of growth for functions in Fock spaces.

Theorem 2.7. *For any $0 < p \le \infty$ and $z \in \mathbb{C}$, we have*

$$\sup\{|f(z)| : \|f\|_{p,\alpha} \le 1\} = e^{\alpha|z|^2/2}.$$

Furthermore, for $0 < p < \infty$, any extremal function is of the form:

$$f(w) = e^{\alpha \bar{z}w - \frac{\alpha}{2}|z|^2 + i\theta},$$

where θ is a real number.

Proof. We first assume that $0 < p < \infty$.

The case $z = 0$ follows from the subharmonicity of the function $|f|^p$ and integration in polar coordinates:

$$|f(0)|^p \le \frac{p\alpha}{2\pi} \int_{\mathbb{C}} \left| f(w) e^{-\frac{\alpha|w|^2}{2}} \right|^p \, dA(w) = \|f\|_{p,\alpha}^p.$$

Equality occurs if and only if f is constant.

More generally, for any $z \in \mathbb{C}$ and $f \in F_\alpha^p$, we consider the function

$$F(w) = f(z - w) e^{\alpha w \bar{z} - (\alpha|z|^2/2)}.$$

From the inequality $|F(0)|^p \le \|F\|_{p,\alpha}^p$ we deduce that

$$
\begin{aligned}
|f(z)|^p e^{-\alpha p|z|^2/2} &\le \frac{p\alpha}{2\pi} \int_{\mathbb{C}} \left| f(z-w) e^{\alpha \bar{z}w} e^{-\alpha|z|^2/2} e^{-\alpha|w|^2/2} \right|^p \, dA(w) \\
&= \frac{p\alpha}{2\pi} \int_{\mathbb{C}} \left| f(z-w) e^{-\alpha|z-w|^2/2} \right|^p \, dA(w) \\
&= \frac{p\alpha}{2\pi} \int_{\mathbb{C}} \left| f(w) e^{-\alpha|w|^2/2} \right|^p \, dA(w) \\
&= \|f\|_{p,\alpha}^p.
\end{aligned}
$$

This shows that

$$|f(z)| \le \|f\|_{p,\alpha} e^{\alpha|z|^2/2}.$$

Furthermore, equality is attained if and only if F is constant. This shows that the extremal functions are of the form

$$f(w) = e^{\alpha \bar{z}w - (\alpha|z|^2/2) + i\theta}.$$

This proves the desired results for $0 < p < \infty$.

If $p = \infty$, it follows from the definition of $\|f\|_{\infty,\alpha}$ that $|f(z)| \le e^{\alpha|z|^2/2}$ for all f with $\|f\|_{\infty,\alpha} \le 1$. Therefore,

$$\sup\{|f(z)| : \|f\|_{\infty,\alpha} \le 1\} \le e^{\alpha|z|^2/2}.$$

On the other hand, the function $f(w) = k_z(w)$ is a unit vector in F_α^∞ and $k_z(z) = e^{\alpha|z|^2/2}$. Thus, we actually have

$$\sup\{|f(z)| : \|f\|_{\infty,\alpha} \le 1\} = e^{\alpha|z|^2/2}.$$

This proves the case for $p = \infty$. $\qquad\qquad\qquad\qquad\qquad\qquad\qquad\qquad\qquad\square$

When $p = \infty$, the extremal functions in Theorem 2.7 consist of more than constant multiples of reproducing kernels. For example, if f is any polynomial normalized so that $\|f\|_{\infty,\alpha} = 1$, then

$$1 = \sup_{z\in\mathbb{C}}|f(z)|e^{-\alpha|z|^2/2} = |f(z_0)|e^{-\alpha|z_0|^2/2}$$

for some $z_0 \in \mathbb{C}$ because in this case we have

$$\lim_{z\to\infty} f(z)e^{-\alpha|z|^2/2} = 0.$$

Therefore, this polynomial f is an extremal function for the extremal problem in Theorem 2.7 when $p = \infty$ and $z = z_0$.

Corollary 2.8. *Let $f \in F_\alpha^p$ and $0 < p \le \infty$. Then*

$$|f(z)| \le \|f\|_{p,\alpha}e^{\alpha|z|^2/2}$$

for all $z \in \mathbb{C}$ and the estimate is sharp.

When $0 < p < \infty$, the estimate above can be somewhat improved. More specifically, we can actually show that

$$\lim_{z\to\infty} f(z)e^{-\alpha|z|^2/2} = 0$$

for every function $f \in F_\alpha^p$. This will follow from the next proposition.

Proposition 2.9. *Suppose $0 < p < \infty$, $f \in F_\alpha^p$, and $f_r(z) = f(rz)$. Then:*

(a) $\|f_r - f\|_{p,\alpha} \to 0$ as $r \to 1^-$.
(b) There is a sequence $\{p_n\}$ of polynomials such that $\|p_n - f\|_{p,\alpha} \to 0$ as $n \to \infty$.

Proof. Suppose $\{g_n\}$ and g are functions in $L^p(X, d\mu)$ such that

$$g_n(x) \to g(x), \qquad n \to \infty,$$

almost everywhere. Then it is well known that

$$\lim_{n\to\infty} \int_X |g_n - g|^p\,d\mu = 0$$

if and only if

$$\lim_{n\to\infty} \int_X |g_n|^p \, d\mu = \int_X |g|^p \, d\mu.$$

This is a simple consequence of Fatou's lemma; see Lemma 3.17 of [119] for example. Given $f \in F_\alpha^p$, we have

$$\|f_r\|_{p,\alpha}^p = \frac{p\alpha}{2\pi} \int_{\mathbb{C}} \left| f(rz)e^{-\alpha|z|^2/2} \right|^p \, dA(z)$$

$$= \frac{p\alpha}{2\pi r^2} \int_{\mathbb{C}} \left| f(z)e^{-\alpha|z|^2/2} \right|^p e^{-p\alpha|z|^2(r^{-2}-1)/2} \, dA(z).$$

Since

$$e^{-p\alpha|z|^2(r^{-2}-1)/2} \le 1$$

for all $z \in \mathbb{C}$ and $0 < r < 1$, an application of the dominated convergence theorem shows that $\|f_r\|_{p,\alpha} \to \|f\|_{p,\alpha}$, and hence $\|f_r - f\|_{p,\alpha} \to 0$ as $r \to 1^-$. This proves part (a).

Part (b) follows from part (a) if we can show that for every $r \in (0,1)$, the function f_r can be approximated by its Taylor polynomials in the norm topology of F_α^p. To this end, we fix some $r \in (0,1)$ and fix some $\beta \in (r^2\alpha, \alpha)$. It follows from Corollary 2.8 that $f_r \in F_\beta^2$. Similarly, it follows from Corollary 2.8 that $F_\beta^2 \subset F_\alpha^p$ and there exists a positive constant C such that $\|g\|_{p,\alpha} \le C\|g\|_{2,\beta}$ for all $g \in F_\beta^2$. Now, if p_n is the nth Taylor polynomial of f_r, then by Proposition 2.1,

$$\|f_r - p_n\|_{p,\alpha} \le C\|f_r - p_n\|_{2,\beta} \to 0$$

as $n \to \infty$. This proves part (b). $\qquad\square$

Let f_α^∞ denote the space of entire functions $f(z)$ such that

$$\lim_{z\to\infty} f(z)e^{-\alpha|z|^2/2} = 0.$$

Obviously, f_α^∞ is a closed subspace of F_α^∞. In fact, f_α^∞ is the closure in F_α^∞ of the set of all polynomials. Thus, the space f_α^∞ is separable while the space F_α^∞ is not.

Theorem 2.10. *If $0 < p < q < \infty$, then $F_\alpha^p \subset F_\alpha^q$, and the inclusion is proper and continuous. Moreover, $F_\alpha^p \subset f_\alpha^\infty$, and the inclusion is proper and continuous.*

Proof. For any entire function f, we consider the integral

$$\|f\|_{q,\alpha}^q = \frac{q\alpha}{2\pi} \int_{\mathbb{C}} |f(z)e^{-\alpha|z|^2/2}|^q \, dA(z).$$

It follows from the pointwise estimate in Corollary 2.8 that

$$\|f\|_{q,\alpha}^q = \frac{q\alpha}{2\pi} \int_{\mathbb{C}} |f(z)|^p |f(z)|^{q-p} e^{-q\alpha|z|^2/2} \, dA(z)$$

$$\leq \frac{q\alpha}{2\pi} \|f\|_{p,\alpha}^{q-p} \int_{\mathbb{C}} |f(z)|^p e^{-p\alpha|z|^2/2} \, dA(z)$$

$$= \frac{q}{p} \|f\|_{p,\alpha}^q.$$

This shows that $F_\alpha^p \subset F_\alpha^q$ with $\|f\|_{q,\alpha} \leq (q/p)^{1/q} \|f\|_{p,\alpha}$ for all $f \in F_\alpha^p$.

To see that the inclusion $F_\alpha^p \subset F_\alpha^q$ is proper, let us assume that $F_\alpha^p = F_\alpha^q$. Then the identity map $I : F_\alpha^p \to F_\alpha^q$ is bounded, one-to-one, and onto. By the open mapping theorem, there must exist a constant $C > 0$ such that

$$C^{-1} \|f\|_{p,\alpha} \leq \|f\|_{q,\alpha} \leq C \|f\|_{p,\alpha}$$

for all $f \in F_\alpha^p$. On the other hand, a computation with Stirling's formula shows that

$$\|z^n\|_{p,\alpha}^p = \alpha p \int_0^\infty r^{np} e^{-p\alpha r^2/2} r \, dr$$

$$= \left(\frac{1}{\alpha p}\right)^{np/2} \Gamma\left(\frac{np}{2} + 1\right)$$

$$\sim \left(\frac{n}{\alpha e}\right)^{np/2} \sqrt{n}.$$

Thus,

$$\|z^n\|_{p,\alpha} \sim \left(\frac{n}{\alpha e}\right)^{n/2} n^{\frac{1}{2p}},$$

and similarly,

$$\|z^n\|_{q,\alpha} \sim \left(\frac{n}{\alpha e}\right)^{n/2} n^{\frac{1}{2q}}.$$

It is then obvious that there is no positive constant C with the property that

$$C^{-1} \|z^n\|_{p,\alpha} \leq \|z^n\|_{q,\alpha} \leq C \|z^n\|_{p,\alpha}$$

for all n. This contradiction shows that the inclusion $F_\alpha^p \subset F_\alpha^q$ must be proper.

To show that $F_\alpha^p \subset f_\alpha^\infty$, observe that for every polynomial f, we have $f \in f_\alpha^\infty$, and it follows from Corollary 2.8 that $\|f\|_{\infty,\alpha} \leq \|f\|_{p,\alpha}$. The desired result then follows from the density of polynomials in F_α^p, the boundedness of the inclusion $F_\alpha^p \subset F_\alpha^\infty$, and the fact that f_α^∞ is closed in F_α^∞.

Finally, by elementary calculations,

$$\|z^n\|_{\infty,\alpha} = \left(\frac{n}{\alpha e}\right)^{n/2}.$$

Another appeal to the open mapping theorem then shows that the inclusion $F_\alpha^p \subset f_\alpha^\infty$ is proper.   □

The next result gives another useful dense subset of F_α^p.

Lemma 2.11. *For any positive parameters α and γ, the set of functions of the form*

$$f(z) = \sum_{k=1}^{n} c_k K_\gamma(z, w_k) = \sum_{k=1}^{n} c_k e^{\gamma z \overline{w}_k},$$

is dense in F_α^p and f_α^∞, where $0 < p < \infty$.

Proof. Since the points w_k are arbitrary, we may assume that $\gamma = \alpha$.

The result is obvious when $p = 2$. In fact, if a function h in F_α^2 is orthogonal to each function $f(z) = K_\alpha(z, w)$, then $h(w) = 0$ for every w.

In general, with the help of Corollary 2.8, we can find a positive parameter β such that $F_\beta^2 \subset F_\alpha^p$ continuously, say $\|f\|_{p,\alpha} \le C\|f\|_{2,\beta}$ for all $f \in F_\beta^2$. In fact, any $\beta \in (0, \alpha)$ works. Now, if f is a polynomial and $\{w_1, \cdots, w_n\}$ are points in the complex plane, then

$$\|f - \sum_{k=1}^{n} c_k K_\alpha(z, w_k)\|_{p,\alpha} \le C\|f - \sum_{k=1}^{n} c_k K_\alpha(z, w_k)\|_{2,\beta}$$

$$= C\|f - \sum_{k=1}^{n} c_k K_\beta(z, \alpha w_k/\beta)\|_{2,\beta}.$$

Combining this with the density of the functions $\sum_{k=1}^{n} c_k K_\beta(z, u_k)$ in F_β^2, we conclude that every polynomial can be approximated in the norm topology of F_α^p by functions of the form $\sum_{k=1}^{n} c_k K_\alpha(z, w_k)$. Since the polynomials are dense in F_α^p, we have proved the result for F_α^p, $0 < p < \infty$.

The proof for f_α^∞ is similar.   □

Finally, in this section, as a consequence of the pointwise estimates, we establish the maximum order and type for functions in the Fock spaces.

Theorem 2.12. *Let $f \in F_\alpha^p$ with $0 < p \le \infty$. Then f is of order less than or equal to 2. When f is of order 2, it must be of type less than or equal to $\alpha/2$.*

Proof. By Corollary 2.8, there exists a positive constant C such that

$$|f(z)| \le C e^{\alpha|z|^2/2}$$

for all $z \in \mathbb{C}$. In particular, $M(r) \leq Ce^{\alpha r^2/2}$ for all $r > 0$. It follows that the order ρ of f satisfies

$$\rho = \limsup_{r \to \infty} \frac{\log \log M(r)}{\log r} \leq 2.$$

Also, if the order of f is actually 2, then its type σ satisfies

$$\sigma = \limsup_{r \to \infty} \frac{\log M(r)}{r^2} \leq \frac{\alpha}{2},$$

completing the proof of the theorem.   □

2.2 Some Integral Operators

In this section, we consider the boundedness of certain integral operators on L^p spaces associated with Gaussian measures. More specifically, for any $\alpha > 0$, we consider the integral operators P_α and Q_α defined by

$$P_\alpha f(z) = \int_{\mathbb{C}} K_\alpha(z,w) f(w) \, d\lambda_\alpha(w), \tag{2.3}$$

and

$$Q_\alpha f(z) = \int_{\mathbb{C}} |K_\alpha(z,w)| f(w) \, d\lambda_\alpha(w), \tag{2.4}$$

respectively.

We need two well-known results from the theory of integral operators. The first one concerns the adjoint of a bounded integral operator.

Lemma 2.13. *Suppose $1 \le p < \infty$ and $1/p + 1/q = 1$. If an integral operator*

$$Tf(x) = \int_X H(x,y) f(y) \, d\mu(y)$$

is bounded on $L^p(X, d\mu)$, then its adjoint

$$T^* : L^q(X, d\mu) \to L^q(X, d\mu)$$

is the integral operator given by

$$T^* f(x) = \int_X \overline{H(y,x)} f(y) \, d\mu(y).$$

Proof. This is a standard result in real analysis. See [113] for example. □

The second result is a useful criterion for the boundedness of integral operators on L^p spaces, which is usually referred to as Schur's test.

Lemma 2.14. *Suppose $H(x,y)$ is a positive kernel and*

$$Tf(x) = \int_X H(x,y) f(y) \, d\mu(y)$$

is the associated integral operator. Let $1 < p < \infty$ with $1/p + 1/q = 1$. If there exist a positive function $h(x)$ and positive constants C_1 and C_2 such that

$$\int_X H(x,y) h(y)^q \, d\mu(y) \le C_1 h(x)^q, \qquad x \in X,$$

and

$$\int_X H(x,y)h(x)^p \, d\mu(x) \le C_2 h(y)^p, \qquad y \in X,$$

then the operator T is bounded on $L^p(X,d\mu)$. Moreover, the norm of T on $L^p(X,d\mu)$ does not exceed $C_1^{1/q} C_2^{1/p}$.

Proof. See [250] for example.　　　　　　　　　　　　　　　　　　□

We now consider the action of the operators P_α and Q_α on the space $L^p(\mathbb{C}, d\lambda_\beta)$. Thus, we fix two positive parameters α and β for the rest of this section and rewrite the integral operators P_α and Q_α as follows:

$$P_\alpha f(z) = \frac{\alpha}{\beta} \int_{\mathbb{C}} e^{\alpha z\bar{w} + \beta|w|^2 - \alpha|w|^2} f(w) \, d\lambda_\beta(w),$$

and

$$Q_\alpha f(z) = \frac{\alpha}{\beta} \int_{\mathbb{C}} |e^{\alpha z\bar{w} + \beta|w|^2 - \alpha|w|^2}| f(w) \, d\lambda_\beta(w).$$

It follows from Lemma 2.13 that the adjoint of P_α and Q_α with respect to the integral pairing

$$\langle f,g \rangle_\beta = \int_{\mathbb{C}} f(z)\overline{g(z)} \, d\lambda_\beta(z)$$

is given, respectively, by

$$P_\alpha^* f(z) = \frac{\alpha}{\beta} e^{(\beta-\alpha)|z|^2} \int_{\mathbb{C}} e^{\alpha z\bar{w}} f(w) \, d\lambda_\beta(w), \tag{2.5}$$

and

$$Q_\alpha^* f(z) = \frac{\alpha}{\beta} e^{(\beta-\alpha)|z|^2} \int_{\mathbb{C}} |e^{\alpha z\bar{w}}| f(w) \, d\lambda_\beta(w). \tag{2.6}$$

We first prove several necessary conditions for the operator P_α to be bounded on $L^p(\mathbb{C}, d\lambda_\beta)$.

Lemma 2.15. *Suppose $0 < p < \infty$, $\alpha > 0$, and $\beta > 0$. If P_α is bounded on $L^p(\mathbb{C}, d\lambda_\beta)$, then $p\alpha \le 2\beta$ and $p \ge 1$.*

Proof. Consider functions of the following form:

$$f_{x,k}(z) = e^{-x|z|^2} z^k, \qquad z \in \mathbb{C},$$

where $x > 0$ and k is a positive integer. We have

$$\int_{\mathbb{C}} |f_{x,k}|^p \, d\lambda_\beta = \frac{\beta}{\pi} \int_{\mathbb{C}} |z|^{pk} e^{-(px+\beta)|z|^2} \, dA(z) = \frac{\beta}{px+\beta} \frac{\Gamma((pk/2)+1)}{(px+\beta)^{pk/2}}.$$

On the other hand, it follows from the reproducing formula in $F^2_{\alpha+x}$ that

$$
\begin{aligned}
P_\alpha(f_{x,k})(z) &= \frac{\alpha}{\pi} \int_{\mathbb{C}} e^{\alpha z \bar{w}} w^k e^{-(\alpha+x)|w|^2} \, dA(w) \\
&= \frac{\alpha}{\alpha+x} \int_{\mathbb{C}} e^{(\alpha+x)[\alpha z/(\alpha+x)\bar{w}]} w^k \, d\lambda_{\alpha+x}(w) \\
&= \frac{\alpha}{\alpha+x} \left(\frac{\alpha z}{\alpha+x} \right)^k = \left(\frac{\alpha}{\alpha+x} \right)^{1+k} z^k.
\end{aligned}
$$

Therefore,

$$
\begin{aligned}
\int_{\mathbb{C}} |P_\alpha(f_{x,k})|^p \, d\lambda_\beta &= \left(\frac{\alpha}{\alpha+x} \right)^{p(1+k)} \int_{\mathbb{C}} |z|^{pk} \, d\lambda_\beta(z) \\
&= \left(\frac{\alpha}{\alpha+x} \right)^{p(1+k)} \frac{\Gamma((pk/2)+1)}{\beta^{pk/2}}.
\end{aligned}
$$

Now, if the integral operator P_α is bounded on $L^p(\mathbb{C}, d\lambda_\beta)$, then there exists a positive constant C (independent of x and k) such that

$$
\left(\frac{\alpha}{\alpha+x} \right)^{p(1+k)} \frac{\Gamma((pk/2)+1)}{\beta^{pk/2}} \leq C \frac{\beta}{px+\beta} \frac{\Gamma((pk/2)+1)}{(px+\beta)^{pk/2}},
$$

or

$$
\left(\frac{\alpha}{\alpha+x} \right)^{p(1+k)} \leq C \left(\frac{\beta}{\beta+px} \right)^{1+(pk/2)}.
$$

Fix any $x > 0$ and look at what happens in the above inequality when $k \to \infty$. We deduce that

$$
\left(\frac{\alpha}{\alpha+x} \right)^2 \leq \frac{\beta}{\beta+px}.
$$

Cross multiply and simplify. The result is

$$
p\alpha^2 \leq 2\alpha\beta + \beta x.
$$

Let $x \to 0$. Then $p\alpha^2 \leq 2\alpha\beta$, or $p\alpha \leq 2\beta$.

Similarly, if we let $k = 0$ and let $x \to \infty$ in the previous paragraph, the result is $p \geq 1$. This completes the proof of the lemma. $\qquad\square$

Since the operator P_α (and hence Q_α) is never bounded on $L^p(\mathbb{C}, d\lambda_\beta)$ when $0 < p < 1$, we need only focus on the case $p \geq 1$.

Lemma 2.16. *Suppose $1 < p < \infty$ and P_α is bounded on $L^p(\mathbb{C}, d\lambda_\beta)$. Then $p\alpha > \beta$.*

Proof. If $p > 1$ and P_α is bounded on $L^p(\mathbb{C}, d\lambda_\beta)$, then P_α^* is bounded on $L^q(\mathbb{C}, d\lambda_\beta)$, where $1/p + 1/q = 1$. Applying the formula for P_α^* from (2.5) to the constant function $f = 1$ shows that the function $e^{(\beta - \alpha)|z|^2}$ is in $L^q(\mathbb{C}, d\lambda_\beta)$. From this, we deduce that

$$q(\beta - \alpha) < \beta,$$

which is easily seen to be equivalent to $\beta < p\alpha$.   □

Lemma 2.17. *If P_α is bounded on $L^1(\mathbb{C}, d\lambda_\beta)$, then $\alpha = 2\beta$.*

Proof. Fix any $a \in \mathbb{C}$ and consider the function

$$f_a(z) = \frac{e^{\alpha z \bar{a}}}{|e^{\alpha z \bar{a}}|}, \qquad z \in \mathbb{C}.$$

Obviously, $\|f_a\|_\infty = 1$ for every $a \in \mathbb{C}$. On the other hand, it follows from (2.5) and Corollary 2.5 that

$$P_\alpha^*(f_a)(a) = \frac{\alpha}{\beta} e^{(\beta - \alpha)|a|^2} \int_{\mathbb{C}} |e^{\alpha w \bar{a}}| \, d\lambda_\beta(w)$$

$$= \frac{\alpha}{\beta} e^{(\beta - \alpha)|a|^2} e^{\alpha^2 |a|^2/(4\beta)}.$$

Since P_α^* is bounded on $L^\infty(\mathbb{C})$, there exists a positive constant C such that

$$\frac{\alpha}{\beta} e^{(\beta - \alpha)|a|^2} e^{\alpha^2 |a|^2/(4\beta)} \leq \|P_\alpha^*(f_a)\|_\infty \leq C\|f_a\|_\infty = C$$

for all $a \in \mathbb{C}$. This clearly implies that

$$\beta - \alpha + \frac{\alpha^2}{4\beta} \leq 0,$$

which is equivalent to $(2\beta - \alpha)^2 \leq 0$. Therefore, we have $\alpha = 2\beta$.   □

Lemma 2.18. *Suppose $1 < p \leq 2$ and P_α is bounded on $L^p(\mathbb{C}, d\lambda_\beta)$. Then $p\alpha = 2\beta$.*

Proof. Once again, we consider functions of the form

$$f_{x,k}(z) = e^{-x|z|^2} z^k, \qquad z \in \mathbb{C},$$

where $x > 0$ and k is a positive integer. It follows from (2.5) and the reproducing property in $F_{\alpha+x}^2$ that

$$P_\alpha^*(f_{x,k})(z) = \frac{\alpha}{\pi} e^{(\beta - \alpha)|z|^2} \int_{\mathbb{C}} e^{\alpha z \bar{w}} w^k e^{-(\beta + x)|w|^2} \, dA(w)$$

$$= \frac{\alpha}{\beta+x} e^{(\beta-\alpha)|z|^2} \int_{\mathbb{C}} e^{(\beta+x)[\alpha z/(\beta+x)]\bar{w}} w^k \, d\lambda_{\beta+x}(w)$$

$$= \frac{\alpha}{\beta+x} e^{(\beta-\alpha)|z|^2} \left(\frac{\alpha z}{\beta+x}\right)^k$$

$$= \left(\frac{\alpha}{\beta+x}\right)^{1+k} e^{(\beta-\alpha)|z|^2} z^k.$$

Suppose $1 < p \le 2$ and $1/p + 1/q = 1$. If the operator P_α is bounded on $L^p(\mathbb{C}, d\lambda_\beta)$, then the operator P_α^* is bounded on $L^q(\mathbb{C}, d\lambda_\beta)$. So there exists a positive constant C, independent of x and k, such that

$$\int_{\mathbb{C}} |P_\alpha^*(f_{x,k})|^q \, d\lambda_\beta \le C \int_{\mathbb{C}} |f_{x,k}|^q \, d\lambda_\beta.$$

We have

$$\int_{\mathbb{C}} |f_{x,k}|^q \, d\lambda_\beta = \frac{\beta}{qx+\beta} \frac{\Gamma((qk/2)+1)}{(qx+\beta)^{qk/2}}.$$

On the other hand, it follows from Lemma 2.16 and its proof that

$$\beta - q(\beta - \alpha) > 0,$$

so the integral

$$I = \int_{\mathbb{C}} |P_\alpha^*(f_{x,k})|^q \, d\lambda_\beta$$

can be evaluated as follows:

$$I = \left(\frac{\alpha}{\beta+x}\right)^{q(1+k)} \frac{\beta}{\pi} \int_{\mathbb{C}} |z|^{qk} e^{-(\beta-q(\beta-\alpha))|z|^2} \, dA(z)$$

$$= \left(\frac{\alpha}{\beta+x}\right)^{q(1+k)} \frac{\beta}{\beta-q(\beta-\alpha)} \int_{\mathbb{C}} |z|^{qk} \, d\lambda_{\beta-q(\beta-\alpha)}(z)$$

$$= \left(\frac{\alpha}{\beta+x}\right)^{q(1+k)} \frac{\beta}{\beta-q(\beta-\alpha)} \frac{\Gamma((qk/2)+1)}{(\beta-q(\beta-\alpha))^{qk/2}}.$$

Therefore,

$$\left(\frac{\alpha}{\beta+x}\right)^{q(1+k)} \frac{\beta}{\beta-q(\beta-\alpha)} \frac{\Gamma((qk/2)+1)}{(\beta-q(\beta-\alpha))^{qk/2}}$$

is less than or equal to

$$\frac{C\beta}{qx+\beta} \frac{\Gamma((qk/2)+1)}{(qx+\beta)^{qk/2}},$$

which easily reduces to

$$\left(\frac{\alpha}{\beta+x}\right)^{q(1+k)} \leq C\left(\frac{\beta-q(\beta-\alpha)}{\beta+qx}\right)^{1+(qk/2)}.$$

Once again, fix $x > 0$ and let $k \to \infty$. We find out that

$$\left(\frac{\alpha}{\beta+x}\right)^2 \leq \frac{\beta-q(\beta-\alpha)}{\beta+qx}.$$

Using the relation $1/p + 1/q = 1$, we can change the right-hand side above to

$$\frac{p\alpha-\beta}{(p-1)\beta+px}.$$

It follows that

$$\alpha^2(p-1)\beta + \alpha^2 px \leq (p\alpha-\beta)(\beta^2+2\beta x+x^2),$$

which can be written as

$$(p\alpha-\beta)x^2 + [2\beta(p\alpha-\beta) - \alpha^2 p]x + \beta^2(p\alpha-\beta) - \alpha^2(p-1)\beta \geq 0.$$

Let $q(x)$ denote the quadratic function on the left-hand side of the above inequality. Since $p\alpha - \beta > 0$ by Lemma 2.16, the function $q(x)$ attains its minimum value at

$$x_0 = \frac{p\alpha^2 - 2\beta(p\alpha-\beta)}{2(p\alpha-\beta)}.$$

Since $2 \geq p$, the numerator above is greater than or equal to

$$p\alpha^2 - 2p\alpha\beta + p\beta^2 = p(\alpha-\beta)^2.$$

It follows that $x_0 \geq 0$ and so $h(x) \geq h(x_0) \geq 0$ for all real x (not just nonnegative x). From this, we deduce that the discriminant of $h(x)$ cannot be positive. Therefore,

$$[2\beta(p\alpha-\beta) - p\alpha^2]^2 - 4(p\alpha-\beta)[\beta^2(p\alpha-\beta) - \alpha^2(p-1)\beta] \leq 0.$$

Elementary calculations reveal that the above inequality is equivalent to

$$(p\alpha-2\beta)^2 \leq 0.$$

Therefore, $p\alpha = 2\beta$. $\square$

Lemma 2.19. *Suppose* $2 < p < \infty$ *and* P_α *is bounded on* $L^p(\mathbb{C}, d\lambda_\beta)$. *Then* $p\alpha = 2\beta$.

Proof. If P_α is a bounded operator on $L^p(\mathbb{C}, d\lambda_\beta)$, then P_α^* is also bounded on $L^q(\mathbb{C}, d\lambda_\beta)$, where $1 < q < 2$ and $1/p + 1/q = 1$. It follows from (2.5) that there exists a positive constant C, independent of f, such that

$$\int_\mathbb{C} \left| e^{(\beta - \alpha)|z|^2} \int_\mathbb{C} e^{\alpha z \bar{w}} \left[f(w) e^{(\alpha - \beta)|w|^2} \right] d\lambda_\alpha(w) \right|^q d\lambda_\beta(z)$$

is less than or equal to

$$C \int_\mathbb{C} |f(w)|^q \, d\lambda_\beta(w),$$

where f is any function in $L^q(\mathbb{C}, d\lambda_\beta)$. Let

$$f(z) = g(z) e^{(\beta - \alpha)|z|^2},$$

where $g \in L^q(\mathbb{C}, d\lambda_{\beta - q(\beta - \alpha)})$. Recall from Lemma 2.16 that

$$\beta - q(\beta - \alpha) > 0.$$

We obtain another positive constant C (independent of g) such that

$$\int_\mathbb{C} |P_\alpha g|^q \, d\lambda_{\beta - q(\beta - \alpha)} \le C \int_\mathbb{C} |g|^q \, d\lambda_{\beta - q(\beta - \alpha)},$$

for all $g \in L^q(\mathbb{C}, d\lambda_{\beta - q(\beta - \alpha)})$. So the operator P_α is bounded on $L^q(\mathbb{C}, d\lambda_{\beta - q(\beta - \alpha)})$. Since $1 < q < 2$, it follows from Lemma 2.18 that

$$q\alpha = 2[\beta - q(\beta - \alpha)].$$

It is easy to check that this is equivalent to $p\alpha = 2\beta$. $\square$

We now prove the main result of this section. Recall that P_α and Q_α are never bounded on $L^p(\mathbb{C}, d\lambda_\beta)$ when $0 < p < 1$.

Theorem 2.20. *Suppose* $\alpha > 0$, $\beta > 0$, *and* $1 \le p < \infty$. *Then the following conditions are equivalent:*

(a) The operator Q_α is bounded on $L^p(\mathbb{C}, d\lambda_\beta)$.
(b) The operator P_α is bounded on $L^p(\mathbb{C}, d\lambda_\beta)$.
(c) The weight parameters satisfy $p\alpha = 2\beta$.

Proof. When $p = 1$, that (b) implies (c) follows from Lemma 2.17, that (c) implies (a) follows from Fubini's theorem and Corollary 2.5, and that (a) implies (b) is obvious.

When $1 < p < \infty$, that (b) implies (c) follows from Lemmas 2.18 and 2.19, and that (a) implies (b) is still obvious.

So we assume $1 < p < \infty$ and proceed to show that condition (c) implies (a). We do this with the help of Schur's test (Lemma 2.14).

Let $1/p + 1/q = 1$ and consider the positive function

$$h(z) = e^{\delta|z|^2}, \qquad z \in \mathbb{C},$$

where δ is a constant to be specified later.

Recall that

$$Q_\alpha f(z) = \int_{\mathbb{C}} H(z,w) f(w) \, d\lambda_\beta(w),$$

where

$$H(z,w) = \frac{\alpha}{\beta} |e^{\alpha z\bar{w}} e^{(\beta-\alpha)|w|^2}|$$

is a positive kernel. We first consider the integrals

$$I(z) = \int_{\mathbb{C}} H(z,w) h(w)^q \, d\lambda_\beta(w), \qquad z \in \mathbb{C}.$$

If δ satisfies

$$\alpha > q\delta, \tag{2.7}$$

then it follows from Corollary 2.5 that

$$\begin{aligned}
I(z) &= \frac{\alpha}{\pi} \int_{\mathbb{C}} |e^{\alpha z\bar{w}}| e^{-(\alpha-q\delta)|w|^2} \, dA(w) \\
&= \frac{\alpha}{\alpha - q\delta} \int_{\mathbb{C}} |e^{\alpha z\bar{w}}| \, d\lambda_{\alpha-q\delta}(w) \\
&= \frac{\alpha}{\alpha - q\delta} e^{\alpha^2 |z|^2/4(\alpha-q\delta)}.
\end{aligned}$$

If we choose δ so that

$$\frac{\alpha^2}{4(\alpha - q\delta)} = q\delta, \tag{2.8}$$

then we obtain

$$\int_{\mathbb{C}} H(z,w) h(w)^q \, d\lambda_\beta(w) \le \frac{\alpha}{\alpha - q\delta} h(z)^q \tag{2.9}$$

for all $z \in \mathbb{C}$.

We now consider the integrals

$$J(w) = \int_{\mathbb{C}} H(z,w) h(z)^p \, d\lambda_\beta(z), \qquad w \in \mathbb{C}.$$

If δ satisfies

$$\beta - p\delta > 0, \tag{2.10}$$

then it follows from Corollary 2.5 that

$$
\begin{aligned}
J(w) &= \frac{\alpha}{\beta} \int_{\mathbb{C}} |e^{\alpha z \bar{w}} e^{(\beta - \alpha)|w|^2} | h(z)^p \, d\lambda_\beta(z) \\
&= \frac{\alpha}{\pi} e^{(\beta - \alpha)|w|^2} \int_{\mathbb{C}} |e^{\alpha z \bar{w}}| e^{-(\beta - p\delta)|z|^2} \, dA(z) \\
&= \frac{\alpha}{\beta - p\delta} e^{(\beta - \alpha)|w|^2} e^{\alpha^2 |w|^2 / 4(\beta - p\delta)} \\
&= \frac{\alpha}{\beta - p\delta} e^{[(\beta - \alpha) + \alpha^2 / 4(\beta - p\delta)]|w|^2}.
\end{aligned}
$$

If we choose δ so that

$$
\beta - \alpha + \frac{\alpha^2}{4(\beta - p\delta)} = p\delta, \tag{2.11}
$$

then we obtain

$$
\int_{\mathbb{C}} H(z, w) h(z)^p \, d\lambda_\beta(z) \le \frac{\alpha}{\beta - p\delta} h(w)^p \tag{2.12}
$$

for all $w \in \mathbb{C}$. In view of Schur's test and the estimates in (2.9) and (2.12), we conclude that the operator Q_α would be bounded on $L^p(\mathbb{C}, d\lambda_\beta)$ provided that we could choose a real δ to satisfy conditions (2.7), (2.8), (2.10), and (2.11) simultaneously.

Under our assumption that $p\alpha = 2\beta$, it is easy to verify that condition (2.8) is the same as condition (2.11). In fact, we can explicitly solve for $q\delta$ and $p\delta$ in (2.8) and (2.11), respectively, to obtain

$$
q\delta = \frac{\alpha}{2}, \quad p\delta = \frac{2\beta - \alpha}{2}.
$$

The relations $p\alpha = 2\beta$ and $1/p + 1/q = 1$ clearly imply that the two resulting δ's above are consistent, namely,

$$
\delta = \frac{\alpha}{2q} = \frac{2\beta - \alpha}{2p}. \tag{2.13}
$$

Also, it is easy to see that the above choice of δ satisfies both (2.7) and (2.10). This completes the proof of the theorem. □

Theorem 2.21. *If $1 \le p < \infty$ and $p\alpha = 2\beta$, then*

$$
\int_{\mathbb{C}} |P_\alpha f|^p \, d\lambda_\beta \le \int_{\mathbb{C}} |Q_\alpha f|^p \, d\lambda_\beta \le 2^p \int_{\mathbb{C}} |f|^p \, d\lambda_\beta
$$

for all $f \in L^p(\mathbb{C}, d\lambda_\beta)$.

Proof. With the choice of δ in (2.13), the constants in (2.9) and (2.12) both reduce to 2. Therefore, Schur's test tells us that, in the case when $1 < p < \infty$, the norm of Q_α on $L^p(\mathbb{C}, d\lambda_\beta)$ does not exceed 2.

When $p = 1$, the desired estimate follows from Fubini's theorem and Corollary 2.5.   □

Corollary 2.22. *For any* $\alpha > 0$ *and* $1 \leq p \leq \infty$, *the operator* P_α *is a bounded projection from* L^p_α *onto* F^p_α. *Furthermore,* $\|P_\alpha f\|_{p,\alpha} \leq 2\|f\|_{p,\alpha}$ *for all* $f \in L^p_\alpha$.

Proof. The case $1 \leq p < \infty$ follows from Theorem 2.21. The case $p = \infty$ follows from Corollary 2.5.   □

2.3 Duality of Fock Spaces

It follows easily from the usual duality of L^p spaces that for any $1 \le p < \infty$, we have $(L_\alpha^p)^* = L_\beta^q$, where $1/p + 1/q = 1$, α and β are any positive parameters, and the duality pairing is given by

$$\langle f, g \rangle_\gamma = \frac{\gamma}{\pi} \int_{\mathbb{C}} f(z)\overline{g(z)} e^{-\gamma|z|^2} \, dA(z).$$

Here, $\gamma = (\alpha + \beta)/2$ is the arithmetic mean of α and β.

In this section, we are going to identify all bounded linear functionals on the Fock space F_α^p, where $0 < p < \infty$. We will also do the same for the space f_α^∞. Somewhat surprisingly, the duality of Fock spaces depends on the geometric mean of α and β instead of their arithmetic mean. Let us begin with the case $p > 1$.

Theorem 2.23. *Suppose $\beta > 0$, $1 < p < \infty$, and $1/p + 1/q = 1$. Then the dual space of F_α^p can be identified with F_β^q under the integral pairing*

$$\langle f, g \rangle_\gamma = \lim_{R \to \infty} \frac{\gamma}{\pi} \int_{|z| < R} f(z)\overline{g(z)} e^{-\gamma|z|^2} \, dA(z),$$

where $\gamma = \sqrt{\alpha\beta}$ is the geometric mean of α and β.

Proof. First, assume that $g \in F_\beta^q$ and F is defined by

$$F(f) = \lim_{R \to \infty} \frac{\gamma}{\pi} \int_{|z| < R} f(z)\overline{g(z)} e^{-\gamma|z|^2} \, dA(z).$$

We proceed to show that F gives rise to a bounded linear functional on F_α^p. To avoid the use of limits all over the place, we appeal to Lemma 2.11 and further assume that g is a finite linear combination of kernel functions.

If $f(z) = e^{\gamma z \bar{a}}$ for some $a \in \mathbb{C}$, then by the reproducing property of the kernel functions $K_\gamma(z, w)$ and $K_\alpha(z, w)$, we have

$$g(a) = \frac{\gamma}{\pi} \int_{\mathbb{C}} \overline{f(z)} g(z) e^{-\gamma|z|^2} \, dA(z),$$

and

$$g(a) = g\left(\sqrt{\frac{\alpha}{\beta}} \frac{\gamma}{\alpha} a \right)$$

$$= \frac{\alpha}{\pi} \int_{\mathbb{C}} e^{\alpha(\gamma a/\alpha)\bar{z}} g\left(\sqrt{\frac{\alpha}{\beta}} z \right) e^{-\alpha|z|^2} \, dA(z)$$

$$= \frac{\alpha}{\pi} \int_{\mathbb{C}} \overline{f(z)} g\left(\sqrt{\frac{\alpha}{\beta}} z \right) e^{-\alpha|z|^2} \, dA(z).$$

Therefore,

$$\int_{\mathbb{C}} f\bar{g} \, d\lambda_\gamma = \frac{\alpha}{\pi} \int_{\mathbb{C}} f(z)\bar{g}\left(\sqrt{\frac{\alpha}{\beta}}z\right) e^{-\alpha|z|^2} \, dA(z). \qquad (2.14)$$

This shows that

$$F(f) = \frac{\alpha}{\pi} \int_{\mathbb{C}} f(z)\bar{g}\left(\sqrt{\frac{\alpha}{\beta}}z\right) e^{-\alpha|z|^2} \, dA(z)$$

$$= \frac{\alpha}{\pi} \int_{\mathbb{C}} \left[f(z)e^{-\frac{\alpha}{2}|z|^2}\right] \left[\bar{g}\left(\sqrt{\frac{\alpha}{\beta}}z\right)e^{-\frac{\alpha}{2}|z|^2}\right] \, dA(z)$$

for all functions f of the form

$$f(z) = \sum_{k=1}^{N} c_k e^{\gamma z \bar{a}_k},$$

which are dense in F_α^p by Lemma 2.11.

It is clear that $g \in L_\beta^q$ is equivalent to the condition that

$$\varphi(z) = g(\sqrt{\alpha/\beta}\, z) \in L_\alpha^q.$$

An application of Hölder's inequality then gives

$$|F(f)| \le C\|f\|_{p,\alpha}\|\varphi\|_{q,\alpha} = C'\|f\|_{p,\alpha}\|g\|_{q,\beta}, \qquad (2.15)$$

where f is any finite linear combination of kernel functions, and C and C' are positive constants. This shows that F defines a bounded linear functional on F_α^p.

Next, assume that $F : F_\alpha^p \to \mathbb{C}$ is a bounded linear functional. Define a function g on the complex plane by

$$\overline{g(w)} = F_z\left(e^{\gamma z \bar{w}}\right).$$

It is easy to show that g is entire. We are going to show that $g \in F_\beta^q$ and $F(f) = \langle f, g \rangle_\gamma$ for all f in a dense subset of F_α^p.

To show that $g \in F_\beta^q$, we need to show that the function $g(w)e^{-\beta|w|^2/2}$ is in $L^q(\mathbb{C}, dA)$. To this end, we consider the integrals

$$\Phi(h) = \int_{\mathbb{C}} h(w)\overline{g(w)}e^{-\beta|w|^2/2} \, dA(w), \qquad h \in L^p(\mathbb{C}, dA).$$

It suffices for us to show that Φ defines a bounded linear functional on the space $L^p(\mathbb{C}, dA)$. Without loss of generality, we may assume that h has compact support in $\mathbb{C}$. In this case, the integral

$$\int_{\mathbb{C}} h(w)e^{\gamma z \bar{w}}e^{-\beta|w|^2/2} \, dA(w)$$

converges in the norm topology of F_α^p, and we have

$$\Phi(h) = \int_{\mathbb{C}} h(w) F_z\left(e^{\gamma z \overline{w}}\right) e^{-\beta|w|^2/2} \, dA(w)$$

$$= F\left(\int_{\mathbb{C}} h(w) e^{\gamma z \overline{w}} e^{-\beta|w|^2/2} \, dA(w)\right)$$

$$= \frac{\alpha}{\beta} F\left(\int_{\mathbb{C}} h\left(\sqrt{\frac{\alpha}{\beta}} w\right) e^{\alpha z \overline{w}} e^{-\alpha|w|^2/2} \, dA(w)\right)$$

$$= \frac{\pi}{\beta} F\left(P_\alpha(\varphi)\right),$$

where

$$\varphi(z) = h\left(\sqrt{\frac{\alpha}{\beta}} z\right) e^{\frac{\alpha}{2}|z|^2}.$$

Since $h \in L^p(\mathbb{C}, dA)$ is equivalent to $\varphi \in L_\alpha^p$ and since the projection P_α maps L_α^p boundedly into F_α^p, we conclude that

$$|\Phi(h)| \le \frac{\pi}{\beta} \|F\| \|P_\alpha(\varphi)\|_{p,\alpha} \le C\|h\|,$$

where $\|h\|$ denotes the usual norm in $L^p(\mathbb{C}, dA)$. This shows that the function g is in F_β^q.

Finally, if $f(z) = e^{\gamma z \overline{a}}$ for some $a \in \mathbb{C}$, then by the remarks immediately following this proof and the reproducing property in F_γ^2,

$$\langle f, g \rangle_\gamma = \lim_{R \to \infty} \frac{\gamma}{\pi} \int_{|z|<R} e^{\gamma z \overline{a}} \overline{g(z)} e^{-\gamma|z|^2} \, dA(z) = \overline{g(a)} = F(f).$$

It follows that $F(f) = \langle f, g \rangle_\gamma$ whenever f is a finite linear combination of kernel functions. This, along with Lemma 2.11, finishes the proof of the theorem. □

Note that (2.14) was proved under the assumption that both f and g are finite linear combinations of kernel functions. By (2.15), the right-hand side of (2.14) converges for all $f \in F_\alpha^p$ and $g \in F_\beta^q$, and the integral is dominated by $\|f\|_{p,\alpha} \|g\|_{q,\beta}$. An approximation argument with the help of Lemma 2.11 then shows that

$$\lim_{R \to \infty} \int_{|z|<R} f(z) \overline{g(z)} \, d\lambda_\gamma(z) = \int_{\mathbb{C}} f(z) \overline{g}\left(\sqrt{\frac{\alpha}{\beta}} z\right) d\lambda_\alpha(z) \qquad (2.16)$$

for all $f \in F_\alpha^p$ and $g \in F_\beta^q$. In particular, the limit on the left-hand side of (2.16) exists for all $f \in F_\alpha^p$ and $g \in F_\beta^q$.

Alternatively, the identity in (2.16) can be proved with the help of Taylor expansions. Details are left to the interested reader. We now consider the case of small exponents.

Theorem 2.24. *Suppose $0 < p \leq 1$ and $\beta > 0$. Then the dual space of F_α^p can be identified with F_β^∞ under the integral pairing*

$$\langle f, g \rangle_\gamma = \lim_{R \to \infty} \frac{\gamma}{\pi} \int_{|z| < R} f(z) \overline{g(z)} e^{-\gamma|z|^2} \, dA(z),$$

where $\gamma = \sqrt{\alpha\beta}$ and the limit above always exists.

Proof. First, assume that $g \in F_\beta^\infty$ and F is defined by $F(f) = \langle f, g \rangle_\gamma$. To show that F extends to a bounded linear functional on F_α^p, we use (2.16) to rewrite

$$\begin{aligned}
F(f) &= \frac{\alpha}{\pi} \int_{\mathbb{C}} f(z) \overline{\varphi(z)} e^{-\alpha|z|^2} \, dA(z) \\
&= \frac{\alpha}{\pi} \int_{\mathbb{C}} \left[f(z) e^{-\frac{\alpha}{2}|z|^2} \right] \overline{\left[\varphi(z) e^{-\frac{\alpha}{2}|z|^2} \right]} \, dA(z),
\end{aligned}$$

where

$$\varphi(z) = g\left(\sqrt{\frac{\alpha}{\beta}} z \right)$$

is in F_α^∞. It follows from this and the embedding in Theorem 2.10 (and its proof) that

$$|F(f)| \leq 2\|\varphi\|_{\infty,\alpha} \|f\|_{1,\alpha} \leq \frac{2}{p} \|\varphi\|_{\infty,\alpha} \|f\|_{p,\alpha}.$$

So F extends to a bounded linear functional on F_α^p, and an approximation argument shows that the limit in the statement of the theorem always exists.

Next, suppose that F is a bounded linear functional on F_α^p. As in the proof of Theorem 2.23, we consider the function g defined on $\mathbb{C}$ by

$$\overline{g(w)} = F_z \left(e^{\gamma z \overline{w}} \right).$$

It follows from the boundedness of F on F_α^p and the integral formula in Corollary 2.5 that

$$\begin{aligned}
|g(w)|^p &\leq \frac{p\alpha\|F\|^p}{2\pi} \int_{\mathbb{C}} |e^{\gamma z \overline{w}} e^{-\alpha|z|^2/2}|^p \, dA(z) \\
&= \frac{p\alpha\|F\|^p}{2\pi} \int_{\mathbb{C}} |e^{p\gamma z \overline{w}}| e^{-p\alpha|z|^2/2} \, dA(z) \\
&= \|F\|^p e^{p\beta|w|^2/2}.
\end{aligned}$$

This shows that $g \in F_\beta^\infty$ with $\|g\|_{\infty,\beta} \leq \|F\|$.

Finally, as in the proof of Theorem 2.23, we have $F(f) = \langle f, g \rangle_\gamma$ for all functions f of the form

$$f(z) = \sum_{k=1}^{N} c_k e^{\gamma z \overline{u_k}}.$$

Since the set of functions of the above form is dense in F_α^p, we have completed the proof of the theorem.   □

Setting $\beta = \alpha$ in Theorems 2.23 and 2.24, we obtain the following special case.

Corollary 2.25. *If $1 \le p < \infty$, then the dual space of F_α^p can be identified with F_α^q under the integral pairing $\langle f, g \rangle_\alpha$, where $1/p + 1/q = 1$. If $0 < p < 1$, then the dual space of F_α^p can be identified with F_α^∞ under the integral pairing $\langle f, g \rangle_\alpha$.*

It is interesting to observe that under the same integral pairing $\langle f, g \rangle_\alpha$, the dual space of each F_α^p, $0 < p \le 1$, can be identified with the same space F_α^∞. This differs from the traditional Hardy and Bergman space theories.

Theorem 2.26. *Suppose $\beta > 0$ and $\gamma = \sqrt{\alpha \beta}$. Then the dual space of f_α^∞ can be identified with F_β^1 under the integral pairing $\langle f, g \rangle_\gamma$.*

Proof. If $g \in F_\beta^1$, then by Theorem 2.24, $F(f) = \langle f, g \rangle_\gamma$ defines a bounded linear functional on f_α^∞.

Now, suppose F is any bounded linear functional on f_α^∞. Since the set of finite linear combinations of kernel functions is dense in f_α^∞ (but not in F_α^∞), we can proceed as in the proof of Theorem 2.23 to obtain

$$F(f) = \lim_{R \to \infty} \frac{\gamma}{\pi} \int_{|w| < R} f(w) \overline{g(w)} e^{-\gamma |w|^2} \, dA(w)$$

for f in a dense subset of f_α^∞, where

$$\overline{g(w)} = F_z \left(e^{\gamma z \overline{w}} \right).$$

It remains for us to show that $g \in F_\beta^1$.

Since the dual space of F_β^1 is identified with F_α^∞ under the integral pairing $\langle f, g \rangle_\gamma$, it suffices to show that there exists a constant $C > 0$ such that

$$|\langle f, g \rangle_\gamma| \le C \|f\|_{\infty, \alpha}$$

for all $f \in F_\alpha^\infty$. For any positive integer n, consider the function:

$$f_n(z) = f \left(\frac{n}{n+1} z \right), \qquad z \in \mathbb{C}.$$

It is clear that $f \in F_\alpha^\infty$ implies that each $f_n \in f_\alpha^\infty$ with $\|f_n\|_{\infty,\alpha} \leq \|f\|_{\infty,\alpha}$ for all n. Now,

$$
\begin{aligned}
\langle f, g \rangle_\gamma &= \lim_{R \to \infty} \frac{\gamma}{\pi} \int_{|w|<R} f(w) F_z \left(e^{\gamma z \overline{w}} \right) e^{-\gamma |w|^2} \, dA(w) \\
&= \lim_{n \to \infty} \lim_{R \to \infty} \frac{\gamma}{\pi} \int_{|w|<R} f_n(w) F_z \left(e^{\gamma z \overline{w}} \right) e^{-\gamma |w|^2} \, dA(w) \\
&= \lim_{n \to \infty} F \left[\frac{\gamma}{\pi} \int_{\mathbb{C}} f_n(w) e^{\gamma z \overline{w}} e^{-\gamma |w|^2} \, dA(w) \right] \\
&= \lim_{n \to \infty} F(f_n).
\end{aligned}
$$

Since $|F(f_n)| \leq \|F\| \|f_n\|_{\infty,\alpha} \leq \|F\| \|f\|_{\infty,\alpha}$ for all n, we conclude that $|\langle f, g \rangle_\gamma| \leq \|F\| \|f\|_{\infty,\alpha}$ for all $f \in F_\alpha^\infty$. This shows that $g \in F_\beta^1$ and completes the proof of the theorem. $\qquad\qquad\square$

2.4 Complex Interpolation

We assume that the reader is familiar with the basic theory of complex interpolation, including the complex interpolation of L^p spaces. The book [250] provides an elementary introduction to the subject. We will begin with the following well-known interpolation theorem of Stein and Weiss.

Theorem 2.27. *Suppose w, w_0, and w_1 are positive weight functions on the complex plane. If $1 \le p_0 \le p_1 \le \infty$ and $0 \le \theta \le 1$, then*

$$[L^{p_0}(\mathbb{C}, w_0 dA), L^{p_1}(\mathbb{C}, w_1 dA)]_\theta = L^p(\mathbb{C}, w dA)$$

with equal norms, where

$$\frac{1}{p} = \frac{1-\theta}{p_0} + \frac{\theta}{p_1}, \qquad w^{\frac{1}{p}} = w_0^{\frac{1-\theta}{p_0}} w_1^{\frac{\theta}{p_1}}.$$

This result is very useful and widely known. See [216] for a proof.

Recall that L_α^p is the space of Lebesgue measurable functions f on the complex plane such that the function $f(z)e^{-\alpha|z|^2/2}$ is in $L^p(\mathbb{C}, dA)$. The norm of f in L_α^p was defined in Sect. 2.1. With the inherited norm, F_α^p is the closed subspace of L_α^p consisting of entire functions.

Specializing to exponential weights, we obtain the following special case of the Stein–Weiss interpolation theorem.

Corollary 2.28. *Suppose $1 \le p_0 \le p_1 \le \infty$ and $0 \le \theta \le 1$. Then for any positive weight parameters α_0 and α_1, we have*

$$\left[L_{\alpha_0}^{p_0}, L_{\alpha_1}^{p_1}\right]_\theta = L_\alpha^p,$$

where

$$\frac{1}{p} = \frac{1-\theta}{p_0} + \frac{\theta}{p_1}, \qquad \alpha = \alpha_0(1-\theta) + \alpha_1\theta.$$

Proof. Since $L_\alpha^p = L^p(\mathbb{C}, d\lambda_{p\alpha/2})$, it follows from the Stein–Weiss interpolation theorem that

$$\left[L_{\alpha_1}^{p_0}, L_{\alpha_2}^{p_1}\right]_\theta = \left[L^{p_0}(\mathbb{C}, d\lambda_{p_0\alpha_1/2}), L^{p_1}(\mathbb{C}, d\lambda_{p_1\alpha_2/2})\right]_\theta$$

$$= L^p(\mathbb{C}, d\lambda_{p\alpha/2}) = L_\alpha^p,$$

where

$$\frac{1}{p} = \frac{1-\theta}{p_0} + \frac{\theta}{p_1}, \qquad \alpha = \alpha_0(1-\theta) + \alpha_1\theta.$$

This proves the desired result.   □

Although F_α^p is a closed subspace of L_α^p, the Fock spaces interpolate in a way that is much different from the containing spaces L_α^p. In some sense, the Lebesgue spaces L_α^p interpolate "arithmetically," while the Fock spaces F_α^p interpolate "geometrically." We begin with the case when the weight parameter α is fixed.

Theorem 2.29. *Suppose* $1 \le p_0 \le p_1 \le \infty$ *and* $0 \le \theta \le 1$. *Then*

$$\left[F_\alpha^{p_0}, F_\alpha^{p_1}\right]_\theta = F_\alpha^p,$$

where

$$\frac{1}{p} = \frac{1-\theta}{p_0} + \frac{\theta}{p_1}.$$

Proof. The inclusion

$$\left[F_\alpha^{p_0}, F_\alpha^{p_1}\right]_\theta \subset F_\alpha^p$$

follows from the definition of complex interpolation, the fact that each $F_\alpha^{p_k}$ is a closed subspace of $L_\alpha^{p_k}$, and the fact that $\left[L_\alpha^{p_0}, L_\alpha^{p_1}\right]_\theta = L_\alpha^p$.

On the other hand, if $f \in F_\alpha^p \subset L_\alpha^p$, then f is entire, and it follows from $\left[L_\alpha^{p_0}, L_\alpha^{p_1}\right]_\theta = L_\alpha^p$ that there exist a function $F(z, \zeta)$ ($z \in \mathbb{C}$ and $0 \le \operatorname{Re} \zeta \le 1$) and a positive constant C such that:

(a) $F(z, \theta) = f(z)$ for all $z \in \mathbb{C}$.
(b) $\|F(\cdot, \zeta)\|_{p_0, \alpha} \le C$ for all $\operatorname{Re} \zeta = 0$.
(c) $\|F(\cdot, \zeta)\|_{p_1, \alpha} \le C$ for all $\operatorname{Re} \zeta = 1$.

Define a function $G(z, \zeta)$ by

$$G(z, \zeta) = \frac{\alpha}{\pi} \int_{\mathbb{C}} F(w, \zeta) e^{\alpha z \bar{w}} e^{-\alpha |w|^2} \, dA(w).$$

Then it follows from Corollary 2.22 that:

(a) $G(z, \theta) = f(z)$.
(b) $\|G(\cdot, \zeta)\|_{p_0, \alpha} \le 2C$ for all $\operatorname{Re} \zeta = 0$.
(c) $\|G(\cdot, \zeta)\|_{p_1, \alpha} \le 2C$ for all $\operatorname{Re} \zeta = 1$.

Since each function $z \mapsto G(z, \zeta)$ is entire, we conclude that $f \in [F_\alpha^{p_0}, F_\alpha^{p_1}]_\theta$. This completes the proof of the theorem. $\qquad \square$

We now consider the case when there are different weight parameters present. Note that α is an arithmetic mean of α_0 and α_1 in Corollary 2.28, but α is a geometric mean of α_0 and α_1 in the following theorem.

Theorem 2.30. *Suppose* $1 \le p_0 \le p_1 \le \infty$ *and* $0 \le \theta \le 1$. *Then for any positive weight parameters* α_0 *and* α_1, *we have*

$$\left[F_{\alpha_0}^{p_0}, F_{\alpha_1}^{p_1}\right]_\theta = F_\alpha^p,$$

where

$$\frac{1}{p} = \frac{1-\theta}{p_0} + \frac{\theta}{p_1}, \qquad \alpha = \alpha_0^{1-\theta} \alpha_1^{\theta}.$$

Proof. For any $\zeta \in \mathbb{C}$, consider the dilation operator S_ζ defined by

$$S_\zeta f(z) = f\left(\left(\frac{\alpha_0}{\alpha_1} \right)^{(\zeta-\theta)/2} z \right).$$

According to Lemma 2.6, S_ζ is an isometry from $F_{\alpha_0}^{p_0}$ onto $F_\alpha^{p_0}$ whenever $\operatorname{Re}\zeta = 0$, and S_ζ is an isometry from $F_{\alpha_1}^{p_1}$ onto $F_\alpha^{p_1}$ whenever $\operatorname{Re}\zeta = 1$. Furthermore, both $S_\zeta f$ and $S_\zeta^{-1} f$ are analytic in ζ when f is analytic. Therefore, by the abstract Stein interpolation theorem (see [215]), the operator S_θ must be an isometry from $[F_{\alpha_0}^{p_0}, F_{\alpha_1}^{p_1}]_\theta$ onto $[F_\alpha^{p_0}, F_\alpha^{p_1}]_\theta$. Since $S_\theta = I$ is the identity operator, we must have

$$\left[F_{\alpha_0}^{p_0}, F_{\alpha_1}^{p_1} \right]_\theta = \left[F_\alpha^{p_0}, F_\alpha^{p_1} \right]_\theta = F_\alpha^p,$$

where the last step follows from Theorem 2.29. $\qquad\square$

As a consequence of the above interpolation theorem, we obtain the following sharp result concerning the action of the Fock projection on L^p spaces.

Theorem 2.31. *Suppose $1 \le p \le \infty$. Then for any positive weight parameters α, β, and γ, we have:*

(a) $P_\alpha L_\beta^p \subset F_\gamma^p$ if and only if $\alpha^2/\gamma \le 2\alpha - \beta$.
(b) $P_\alpha L_\beta^p = F_\gamma^p$ if and only if $\alpha^2/\gamma = 2\alpha - \beta$.

Proof. It is easy to see that a necessary condition for $P_\alpha L_\beta^p \subset F_\gamma^p$, $1 \le p \le \infty$, is that $2\alpha > \beta$. So for the rest of the proof, we always assume that $2\alpha > \beta$.

If $\alpha^2/\gamma \le 2\alpha - \beta$, it follows from Corollary 2.5 that P_α maps L_β^∞ into F_γ^∞. Similarly, it follows from Fubini's theorem and Corollary 2.5 that P_α maps L_β^1 into F_γ^1. By complex interpolation, P_α maps L_β^p into F_γ^p for all $1 \le p \le \infty$.

If $\alpha^2/\gamma = 2\alpha - \beta$ and $f \in F_\gamma^p$, then the function

$$g(z) = \frac{\alpha}{\gamma} f\left(\frac{\alpha}{\gamma} z \right) e^{(\beta-\alpha)|z|^2}$$

belongs to L_β^p and $P_\alpha g = f$. Therefore, $P_\alpha L_\beta^p = F_\gamma^p$ for $1 \le p \le \infty$.

If $\alpha^2/\gamma > 2\alpha - \beta$, then there exists some $\gamma' > \gamma$ such that $\alpha^2/\gamma' = 2\alpha - \beta$ (here, we used the assumption that $2\alpha > \beta$). By what was proved in the previous paragraph, $P_\alpha L_\beta^p = F_{\gamma'}^p$. Since F_γ^p is strictly contained in $F_{\gamma'}^p$, we see that P_α cannot possibly map L_β^p into F_γ^p. A similar argument shows that if $\alpha^2/\gamma < 2\alpha - \beta$, then $P_\alpha L_\beta^\infty \ne F_\gamma^\infty$. This completes the proof of the theorem. $\qquad\square$

2.5 Atomic Decomposition

Recall from Lemma 2.11 that the set of finite linear combinations of kernel functions is dense in F_α^p, $0 < p < \infty$. In this section, we improve upon this result. We show that every function in F_α^p can actually be decomposed into an infinite series of kernel functions.

We begin with a basic estimate for integral averages of functions in Fock spaces.

Lemma 2.32. *For any positive parameters α, p, and R, there exists a positive constant $C = C(p, \alpha, R)$ such that*

$$\left| f(a)e^{-\alpha|a|^2/2} \right|^p \leq \frac{C}{r^2} \int_{B(a,r)} \left| f(z)e^{-\alpha|z|^2/2} \right|^p dA(z)$$

for all entire functions f, all complex numbers a, and all $r \in (0,R]$. Here, $B(a,r)$ is the Euclidean disk centered at a with radius r.

Proof. Let I denote the integral above. Then

$$I = \int_{B(a,r)} |f(z)|^p e^{-p\alpha|z|^2/2} dA(z)$$

$$= \int_{|w|<r} |f(w+a)|^p e^{-p\alpha|w+a|^2/2} dA(w)$$

$$= \int_{|w|<r} |f(w+a)e^{-\alpha w\bar{a}}|^p e^{-p\alpha(|w|^2+|a|^2)/2} dA(w).$$

Writing the integral in polar coordinates and using the subharmonicity of the function $|f(w+a)e^{-\alpha w\bar{a}}|^p$, we obtain

$$I \geq |f(a)|^p \int_{|w|<r} e^{-p\alpha(|w|^2+|a|^2)/2} dA(w)$$

$$= 2\pi |f(a)|^p \int_0^r t e^{-p\alpha(t^2+|a|^2)/2} dt$$

$$= \pi |f(a)e^{-\alpha|a|^2/2}|^p \int_0^{r^2} e^{-p\alpha s/2} ds$$

$$= \frac{2\pi}{p\alpha}(1 - e^{-p\alpha r^2/2})|f(a)e^{-\alpha|a|^2/2}|^p.$$

This proves the desired estimate.   □

Recall that for any positive number r,

$$r\mathbb{Z}^2 = \{nr + imr : n \in \mathbb{Z}, m \in \mathbb{Z}\}$$

is a square lattice in the complex plane. The fundamental region of $r\mathbb{Z}^2$, if we ignore the boundary points, is the square

$$S_r = \{z = x + iy : -r/2 \leq x < r/2, -r/2 \leq y < r/2\}.$$

We also consider the square

$$Q_r = \{z = x + iy : -r \leq x < r, -r \leq y < r\}.$$

It is clear that the complex plane admits the following decomposition:

$$\mathbb{C} = \bigcup\{S_r + z : z \in r\mathbb{Z}^2\}.$$

Moreover, the use of half-open and half-closed squares makes the decomposition above a disjoint union. Thus,

$$\int_{\mathbb{C}} f(z)\,d\mu(z) = \sum_{w \in r\mathbb{Z}^2} \int_{S_r+w} f(z)\,d\mu(z),$$

whenever $f \in L^1(\mathbb{C}, d\mu)$. Furthermore, there exists a positive integer N such that every point in the complex plane belongs to at most N of the squares $Q_r + w$. Therefore,

$$\int_{\mathbb{C}} f(z)\,d\mu(z) \leq \sum_{w \in r\mathbb{Z}^2} \int_{Q_r+w} f(z)\,d\mu(z) \leq N \int_{\mathbb{C}} f(z)\,d\mu(z)$$

whenever f is a nonnegative measurable function.

Also, recall that for each $a \in \mathbb{C}$, the normalized reproducing kernel of F_α^2 at the point a is given by

$$k_a(z) = K(z,a)/\sqrt{K(a,a)} = e^{\alpha z\bar{a} - \frac{1}{2}\alpha|a|^2}.$$

This is of course a unit vector in F_α^2. The following result is a pleasant surprise.

Lemma 2.33. *Each k_a is also a unit vector in F_α^p, where $0 < p \leq \infty$.*

Proof. It follows from the definition of the norm in F_α^p and the reproducing formula in $F_{p\alpha/2}^2$ that

$$\|k_a\|_{p,\alpha}^p = \frac{p\alpha}{2\pi} \int_{\mathbb{C}} \left|k_a(z)e^{-\frac{1}{2}\alpha|z|^2}\right|^p dA(z)$$

$$= \frac{p\alpha}{2\pi} e^{-\frac{1}{2}p\alpha|a|^2} \int_{\mathbb{C}} \left|e^{\frac{p\alpha}{2}z\bar{a}}\right|^2 e^{-\frac{p\alpha}{2}|z|^2} dA(z)$$

$$= e^{-\frac{p\alpha}{2}|a|^2} e^{\frac{p\alpha}{2}|a|^2} = 1,$$

which proves the desired result for $0 < p < \infty$. For $p = \infty$, observe that

$$|k_a(z)|e^{-\frac{\alpha}{2}|z|^2} = e^{-\frac{\alpha}{2}|z-a|^2}.$$

It follows that

$$\sup_{z \in \mathbb{C}} |k_a(z)|e^{-\frac{\alpha}{2}|z|^2} = 1,$$

and the proof of the lemma is complete. $\square$

The main result of this section is the following:

Theorem 2.34. *Let $0 < p \le \infty$. There exists a positive constant r_0 such that for any $0 < r < r_0$, the space F_α^p consists exactly of the following functions:*

$$f(z) = \sum_{w \in r\mathbb{Z}^2} c_w k_w(z), \tag{2.17}$$

where $\{c_w : w \in r\mathbb{Z}^2\} \in l^p$. Moreover, there exists a positive constant C (independent of f) such that

$$C^{-1}\|f\|_{p,\alpha} \le \inf \|\{c_w\}\|_{l^p} \le C\|f\|_{p,\alpha}$$

for all $f \in F_\alpha^p$, where the infimum is taken over all sequences $\{c_w\}$ that give rise to the decomposition (not unique) in (2.17).

Proof. If $0 < p \le 1$ and f is given by (2.17) with $\{c_w\} \in l^p$, then by Hölder's inequality,

$$|f(z)e^{-\alpha|z|^2/2}|^p \le \sum_{w \in r\mathbb{Z}^2} |c_w|^p |k_w(z)e^{-\alpha|z|^2/2}|^p.$$

It follows from this and Lemma 2.33 that

$$\|f\|_{p,\alpha}^p \le \sum_{w \in r\mathbb{Z}^2} |c_w|^p.$$

Thus, $f \in F_\alpha^p$ and

$$\|f\|_{p,\alpha}^p \le \inf \sum_{w \in r\mathbb{Z}^2} |c_w|^p.$$

If $\{c_w\} \in l^\infty$ and f is given by (2.17), then

$$|f(z)|e^{-\frac{\alpha}{2}|z|^2} \le \|\{c_w\}\|_\infty \sum_{w \in r\mathbb{Z}^2} e^{-\frac{\alpha}{2}|z-w|^2}.$$

By Lemma 1.12, there exists a positive constant C such that

$$\|f\|_{\infty,\alpha} \le C \inf \|\{c_w\}\|_\infty,$$

where the infimum is taken over all sequences $\{c_w\}$ in (2.17).

After interpolating between $p = 1$ and $p = \infty$, we have now shown that, for all $p \in (0,\infty]$ and $\{c_w\} \in l^p$, the function f given by (2.17) is in F_α^p. Furthermore,

$$\|f\|_{p,\alpha} \le C \inf \|\{c_w\}\|_{l^p},$$

where $C = C(p,\alpha,r)$ is a positive constant and the infimum is taken over all sequences $\{c_w\}$ that give rise to the representation of f in (2.17). It is interesting to note that this part of the proof works for any positive r.

To prove the other part of the theorem, we assume that $0 < r < 1$ and consider the linear operator T_r defined on the space of entire functions as follows:

$$T_r f(z) = \frac{\alpha}{\pi} \sum_{w \in r\mathbb{Z}^2} e^{\alpha z \bar{w} - \frac{\alpha}{2}|w|^2} \int_{S_r+w} f(u) e^{-\frac{\alpha}{2}|u|^2 + \alpha i \mathrm{Im}\,(w\bar{u})} \, dA(u).$$

We proceed to show that T_r is a bounded linear operator on F_α^p and to estimate $\|I - T_r\|$, the norm of $I - T_r$ on F_α^p, in terms of r, where I is the identity operator.

Let $D_r = I - T_r$. If f is in F_α^p, then

$$f(z) = \int_{\mathbb{C}} f(u) e^{\alpha z \bar{u}} \, d\lambda_\alpha(u)$$

$$= \frac{\alpha}{\pi} \sum_{w \in r\mathbb{Z}^2} \int_{S_r+w} f(u) e^{\alpha z \bar{u} - \frac{\alpha}{2}|u|^2 - \alpha i \mathrm{Im}\,(w\bar{u})} e^{-\frac{\alpha}{2}|u|^2 + \alpha i \mathrm{Im}\,(w\bar{u})} \, dA(u).$$

It follows that

$$D_r f(z) = \frac{\alpha}{\pi} \sum_{w \in r\mathbb{Z}^2} \int_{S_r+w} f(u) H(z,w,u) \, dA(u), \tag{2.18}$$

where

$$H(z,w,u) = \left[e^{\alpha z \bar{w} - \frac{\alpha}{2}|w|^2} - e^{\alpha z \bar{u} - \frac{\alpha}{2}|u|^2 - \alpha i \mathrm{Im}\,(w\bar{u})} \right] e^{-\frac{\alpha}{2}|u|^2 + \alpha i \mathrm{Im}\,(w\bar{u})}.$$

We now estimate the norm of the operator D_r on F_α^∞ and on F_α^1.

By (2.18),

$$|D_r(z)| e^{-\frac{\alpha}{2}|z|^2} \le \frac{\alpha}{\pi} \|f\|_{\infty,\alpha} J_r(z),$$

where

$$J_r(z) = \sum_{w \in r\mathbb{Z}^2} \int_{S_r+w} \left| e^{\alpha z \bar{u} - \frac{\alpha}{2}|u|^2 - \alpha i \mathrm{Im}\,(w\bar{u})} - e^{\alpha z \bar{w} - \frac{\alpha}{2}|w|^2} \right| e^{-\frac{\alpha}{2}|z|^2} \, dA(u).$$

Elementary calculations show that

$$J_r(z) = \sum_{w \in r\mathbb{Z}^2} \int_{S_r+w} \left| e^{-\frac{\alpha}{2}|z-w|^2} - e^{-\frac{\alpha}{2}|z-u|^2+\alpha i \operatorname{Im}(z-w)(\bar{u}-\bar{w})} \right| dA(u)$$

$$= \sum_{w \in r\mathbb{Z}^2} e^{-\frac{\alpha}{2}|z-w|^2} \int_{S_r+w} \left| 1 - e^{-\frac{\alpha}{2}|u-w|^2+\alpha(z-w)(\bar{u}-\bar{w})} \right| dA(u)$$

$$= \sum_{w \in r\mathbb{Z}^2} e^{-\frac{\alpha}{2}|z-w|^2} \int_{S_r} \left| 1 - e^{-\frac{\alpha}{2}|u|^2+\alpha(z-w)\bar{u}} \right| dA(u).$$

Since $|u| < r$ for all $u \in S_r$ and

$$|1 - e^\zeta| = \left| \sum_{k=1}^\infty \frac{\zeta^k}{k!} \right| \leq \sum_{k=1}^\infty \frac{|\zeta|^k}{k!} = e^{|\zeta|} - 1$$

for all complex numbers ζ, we have

$$\left| 1 - e^{-\frac{\alpha}{2}|u|^2+\alpha(z-w)\bar{u}} \right| \leq e^{\alpha|z-w|r+\frac{\alpha}{2}r^2} - 1 \leq r(e^{\alpha|z-w|+\frac{\alpha}{2}} - 1)$$

$$\leq C r e^{\frac{\alpha}{4}|z-w|^2}$$

for all $u \in S_r$, where C is a positive constant that only depends on α. Here, we used the additional assumption that $0 < r < 1$. It follows that there exists another positive constant C, independent of r and z, such that

$$J_r(z) \leq C r^3 \sum_{w \in r\mathbb{Z}^2} e^{-\frac{\alpha}{4}|z-w|^2}$$

for all $z \in \mathbb{C}$ and $0 < r < 1$. Since

$$e^{-\frac{\alpha}{4}|z-w|^2} = e^{-\frac{\alpha}{4}|z|^2} \left| e^{\frac{\alpha}{4}w\bar{z}} e^{-\frac{\alpha}{8}|w|^2} \right|^2,$$

an application of Lemma 2.32 shows that there is yet another positive constant C, independent of z and r, such that

$$J_r(z) \leq C r \sum_{w \in r\mathbb{Z}^2} \int_{S_r+w} e^{-\frac{\alpha}{4}|z-u|^2} dA(u)$$

$$= C r \int_{\mathbb{C}} e^{-\frac{\alpha}{4}|z-u|^2} dA(u)$$

$$= C r \int_{\mathbb{C}} e^{-\frac{\alpha}{4}|u|^2} dA(u) = \frac{4\pi C r}{\alpha}.$$

This shows that there exists another positive constant C, independent of r, such that

$$\|D_r f\|_{\infty,\alpha} \le Cr \|f\|_{\infty,\alpha}.$$

Consequently, the norm of D_r on F_α^∞ satisfies

$$\|D_r\|_{\infty,\alpha} \le Cr, \qquad 0 < r < 1. \tag{2.19}$$

To estimate the norm of D_r on F_α^1, first note that $|D_r f(z)|$ is less than or equal to

$$\frac{\alpha}{\pi} \sum_{w \in r\mathbb{Z}^2} \int_{S_r+w} \left| e^{\alpha z \bar{w} - \frac{\alpha}{2}|w|^2} - e^{\alpha z \bar{u} - \frac{\alpha}{2}|u|^2 - \alpha i \operatorname{Im}(w\bar{u})} \right| \, |f(u)| e^{-\frac{\alpha}{2}|u|^2} \, dA(u).$$

By Fubini's theorem, the integral

$$\int_{\mathbb{C}} |D_r f(z)| e^{-\frac{\alpha}{2}|z|^2} \, dA(z)$$

is less than or equal to

$$\frac{\alpha}{\pi} \sum_{w \in r\mathbb{Z}^2} \int_{S_r+w} |f(u)| e^{-\frac{\alpha}{2}|u|^2} H(w,u) \, dA(u),$$

where

$$\begin{aligned}
H(w,u) &= \int_{\mathbb{C}} e^{-\frac{\alpha}{2}|z|^2} \left| e^{\alpha z \bar{w} - \frac{\alpha}{2}|w|^2} - e^{\alpha z \bar{u} - \frac{\alpha}{2}|u|^2 - \alpha i \operatorname{Im}(w\bar{u})} \right| \, dA(z) \\
&= \int_{\mathbb{C}} \left| e^{-\frac{\alpha}{2}|z-w|^2} - e^{-\frac{\alpha}{2}|z-u|^2 + \alpha i \operatorname{Im}(z-w)(\bar{u}-\bar{w})} \right| \, dA(z) \\
&= \int_{\mathbb{C}} \left| e^{-\frac{\alpha}{2}|z|^2} - e^{-\frac{\alpha}{2}|z-(u-w)|^2 + \alpha i \operatorname{Im} z(\bar{u}-\bar{w})} \right| \, dA(z) \\
&= \int_{\mathbb{C}} e^{-\frac{\alpha}{2}|z|^2} \left| 1 - e^{\alpha z(\bar{u}-\bar{w}) - \frac{\alpha}{2}|u-w|^2} \right| \, dA(z).
\end{aligned}$$

Since $|u - w| < r$ for $u \in S_r + w$ and $|1 - e^\zeta| \le e^{|\zeta|} - 1$ for all complex numbers ζ, we have

$$\left| 1 - e^{\alpha z(\bar{u}-\bar{w}) - \frac{\alpha}{2}|u-w|^2} \right| \le e^{\alpha|z|r + \frac{\alpha}{2}r^2} - 1 \le r \left(e^{\alpha|z| + \frac{\alpha}{2}} - 1 \right).$$

It is now clear that we can find a positive constant $C = C(\alpha)$ such that $H(w,u) \le Cr$ for all w and u. It follows that

$$\int_{\mathbb{C}} |D_r f(z)| e^{-\frac{\alpha}{2}|z|^2} \, dA(z) \le Cr \int_{\mathbb{C}} |f(u)| e^{-\frac{\alpha}{2}|u|^2} \, dA(u)$$

for all $f \in F_\alpha^1$. Thus, the norm of D_r on F_α^1 satisfies

$$\|D_r\|_{1,\alpha} \leq Cr, \qquad 0 < r < 1. \tag{2.20}$$

By (2.19) and (2.20), if r is sufficiently small, then $\|D_r\|_{\infty,\alpha} < 1$ and $\|D_r\|_{1,\alpha} < 1$. By complex interpolation, we also have $\|D_r\|_{p,\alpha} < 1$ for all $1 \leq p \leq \infty$. This shows that if r is small enough, the operator T_r is invertible on F_α^p for all $1 \leq p \leq \infty$. When T_r is invertible and $f \in F_\alpha^p$, we can write $f = T_r g$ with $g = T_r^{-1} f$ and obtain the atomic decomposition (2.17) with

$$c_w = \frac{\alpha}{\pi} \int_{S_r+w} g(u) e^{-\frac{\alpha}{2}|u|^2 + \alpha i \mathrm{Im}\,(w\bar{u})} \, dA(u).$$

A simple argument with the help of Lemma 2.32 shows that the above sequence $\{c_w\}$ is in l^p whenever $g \in F_\alpha^p$. This completes the proof of the theorem in the case $1 \leq p \leq \infty$.

We will complete the proof of the case $0 < p < 1$ after we have proved the following three lemmas.   □

Lemma 2.35. *Suppose $0 < r < 1$, $0 < p \leq 1$, and m is a nonnegative integer. For any entire function f, we define a sequence*

$$\{(Sf)_{w,k} : w \in r\mathbb{Z}^2, 0 \leq k \leq m\}$$

by

$$(Sf)_{w,k} = \frac{\alpha}{\pi} \int_{S_r+w} e^{\alpha i \mathrm{Im}\, z(\bar{z}-\overline{w}) - \frac{\alpha}{2}|z-w|^2} \frac{(\bar{z} - \overline{w})^k}{k!} f(z) e^{-\frac{\alpha}{2}|z|^2} \, dA(z).$$

Then S maps F_α^p boundedly into l^p.

Proof. For any $w \in r\mathbb{Z}^2$, $z \in S_r + w$, and $1 \leq k \leq m$, we have

$$|(Sf)_{w,k}|^p = \frac{\alpha^p}{\pi^p} \left| \int_{S_r+w} e^{\alpha i \mathrm{Im}\, z(\bar{z}-\overline{w}) - \frac{\alpha}{2}|z-w|^2} \frac{(\bar{z} - \overline{w})^k}{k!} f(z) e^{-\alpha(z-w)\overline{w}} \right.$$
$$\left. e^{-\frac{\alpha}{2}|z-w|^2 - \frac{\alpha}{2}|w|^2 + \alpha i \mathrm{Im}\, (z-w)\overline{w}} \, dA(z) \right|^p$$

$$\leq C_1 r^{pk} \left[e^{-\frac{\alpha}{2}|w|^2} \int_{S_r+w} \left| f(z) e^{-\alpha(z-w)\overline{w}} \right| dA(z) \right]^p$$

$$\leq C_1 r^{p(2+k)} e^{-\frac{p\alpha}{2}|w|^2} \sup\{|f(z) e^{-\alpha(z-w)\overline{w}}|^p : z \in S_r + w\}$$

$$\leq C_2 r^{p(2+k)-2} e^{-\frac{p\alpha}{2}|w|^2} \int_{Q_r+w} \left| f(z) e^{-\alpha(z-w)\overline{w}} \right|^p dA(z)$$

$$= C_2 r^{p(2+k)-2} \int_{Q_r+w} \left| f(z) e^{\frac{\alpha}{2}|z-w|^2 - \frac{\alpha}{2}|z|^2} \right|^p dA(z)$$

$$\leq C_3 r^{p(2+k)-2} \int_{Q_r+w} \left| f(z) e^{-\frac{\alpha}{2}|z|^2} \right|^p dA(z).$$

Let $C_4 = C_3(m+1)$. Then

$$\sum_{w\in r\mathbb{Z}^2}\sum_{k=0}^{m}|(Sf)_{w,k}|^p \le C_4 r^{2(p-1)}\sum_{w\in r\mathbb{Z}^2}\int_{Q_{r+w}}\left|f(z)e^{-\frac{\alpha}{2}|z|^2}\right|^p dA(z)$$

$$\le C_5 r^{2(p-1)}\int_{\mathbb{C}}\left|f(z)e^{-\frac{\alpha}{2}|z|^2}\right|^p dA(z).$$

This proves the desired result. $\qquad\qquad\qquad\qquad\qquad\qquad\qquad\qquad\square$

Lemma 2.36. *Suppose $0 < r < 1$, $0 < p \le 1$, and m is a nonnegative integer. For every sequence*

$$c = \{c_{w,k} : w \in r\mathbb{Z}^2, 0 \le k \le m\},$$

define a function Tc by

$$Tc(z) = \sum_{w\in r\mathbb{Z}^2}\sum_{k=0}^{m}c_{w,k}[\alpha(z-w)]^k e^{\alpha z\bar{w}-\frac{\alpha}{2}|w|^2}.$$

Then T is a bounded linear operator from l^p into F_α^p.

Proof. It is obvious that the series converges to an entire function $f(z)$ uniformly on compact subsets of $\mathbb{C}$. Since $0 < p < 1$, it follows from Hölder's inequality that

$$|f(z)|^p \le \sum_{w\in r\mathbb{Z}^2}\sum_{k=0}^{m}|c_{w,k}|^p[\alpha|z-w|]^{pk}\left|e^{\alpha z\bar{w}-\frac{\alpha}{2}|w|^2}\right|^p.$$

Thus

$$\left|f(z)e^{-\frac{\alpha}{2}|z|^2}\right|^p \le \sum_{w\in r\mathbb{Z}^2}\sum_{k=0}^{m}|c_{w,k}|^k[\alpha|z-w|]^{pk}\left|e^{-\frac{\alpha}{2}|z-w|^2}\right|^p,$$

and hence

$$\int_{\mathbb{C}}\left|f(z)e^{-\frac{\alpha}{2}|z|^2}\right|^p dA(z) \le \sum_{w\in r\mathbb{Z}^2}\sum_{k=0}^{m}|c_{w,k}|^p\int_{\mathbb{C}}\left[|\alpha z|^k e^{-\frac{\alpha}{2}|z|^2}\right]^p dA(z)$$

$$\le C\sum_{w\in r\mathbb{Z}^2}\sum_{k=0}^{m}|c_{w,k}|^p.$$

This proves the desired result. $\qquad\qquad\qquad\qquad\qquad\qquad\qquad\qquad\square$

Lemma 2.37. *Let r_0 be the number from Theorem 2.34 in the case $p = \infty$. Suppose $0 < r < r_0$ and $0 < p \le 1$. Then every monomial z^k can be represented as*

$$z^k = \sum_{w\in r\mathbb{Z}^2}c_w k_w(z),$$

where $\{c_w\} \in l^p$.

Proof. Fix $\rho \in (r, r_0)$. By the already-proved case $p = \infty$ of Theorem 2.34, every monomial z^k can be represented as

$$z^k = \sum_{w \in \rho \mathbb{Z}^2} c_w k_w(z),$$

where $\{c_w\} \in l^\infty$. For $w \in \rho \mathbb{Z}^2$, we can write $w = \rho(m + in)$ for some integers m and n. Since

$$k_w(z) = e^{\alpha z \overline{w} - \frac{\alpha}{2}|w|^2},$$

we have for $w' = r(m + in)$ that

$$k_w((r/\rho)z) = e^{\alpha z \overline{w'} - \frac{\alpha}{2}|w'|^2} e^{\frac{\alpha}{2}(|w'|^2 - |w|^2)}.$$

It follows that

$$\left(\frac{r}{\rho}z\right)^k = \sum_{w' \in r\mathbb{Z}^2} c'_{w'} k_{w'}(z),$$

where

$$c'_{w'} = c_w e^{-(\rho^2 - r^2)(n^2 + m^2)}$$

is clearly a sequence in l^p. This proves the desired decomposition for monomials. $\qquad\square$

We can now finish the proof of Theorem 2.34 in the case $0 < p < 1$.

Fix a sufficiently small $r \in (0, 1)$, let m be the integer part of $2(1 - p)/p$, and let S and T be the operators defined in the previous two lemmas. We have

$$(I - TS)f(z) = \frac{\alpha}{\pi} \sum_{w \in r\mathbb{Z}^2} \int_{S_r + w} G(z, w, u) f(u) e^{-\frac{\alpha}{2}|u|^2} \, dA(u),$$

where

$$G = k_u(z) - e^{\alpha i \mathrm{lm}\, u(\overline{u} - \overline{w}) - \frac{\alpha}{2}|u - w|^2} \left[\sum_{k=0}^{m} \frac{[\alpha(z - w)(\overline{u} - \overline{w})]^k}{k!} \right] k_w(z).$$

It is elementary to check that

$$G = e^{\alpha i \mathrm{lm}\, u(\overline{u} - \overline{w}) - \frac{\alpha}{2}|u - w|^2} \left[\sum_{k=m+1}^{\infty} \frac{[\alpha(z - w)(\overline{u} - \overline{w})]^k}{k!} \right] k_w(z).$$

For $u \in S_r + w$, we have $|u - w| < r$. Therefore,

$$|G| \le |k_w(z)| \sum_{k=m+1}^{\infty} \frac{(\alpha r|z - w|)^k}{k!},$$

and so by Hölder's inequality, $|(I - TS)f(z)|^p$ is less than or equal to

$$\frac{\alpha^p}{\pi^p} \sum_{w \in r\mathbb{Z}^2} |k_w(z)|^p \left[\sum_{k=m+1}^{\infty} \frac{(\alpha r|z - w|)^k}{k!} \right]^p \left[\int_{S_r+w} |f(u)|e^{-\frac{\alpha}{2}|u|^2} dA(u) \right]^p.$$

It follows from this and Fubini's theorem that

$$\int_{\mathbb{C}} \left| (I - TS)f(z)e^{-\frac{\alpha}{2}|z|^2} \right|^p dA(z)$$

is less than or equal to

$$\frac{\alpha^p}{\pi^p} \sum_{w \in r\mathbb{Z}^2} C(w) \left[\int_{S_r+w} |f(u)|e^{-\frac{\alpha}{2}|u|^2} dA(u) \right]^p,$$

where

$$C(w) = \int_{\mathbb{C}} e^{-\frac{p\alpha}{2}|z-w|^2} \left[\sum_{k=m+1}^{\infty} \frac{(\alpha r|z - w|)^k}{k!} \right]^p dA(z)$$

$$= \int_{\mathbb{C}} e^{-\frac{p\alpha}{2}|z|^2} \left[\sum_{k=m+1}^{\infty} \frac{(\alpha r|z|)^k}{k!} \right]^p dA(z)$$

$$\leq r^{(m+1)p} \int_{\mathbb{C}} e^{-\frac{p\alpha}{2}|z|^2} \left[\sum_{k=m+1}^{\infty} \frac{(\alpha|z|)^k}{k!} \right]^p dA(z)$$

$$\leq r^{(m+1)p} \int_{\mathbb{C}} e^{-\frac{p\alpha}{2}|z|^2 + p\alpha|z|} dA(z).$$

So there is a constant $C > 0$ such that

$$\int_{\mathbb{C}} \left| (I - TS)f(z)e^{-\frac{\alpha}{2}|z|^2} \right|^p dA(z)$$

is less than or equal to

$$Cr^{(m+1)p} \sum_{w \in r\mathbb{Z}^2} \left[\int_{S_r+w} |f(u)|e^{-\frac{\alpha}{2}|u|^2} \right]^p.$$

On the other hand,

$$\left[\int_{S_r+w} |f(u)|e^{-\frac{\alpha}{2}|u|^2} \right]^p \leq r^{2p} \sup_{u \in S_r+w} \left| f(u)e^{-\frac{\alpha}{2}|u|^2} \right|^p,$$

and an application of Lemma 2.32 produces another constant $C > 0$ (independent of $r \in (0,1)$) such that

$$\left[\int_{S_r+w} |f(u)| e^{-\frac{\alpha}{2}|u|^2} \right]^p \leq Cr^{2p-2} \int_{Q_r+w} \left| f(u) e^{-\frac{\alpha}{2}|u|^2} \right|^p dA(z).$$

Thus,

$$\int_{\mathbb{C}} \left| (I - TS)f(z) e^{-\frac{\alpha}{2}|z|^2} \right|^p dA(z)$$

is less than or equal to

$$Cr^{(m+1)p+2p-2} \sum_{w \in r\mathbb{Z}^2} \int_{Q_r+w} \left| f(u) e^{-\frac{\alpha}{2}|u|^2} \right|^p dA(u).$$

So we can find another constant $C > 0$, independent of $r \in (0,1)$, such that

$$\|I - TS\|_{p,\alpha} \leq Cr^{(m+3)-(2/p)}, \qquad 0 < r < 1.$$

Since

$$m+3 - \frac{2}{p} > \frac{2(1-p)}{p} - 1 + 3 - \frac{2}{p} = 0,$$

we see that there exists some $r_0 \in (0,1)$ such that $\|I - TS\|_{p,\alpha} < 1$ whenever $r \in (0, r_0)$. This shows that the operator TS is invertible on F_α^p whenever $r \in (0, r_0)$.

Consequently, for any $r \in (0, r_0)$, the operator T is onto, and so every function $f \in F_\alpha^p$ can be written as

$$f(z) = \sum_{w \in r\mathbb{Z}^2} \sum_{k=0}^{m} c_{w,k}(z-w)^k e^{-\alpha z \bar{w} - \frac{\alpha}{2}|w|^2}. \tag{2.21}$$

Furthermore, the coefficients $c_{w,k}$ in (2.21) all depend on f linearly, and

$$\sum_{w \in r\mathbb{Z}^2} \sum_{k=0}^{m} |c_{w,k}|^p \leq C \|f\|_{p,\alpha}^p,$$

where C is a positive constant independent of f.

Given any $\delta > 0$ and any $r \in (0, r_0)$, it follows from Lemma 2.37 that there exist coefficients $c'_{w,k}$, $0 \leq k \leq m$, $w \in r\mathbb{Z}^2$, $|w| \leq N$, such that

$$\left\| z^k - \sum_{u \in r\mathbb{Z}^2, |u| \leq N} c'_{u,k} e^{\alpha z \bar{u} - \frac{\alpha}{2}|u|^2} \right\|_{p,\alpha} < \delta$$

for all $0 \leq k \leq m$. By a change of variables, the norm of

$$(z-w)^k e^{\alpha z \bar{w} - \frac{\alpha}{2}|w|^2} - \sum_{u \in r\mathbb{Z}^2, |u| \leq N} c'_{u,k} e^{\alpha(z-w)\bar{u} - \frac{\alpha}{2}|u|^2 + \alpha z \bar{w} - \frac{\alpha}{2}|w|^2}$$

in F_α^p is less than δ for all $0 \le k \le m$ and $w \in r\mathbb{Z}^2$. Define an operator A_r on F_α^p by

$$A_r f(z) = \sum_{w \in r\mathbb{Z}^2, 0 \le k \le m} c_{w,k} \sum_{u \in r\mathbb{Z}^2, |u| \le N} c'_{u,k} e^{\alpha \mathrm{iIm}\,(w\bar{u})} e^{\alpha z(\bar{w}+\bar{u}) - \frac{\alpha}{2}|w+u|^2}$$

and observe that

$$\alpha(z-w)\bar{u} - \frac{\alpha}{2}|u|^2 + \alpha z\bar{w} - \frac{\alpha}{2}|w|^2 = \alpha \mathrm{iIm}\,(w\bar{u}) + \alpha z(\bar{w}+\bar{u}) - \frac{\alpha}{2}|w+u|^2.$$

It then follows from Hölder's inequality that

$$\|f - A_r f\|_{p,\alpha}^p \le \sum_{w \in r\mathbb{Z}^2, 0 \le k \le m} |c_{w,k}|^p \delta^p \le C\delta^p \|f\|_{p,\alpha}^p$$

for all $f \in F_\alpha^p$. If we choose δ such that $C\delta^p < 1$, then $\|I - A_r\|_{p,\alpha} < 1$, and so the operator A_r is surjective on F_α^p. Since $w + u \in r\mathbb{Z}^2$ whenever $w \in r\mathbb{Z}^2$ and $u \in r\mathbb{Z}^2$, the proof of Theorem 2.34 is now complete.

2.6   Translation Invariance

In this section, we consider the action of translations on Fock spaces and determine three spaces that are unique under such actions: the space F_α^∞ is maximal among translation invariant Banach spaces of entire functions, the space F_α^1 is minimal among translation invariant Banach spaces of entire functions, and the space F_α^2 is the only Hilbert space of entire functions invariant under translations.

For any point $a \in \mathbb{C}$, we define three analytic self-maps of the complex plane as follows:

$$t_a(z) = z + a, \qquad \tau_a(z) = z - a, \qquad \varphi_a(z) = a - z.$$

The map t_a is naturally called the translation by a, and it is clear that $\tau_a = t_{-a} = t_a^{-1}$. The map φ_a is the composition of the translation t_a with the reflection $z \mapsto -z$. Note that φ_a is its own inverse.

When making a change of variables, observe that

$$\int_{\mathbb{C}} f \circ t_a(z)\, d\lambda_\alpha(z) = \int_{\mathbb{C}} f \circ \varphi_a(z)\, d\lambda_\alpha(z)$$

$$= \frac{\alpha}{\pi} \int_{\mathbb{C}} f(w) e^{-\alpha|a-w|^2}\, dA(w)$$

$$= \int_{\mathbb{C}} f(w) |k_a(w)|^2\, d\lambda_\alpha(w).$$

On the other hand,

$$\int_{\mathbb{C}} f \circ \tau_a(z)\, d\lambda_\alpha(z) = \frac{\alpha}{\pi} \int_{\mathbb{C}} f(w) e^{-\alpha|w+a|^2}\, dA(w)$$

$$= \int_{\mathbb{C}} f(w) |k_{-a}(w)|^2\, d\lambda_\alpha(w).$$

Similarly,

$$\int_{\mathbb{C}} f \circ \tau_a(z) |k_a(z)|^2\, d\lambda_\alpha(z) = \int_{\mathbb{C}} f \circ \varphi_a(z) |k_a(z)|^2\, d\lambda_\alpha(z)$$

$$= \int_{\mathbb{C}} f(z)\, d\lambda_\alpha(z),$$

while

$$\int_{\mathbb{C}} f \circ t_a(z) |k_a(z)|^2\, d\lambda_\alpha(z) = \int_{\mathbb{C}} f(z + 2a)\, d\lambda_\alpha(z).$$

See Corollary 2.4. These are some of the subtle differences that can easily be overlooked.

We can use τ_a and φ_a to define certain unitary operators on F_α^2. Although there is an obvious temptation to use only one of these maps in the book, we have found that there are situations in which one choice is more convenient than the other. Therefore, we are going to use both in the book.

For a fixed weight parameter α and $a \in \mathbb{C}$, we define two operators W_a and U_a as follows:

$$W_a f = f \circ \tau_a k_a, \qquad U_a f = f \circ \varphi_a k_a,$$

where k_a is the normalized reproducing kernel of F_α^2 at a. We will consider the action of these operators on both L_α^p and F_α^p. The focus in this section is their action on Fock spaces.

These are weighted translation operators. In some of the literature, the operators W_a are called Weyl (unitary) operators. We first show that both W_a and U_a are isometries on the Fock spaces F_α^p.

Proposition 2.38. *Let* $0 < p \le \infty$. *We have*

$$\|W_a f\|_{p,\alpha} = \|U_a f\|_{p,\alpha} = \|f\|_{p,\alpha}$$

for all $a \in \mathbb{C}$ *and* $f \in F_\alpha^p$. *Furthermore, both* W_a *and* U_a *are invertible on* F_α^p *with* $W_a^{-1} = W_{-a}$ *and* $U_a^{-1} = U_a$. *Consequently,* W_a *and* U_a *are both unitary operators on* F_α^2 *with* $W_a^* = W_{-a}$ *and* $U_a^* = U_a$.

Proof. It is easy to check that

$$e^{-\frac{\alpha}{2}|z|^2}|W_a f(z)| = e^{-\frac{\alpha}{2}|z-a|^2}|f(z-a)|,$$

and

$$e^{-\frac{\alpha}{2}|z|^2}|U_a f(z)| = e^{-\frac{\alpha}{2}|a-z|^2}|f(a-z)|.$$

The identities

$$\|W_a f\|_{p,\alpha} = \|U_a f\|_{p,\alpha} = \|f\|_{p,\alpha}$$

then follow from a change of variables. See Corollary 2.4.

To see that W_a is invertible with $W_a^{-1} = W_{-a}$, take any $f \in F_\alpha^p$ and note that

$$
\begin{aligned}
W_{-a} W_a f(z) &= e^{-\alpha \bar{a} z - \frac{\alpha}{2}|a|^2}(W_a f)(z+a) \\
&= e^{-\alpha \bar{a} z - \frac{\alpha}{2}|a|^2} e^{\alpha \bar{a}(z+a) - \frac{\alpha}{2}|a|^2} f(z+a-a) \\
&= f(z).
\end{aligned}
$$

A similar argument shows that U_a is invertible with $U_a^{-1} = U_a$. This completes the proof of the proposition. $\square$

Although the operators W_a and U_a behave similarly in many situations, there are sometimes reasons to pick one over the other. For example, the operators W_a almost have a semigroup property with respect to a, while the operators U_a are all self-adjoint. In particular, we can use the Weyl operators to obtain the following unitary representation of the Heisenberg group. Recall that another unitary representation was given in Chap. 1 based on Weyl pseudodifferential operators.

Theorem 2.39. *The mapping* $(a, \theta) \mapsto e^{i\theta} W_a$ *is a unitary representation of the Heisenberg group* $\mathbb{H}$ *on the Fock space* F_α^2.

Proof. For any two points a and b in $\mathbb{C}$, we easily check that

$$W_a W_b = e^{-\alpha i \operatorname{Im}(a\bar{b})} W_{a+b} = e^{\alpha i \operatorname{Im}(\bar{a}b)} W_{a+b}. \tag{2.22}$$

This shows that $(a, \theta) \mapsto e^{i\theta} W_a$ is a group embedding of $\mathbb{H}$ into the group of unitary operators on F_α^2. □

In the rest of this section, we work with the Weyl unitary operators W_a. A similar theory can be developed with the unitary operators U_a, which is left to the reader as an exercise.

Proposition 2.40. *The Fock space* F_α^∞ *is maximal in the sense that if* X *is any Banach space of entire functions with the following properties:*

(a) $\|W_a f\|_X = \|f\|_X$ *for all* $a \in \mathbb{C}$ *and* $f \in X$,
(b) *the point evaluation* $f \mapsto f(0)$ *is a bounded linear functional on* X,

then $X \subset F_\alpha^\infty$ *and the inclusion is continuous.*

Proof. Condition (a) implies that $W_a f \in X$ for every $f \in X$ and every $a \in \mathbb{C}$. Combining this with condition (b), we see that for every $a \in \mathbb{C}$, the point evaluation $f \mapsto f(a)$ is also a bounded linear functional on X, and

$$e^{-\frac{\alpha}{2}|a|^2} |f(a)| = |W_{-a} f(0)| \le C \|W_{-a} f\|_X = C \|f\|_X,$$

where C is a positive constant that is independent of $a \in \mathbb{C}$ and $f \in X$. Since a is arbitrary, we conclude that $f \in F_\alpha^\infty$ with $\|f\|_{\infty,\alpha} \le C \|f\|_X$ for all $f \in X$. □

Proposition 2.41. *The Fock space* F_α^1 *is minimal in the sense that if* X *is a Banach space of entire functions with the following properties:*

(a) $\|W_a f\|_X = \|f\|_X$ *for all* $a \in \mathbb{C}$ *and* $f \in X$,
(b) X *contains all constant functions,*

then $F_\alpha^1 \subset X$ *and the inclusion is continuous.*

Proof. Since X contains all constant functions, applying W_a to the constant function 1 shows that for each $a \in \mathbb{C}$, the function

$$k_a(z) = e^{\alpha \bar{a} z - \frac{\alpha}{2}|a|^2}$$

belongs to X. Furthermore, $\|k_a\|_X = \|W_a 1\|_X = \|1\|_X$ for all $a \in \mathbb{C}$.

Let $\{z_n\}$ denote a sequence in $\mathbb{C}$ on which we have atomic decomposition for F_α^1. If $f \in F_\alpha^1$, there exists a sequence $\{c_n\} \in l^1$ such that

$$f = \sum_{n=1}^{\infty} c_n k_{z_n}. \tag{2.23}$$

Since each k_{z_n} belongs to X and $\sum |c_n| < \infty$, we conclude that $f \in X$ with

$$\|f\|_X \leq \sum_{n=1}^{\infty} |c_n| \|k_{z_n}\|_X = C \sum_{n=1}^{\infty} |c_n|,$$

where $C = \|1\|_X > 0$. Taking the infimum over all sequences $\{c_n\}$ satisfying (2.23), we obtain another constant $C > 0$ such that

$$\|f\|_X \leq C\|f\|_{F_\alpha^1}, \qquad f \in F_\alpha^1.$$

This proves the desired result.   □

Proposition 2.42. *Suppose H is a nontrivial separable Hilbert space of entire functions with the following properties:*

(a) $\|W_a f\|_H = \|f\|_H$ for all $a \in \mathbb{C}$ and $f \in H$.
(b) $f \mapsto f(0)$ is a bounded linear functional on H.

Then $H = F_\alpha^2$ and there exists a positive constant c such that $\langle f, g \rangle_H = c\langle f, g \rangle_\alpha$ for all f and g in H.

Proof. Since H contains at least one function that is not identically zero, it follows from conditions (a) and (b) that for any $z \in \mathbb{C}$, the mapping $f \mapsto f(z)$ is a nonzero bounded linear functional on H. Furthermore, for any compact subset S of $\mathbb{C}$, there exists a positive constant C such that $|f(z)| \leq C\|f\|_H$ for all $f \in H$ and all $z \in S$.

Consequently, the space H possesses a reproducing kernel $K_H(z, w)$. Moreover, if $\{e_n\}$ is an orthonormal basis of H, then

$$K_H(z, w) = \sum_{n=0}^{\infty} e_n(z) \overline{e_n(w)}, \qquad (2.24)$$

and the convergence is uniform when z and w are restricted to compact subsets of $\mathbb{C}$. In particular, the series representation for $K_H(z, w)$ in (2.24) is independent of the choice of the orthonormal basis $\{e_n\}$.

It is easy to see from condition (a) and the proof of Proposition 2.38 that each W_a is a unitary operator on H. Fix any $a \in \mathbb{C}$ and let $\sigma_n = W_a e_n$, $n \geq 1$. Then $\{\sigma_n\}$ is also an orthonormal basis of H. Therefore, by (2.24), we have

$$K_H(z, w) = \sum_{n=1}^{\infty} \sigma_n(z) \overline{\sigma_n(w)}$$

$$= k_a(z) \overline{k_a(w)} \sum_{n=1}^{\infty} e_n(z - a) \overline{e_n(w - a)}$$

$$= k_a(z) \overline{k_a(w)} K_H(z - a, w - a),$$

where k_a is the normalized reproducing kernel of F_α^2 at a. Let $z = w = a$. We obtain

$$K_H(z,z) = e^{\alpha |z|^2} K_H(0,0) = K_\alpha(z,z) K_H(0,0), \qquad z \in \mathbb{C}^n,$$

where $K_\alpha(z,w)$ is the reproducing kernel of F_α^2.

By a well-known result in the function theory of several complex variables, any reproducing kernel is uniquely determined by its values on the diagonal. See [142]. Therefore, we must have $K_H(z,w) = cK(z,w)$ for all z and w, where $c = K_H(0,0) > 0$ as H contains functions that do not vanish at the origin. This shows that, after an adjustment of the inner product by a positive scalar, the two spaces H and F_α^2 have the same reproducing kernel, from which it follows that $H = F_\alpha^2$. This completes the proof of the proposition. $\square$

2.7 A Maximum Principle

The classical maximum principle asserts that if f is an entire function and $|f(z)| \leq M$ for all $|z| = R$, then $|f(z)| \leq M$ for all $|z| \leq R$. The purpose of this section is to prove the following version of the maximum principle for Fock spaces.

Theorem 2.43. *For any $\alpha > 0$ and $p \geq 1$, there exists a positive radius $R = R(\alpha, p)$ such that $\|f\|_{p,\alpha} \leq \|g\|_{p,\alpha}$ for all entire functions f and g satisfying*

$$|f(z)| \leq |g(z)|, \qquad |z| \geq R.$$

Proof. Without loss of generality, we may assume that $g \in F_\alpha^p$. Otherwise, the desired result is obvious. Under this assumption, we also have

$$\int_{|z| \geq R} |f|^p \, d\lambda_\alpha \leq \int_{|z| \geq R} |g|^p \, d\lambda_\alpha < \infty,$$

which easily implies that $f \in F_\alpha^p$ as well.

For any positive radius r and any function F in the complex plane, we write

$$I(r, F) = \int_0^{2\pi} F(re^{i\theta}) \, d\theta.$$

We fix some $R > 0$ and assume that $|f(z)| \leq |g(z)|$ for all $R \leq |z| < \infty$.

We will try to compare $I(r, |f|^p - |g|^p)$ for $0 < r < R$ to $I(\rho, |g|^p - |f|^p)$ for $R < \rho < \infty$. To this end, we let $\omega(z) = f(z)/g(z)$, which is analytic and has modulus less than or equal to 1 in the region $R < |z| < \infty$. We may assume that $|\omega(z)| < 1$ for all $R < |z| < \infty$. In fact, if $|\omega(z_0)| = 1$ for some $R < |z_0| < \infty$, then by the classical maximum modulus principle, the analytic function ω on $R < |z| < \infty$ must be constant, which would then imply that f and g differ by a constant multiple in the whole complex plane, from which the desired result clearly follows.

For any $\rho \in (R, \infty)$, pick a point $\zeta(\rho)$ such that $|\zeta(\rho)| = \rho$ and

$$|\omega(\zeta(\rho))| = \max\{|\omega(z)| : |z| = \rho\}.$$

We may assume that f is not identically 0, for otherwise the desired result is trivial. Thus, $0 < |\omega(\zeta(\rho))| < 1$ for all $\rho \in (R, \infty)$. To simplify notation, let us write $\omega_\rho = \omega(\zeta(\rho))$.

Since $p \geq 1$, it follows from elementary calculus that

$$py^{p-1}(x - y) \leq x^p - y^p \leq px^{p-1}(x - y), \tag{2.25}$$

for all $x \geq 0$ and $y \geq 0$. We deduce from the second inequality in (2.25) that for any $0 \leq r \leq R < \rho < \infty$, we have

$$I(r, |f|^p - |g|^p) \leq I(r, |f|^p - |\omega_\rho g|^p)$$
$$\leq I(r, p|f|^{p-1}(|f| - |\omega_\rho g|))$$
$$\leq I(r, p|f|^{p-1}|f - \omega_\rho g|).$$

The function $p|f|^{p-1}|f - \omega_\rho g|$ is subharmonic on the complex plane, so its integral mean on $|z| = r$ is an increasing function of r (see [76] for example). Thus,

$$I(r, |f|^p - |g|^p) \leq I(\rho, p|f|^{p-1}|f - \omega_\rho g|)$$
$$= I(\rho, p|\omega|^{p-1}|\omega - \omega_\rho|(|g|^p - |f|^p)/(1 - |\omega|^p)).$$

Taking $x = 1$ and $y = |\omega|$ in the first inequality of (2.25), we get

$$\frac{p|\omega|^{p-1}}{1 - |\omega|^p} \leq \frac{1}{1 - |\omega|} = \frac{1 + |\omega|}{1 - |\omega|^2} < \frac{2}{1 - |\omega|^2}.$$

Therefore,

$$I(r, |f|^p - |g|^p) \leq 2I(\rho, |\omega - \omega_\rho|(|g|^p - |f|^p)/(1 - |\omega|^2)) \qquad (2.26)$$

for all $0 \leq r \leq R < \rho < \infty$.
     Set

$$\gamma(\rho) = \max\left\{ \frac{|\omega(z) - \omega_\rho|}{1 - |\omega(z)|^2} : |z| = \rho \right\}$$

for $\rho \in (R, \infty)$. By (2.26),

$$I(r, |f|^p - |g|^p) \leq 2\gamma(\rho)I(\rho, |g|^p - |f|^p) \qquad (2.27)$$

for all $0 \leq r \leq R < \rho < \infty$. Fix ρ and integrate both sides of (2.27) over $[0, R]$ against the measure $re^{-\alpha r^2} dr$. The result is

$$\int_{|z| \leq R} (|f|^p - |g|^p) \, d\lambda_\alpha \leq \frac{1 - e^{-\alpha R^2}}{\pi} \gamma(\rho)I(\rho, |g|^p - |f|^p) \qquad (2.28)$$

for all $R < \rho < \infty$. Divide both sides of (2.28) by $\gamma(\rho)$ and integrate both sides over (R, ∞) against the measure $\rho e^{-\alpha \rho^2} d\rho$. The result is

$$\int_{|z| \leq R} (|f|^p - |g|^p) \, d\lambda_\alpha \leq C_R \int_{|z| \geq R} (|g|^p - |f|^p) \, d\lambda_\alpha,$$

where

$$C_R = \frac{1 - e^{-\alpha R^2}}{\alpha \int_R^\infty \dfrac{\rho e^{-\alpha \rho^2}}{\gamma(\rho)} \, d\rho}.$$

If R is a positive radius such that $C_R < 1$, then the integral

$$J = \int_{\mathbb{C}} (|f|^p - |g|^p) \, d\lambda_\alpha$$

satisfies the following estimates:

$$J = \left(\int_{|z| \le R} + \int_{|z| \ge R} \right) (|f|^p - |g|^p) \, d\lambda_\alpha$$

$$\le C_R \int_{|z| \ge R} (|g|^p - |f|^p) \, d\lambda_\alpha + \int_{|z| \ge R} (|f|^p - |g|^p) \, d\lambda_\alpha$$

$$\le \int_{|z| \ge R} (|g|^p - |f|^p) \, d\lambda_\alpha + \int_{|z| \ge R} (|f|^p - |g|^p) \, d\lambda_\alpha$$

$$= 0,$$

which proves the desired result.

We will actually show that $C_R < 1$ for all sufficiently small positive radius R. To this end, let d denote the pseudohyperbolic metric in the unit disk $\mathbb{D}$, namely,

$$d(z, w) = \left| \frac{z - w}{1 - \bar{z}w} \right|.$$

Since

$$\frac{|a - b|}{1 - |a|^2} = \frac{d(a, b)}{\sqrt{1 - d^2(a, b)}} \frac{\sqrt{1 - |b|^2}}{\sqrt{1 - |a|^2}}$$

for all a and b in the unit disk, we see that for all z with $|z| = \rho$,

$$\frac{|\omega(z) - \omega_\rho|}{1 - |\omega(z)|^2} = \frac{d(\omega(z), \omega_\rho)}{\sqrt{1 - d^2(\omega(z), \omega_\rho)}} \frac{\sqrt{1 - |\omega_\rho|^2}}{\sqrt{1 - |\omega(z)|^2}}$$

$$\le \frac{d(\omega(z), \omega(\zeta(\rho)))}{\sqrt{1 - d^2(\omega(z), \omega(\zeta(\rho)))}}.$$

It follows that

$$\gamma(\rho) \le \sup_{|z| = \rho} \frac{d(\omega(z), \omega(\zeta(\rho)))}{\sqrt{1 - d^2(\omega(z), \omega(\zeta(\rho)))}}, \qquad R < \rho < \infty. \qquad (2.29)$$

The function $H(z) = \omega(R/z)$ is analytic from the punctured disk $0 < |z| < 1$ into the unit disk. Since H is bounded near $z = 0$, it has a removable singularity at $z = 0$. Thus, we can think of H as analytic self-maps of the unit disk. By the classical Schwarz lemma, we have

$$d(H(z), H(w)) \leq d(z, w), \qquad z, w \in \mathbb{D}.$$

It follows that

$$d(\omega(z), \omega(\zeta(\rho))) = d(H(R/z), H(R/\zeta(\rho))) \leq d(R/z, R/\zeta(\rho))$$

for all $|z| = \rho$. Combining this with (2.29), we obtain

$$\gamma(\rho) \leq \sup_{|z|=\rho} \frac{d(R/z, R/\zeta(\rho))}{\sqrt{1 - d^2(R/z, R/\zeta(\rho))}}.$$

By symmetry of the unit disk,

$$\sup_{|z|=\rho} d(R/z, R/\zeta(\rho)) = d(-R/\zeta(\rho), R/\zeta(\rho)) = \frac{2R\rho}{\rho^2 + R^2}.$$

From this, we deduce that

$$\gamma(\rho) \leq \frac{2R\rho}{\rho^2 - R^2}, \qquad R < \rho < \infty.$$

Plugging this into the formula for C_R, we obtain the estimate

$$C_R \leq \frac{2R(1 - e^{-\alpha R^2})}{\alpha \displaystyle\int_R^\infty (\rho^2 - R^2) e^{-\alpha \rho^2}\, d\rho}.$$

The quotient above tends to 0 as $R \to 0^+$. Therefore, $C_R < 1$ for all sufficiently small positive radius R. This completes the proof of the theorem. $\qquad \square$

If $0 < p < 1$, the inequalities in (2.25) are replaced by

$$px^{p-1}(x - y) \leq x^p - y^p \leq py^{p-1}(x - y), \tag{2.30}$$

and a similar sequence of estimates leads to

$$I(r, |f|^p - |g|^p) \leq I(\rho, p|\omega_\rho|^{p-1}|\omega - \omega_\rho|(|g|^p - |f|^p)/(1 - |\omega|^p))$$

for all $0 \leq r \leq R < \rho < \infty$. So in this case, we need to consider the function

$$\gamma(\rho) = \max \left\{ \frac{p|\omega_\rho|^{p-1}|\omega(z) - \omega_\rho|}{1 - |\omega(z)|^p} : |z| = \rho \right\}, \qquad R < \rho < \infty.$$

Note that the function γ depends on f and g. We just need to bound γ from above by a function that is independent of f and g. By the left inequality in (2.30), we have

$$\gamma(\rho) \leq \sup_{|z|=\rho} \frac{|\omega_\rho|^{p-1}|\omega(z) - \omega_\rho|}{1 - |\omega(z)|} \leq 2|\omega_\rho|^{p-1} \sup_{|z|=\rho} \frac{|\omega(z) - \omega_\rho|}{1 - |\omega(z)|^2}.$$

Therefore, we just need to bound $|\omega_\rho|$ from below by a positive function that is independent of f and g. But this is impossible, for we may have a situation like $f(z) = g(z)/N$, where N is large; in this case, we have $H = 1/N$, and we can choose N to be arbitrarily large.

2.8 Notes

There are two ways to define the Fock spaces. One way is to consider subspaces $L^p(\mathbb{C}, d\lambda_\alpha)$ consisting of entire functions. This would be similar to the definitions of the more classical Hardy and Bergman spaces. It turns out that this is not a good way to define the Fock spaces. The seemingly cumbersome definition of F_α^p as the space of entire functions f such that $f(z)e^{-\alpha|z|^2/2}$ belongs to $L^p(\mathbb{C}, dA)$ will make the statements and proofs of many results much easier and more convenient later on.

The constant α in F_α^p is not essential in our theory. No generality is lost if we choose to develop the theory with a particular choice of α, say $\alpha = 1$. This weight parameter plays the role of Planck's constant in mathematical physics, and it provides us with an extra level of freedom that is useful in several situations.

Although the Fock space F_α^2 is a central subject in quantum physics, this book is focused on purely mathematical analysis on Fock spaces. No serious effort is made to show any connections or applications to physics. We refer the interested reader to books such as [177] for applications of the Fock space in physics.

The characterization of the boundedness of P_α and Q_α on L^p spaces was obtained in [74], where more precise norm estimates can also be found. See [96] for an even more elaborate study of similar integral operators. The boundedness of the projection P_α on L_α^p for $1 \le p \le \infty$ can be found in [138]. The papers [214] and [217] also study the boundedness of P_α and Q_α on L^p spaces.

The study of the Heisenberg group is a small industry by itself. This is especially so in quantum physics and harmonic analysis, where the connection of Fock spaces to the Heisenberg group is evident. But we will not use the Heisenberg group in any way other than the special elements W_a in it.

The paper [138] by Janson, Peetre, and Rochberg is a key reference throughout this book. In particular, the duality, atomic decomposition, and complex interpolation for the Fock spaces F_α^p, where $1 \le p \le \infty$, were proved in [138]. Our presentation of the case $0 < p < 1$ follows [231] very closely.

The translation invariance of the Fock spaces was first considered in [138], where it was shown that F_α^1 is minimal and F_α^∞ is maximal among Banach spaces of entire functions whose norm is invariant under the action of the Heisenberg group. The uniqueness of F_α^2 among Hilbert spaces of entire functions whose norm is invariant under the action of W_a was proved in [255].

The version of the maximum modulus principle in Sect. 2.7 was first proved in [194], based on a technique introduced in [122] to tackle the corresponding problem for Bergman spaces on the unit disk. That such a maximum modulus principle might be true for the Bergman space was first conjectured by Korenblum in [141] and was proved in [117] in the case $p = 2$ and in [122] when $1 \le p < \infty$. See [232–239] for other work concerning Korenblum's maximum principle.

2.9 Exercises

1. Show that the Fock space F_α^p is a closed subspace of $L^p(\mathbb{C}, d\lambda_{p\alpha/2})$.
2. Show that M_z, the operator of multiplication by the coordinate function z, is a densely defined unbounded linear operator on F_α^2. Show that the adjoint of M_z on F_α^2 is essentially the operator of differentiation. More specifically, $M_z^* f(z) = (1/\alpha) f'(z)$ for all $f \in F_\alpha^2$.
3. Let $0 < p \le \infty$, S be a compact subset of $\mathbb{C}$, and k be a positive integer. Show that there exists a positive constant C such that

$$|f^{(k)}(z)| \le C\|f\|_{p,\alpha}$$

   for all $z \in S$ and $f \in F_\alpha^p$.
4. Let φ be an entire function. Show that the composition operator C_φ defined by $C_\varphi f = f \circ \varphi$ is bounded on F_α^p if and only if $\varphi(z) = az + b$, where $|a| < 1$ or $|a| = 1$ and $b = 0$. Characterize compact composition operators on F_α^p. See [46] and [110].
5. Suppose $1 < p < \infty$ and $f \in F_\alpha^p$. Show that the Taylor polynomials of f converge to f in the norm topology of F_α^p.
6. Suppose $0 < p \le 1$. Are there functions $f \in F_\alpha^p$ such that the Taylor polynomials of f do not converge to f in the norm topology of F_α^p? See [256] for the corresponding problem in the context of Hardy and Bergman spaces.
7. Show that f_α^∞ is a closed subspace of F_α^∞.
8. Show that the set of polynomials is dense in f_α^∞.
9. Characterize the space $P_\alpha C_0(\mathbb{C})$, where $C_0(\mathbb{C})$ is the space of continuous functions on $\mathbb{C}$ that vanish at ∞.
10. If $1 < p < \infty$ and $f(z) = \sum a_n z^n$ is a function in F_α^p, then

$$a_n = o\left(\sqrt{\frac{\alpha^n}{n!}}\, n^{\frac{1}{4} - \frac{1}{2p}}\right), \qquad n \to \infty.$$

    See [224] for this and the next few problems.
11. If $f(z) = \sum a_n z^n$ is a function in F_α^1, then

$$a_n = O\left(\sqrt{\frac{\alpha^n}{n!}}\, n^{-\frac{1}{4}}\right), \qquad n \to \infty.$$

12. Let $1 \le p < \infty$ and let $\{\delta_n\}$ be any sequence of positive numbers decreasing to 0. Then there exists a function $f(z) = \sum a_n z^n$ in F_α^p such that

$$a_n \ne O\left(\sqrt{\frac{\alpha^n}{n!}}\, n^{\frac{1}{4} - \frac{1}{2p}} \delta_n\right), \qquad n \to \infty.$$

13. If $0 < p \leq 2$ and

$$\sum_{n=0}^{\infty} |a_n|^p \left(\frac{n!}{\alpha^n} \right)^{\frac{p}{2}} n^{-\frac{p}{4}+\frac{1}{2}} < \infty,$$

then the function $f(z) = \sum a_n z^n$ belongs to F_α^p.

14. If $0 < p \leq 2$ and the function $f(z) = \sum a_n z^n$ belongs to F_α^p, then

$$\sum_{n=0}^{\infty} |a_n|^p \left(\frac{n!}{\alpha^n} \right)^{\frac{p}{2}} n^{\frac{3p}{4}-\frac{3}{2}} < \infty.$$

15. If $2 \leq p < \infty$ and

$$\sum_{n=0}^{\infty} |a_n|^p \left(\frac{n!}{\alpha^n} \right)^{\frac{p}{2}} n^{\frac{3p}{4}-\frac{3}{2}} < \infty,$$

then the function $f(z) = \sum a_n z^n$ belongs to F_α^p.

16. If $2 \leq p < \infty$ and the function $f(z) = \sum a_n z^n$ is in F_α^p, then

$$\sum_{n=0}^{\infty} |a_n|^p \left(\frac{n!}{\alpha^n} \right)^{\frac{p}{2}} n^{-\frac{p}{4}+\frac{1}{2}} < \infty.$$

17. Let $1 < p \leq 2$ with $1/p + 1/q = 1$ and $f(z) = \sum a_n z^n$ is in F_α^p. Then

$$\sum_{n=0}^{\infty} |a_n|^q \left(\frac{n!}{\alpha^n} \right)^{\frac{q}{2}} n^{\frac{q}{4}-\frac{1}{2}} < \infty.$$

18. If $f(z) = \sum a_n z^n$ is in F_α^p, where $0 < p < \infty$, then

$$|a_n| \leq \left(\frac{\alpha e}{n} \right)^{\frac{n}{2}} \|f\|_{p,\alpha},$$

for all $n \geq 1$.

19. Suppose $2 \leq p \leq \infty$, $1/p + 1/q = 1$, and

$$\sum_{n=0}^{\infty} |a_n|^q \left(\frac{n!}{\alpha^n} \right)^{\frac{q}{2}} n^{\frac{q}{4}-\frac{1}{2}} < \infty,$$

then the function $f(z) = \sum a_n z^n$ is in F_α^p.

20. Suppose (X, μ) is a measure space and $f_n \in L^p(X, d\mu)$ for $n \geq 0$, where $0 < p < \infty$. Show that

$$\lim_{n \to \infty} \int_X |f_n - f_0|^p \, d\mu = 0$$

if and only if $f_n \to f_0$ pointwise and

$$\lim_{n\to\infty} \int_X |f_n|^p \, d\mu = \int_X |f_0|^p \, d\mu.$$

21. Let R be a positive radius and let

$$dA_R(z) = \frac{\alpha R^2}{\pi R^{2\alpha R^2}} (R^2 - |z|^2)^{\alpha R^2 - 1} \, dA(z)$$

denote the normalized weighted area measure on the disk $B(0,R)$, where dA is area measure. For any entire function f, show that

$$\lim_{R\to\infty} \int_{B(0,R)} |f(z)|^p \, dA_R(z) = \int_{\mathbb{C}} |f(z)|^p \, d\lambda_\alpha(z).$$

Therefore, we can think of the Fock space as a certain limit of weighted Bergman spaces.

22. Show that the norm of the operator Q_α on L_α^p is exactly 2, where $1 \le p \le \infty$. See [74].

23. If we define

$$T_r f(z) = \frac{\alpha}{\pi} \sum_{w \in r\mathbb{Z}^2} e^{\alpha z \bar{w} - \frac{\alpha}{2}|w|^2} \int_{S_r + w} f(u) e^{-\frac{\alpha}{2}|u|^2} \, dA(u),$$

show that $\|T_r - I\|_{\infty,\alpha} \to 0$ as $r \to 0$ and $\|(T_r - I)f\|_{1,\alpha} \to 0$ as $r \to 0$ for any $f \in F_\alpha^1$. Do we have $\|T_r - I\|_{1,\alpha} \to 0$ as $r \to 0$?

24. Determine the interpolation space $[f_\alpha^\infty, F_\beta^p]_\theta$, where $1 \le p \le \infty$.

25. Use the mean value theorem and Hölder's inequality to show that there exists a positive constant $C = C(\alpha, p)$ such that $\|D_r\|_{p,\alpha} \le Cr^{3-(2/p)}$ for all $0 < p \le 1$. This shows that the method employed to prove atomic decomposition for F_α^1 can be extended to the range $2/3 < p < 1$.

26. Let f be an entire function and $0 < p \le \infty$. Show that $f \in F_\alpha^p$ if and only if there exists a complex Borel measure μ such that

$$f(z) = \int_{\mathbb{C}} e^{\alpha \bar{a} z - \frac{\alpha}{2}|a|^2} \, d\mu(a)$$

and $\{|\mu|(S_r + w) : w \in r\mathbb{Z}^2\} \in l^p$.

27. If μ is a positive Borel measure on $\mathbb{C}$ and $0 < p \le \infty$, show that the condition $\{\mu(S_r + w) : w \in r\mathbb{Z}^2\} \in l^p$ is equivalent to the condition that the function $z \mapsto \mu(B(z,r))$ is in $L^p(\mathbb{C}, dA)$.

28. Suppose $f \in F_\alpha^p$. Then there are constants a, b, and c such that $f(z) = z^k P(z) e^{az^2 + bz + c}$, where k is the order of zero of f at the origin and $P(z)$ is the Weierstrass product associated with the zeros (excluding the origin) of f.

29. Suppose T is a bounded linear operator on F_α^2 and it commutes with every operator W_a. Show that T is a constant multiple of the identity operator. This result is called Schur's lemma in mathematical physics. See [177] for example.

30. Show that the main atomic decomposition theorem remains valid if we replace the square lattice $r\mathbb{Z}^2$ by any sequence $\{w_k\}$ in the complex plane with the following properties: $\mathbb{C} = \cup_k B(w_k, r)$, $|w_k - w_j| \le r$, and $|w_k - w_j| \ge r/4$ whenever $k \ne j$. Here, r is any sufficiently small positive radius.

31. Show that harmonic conjugation is a bounded linear operator on L_α^p for $1 \le p \le \infty$.

32. Characterize lacunary series in F_α^p. See [226].

33. Prove an atomic decomposition for the space f_α^∞.

34. Suppose $f \in F_\alpha^p$ and $f(a) \ne 0$. Show that there exists a positive integer N and at most one more point b such that

$$\|f\|_{p,\alpha}^p = N\left|f(a)e^{-\frac{\alpha}{2}|a|^2}\right|^p + \left|f(b)e^{-\frac{\alpha}{2}|b|^2}\right|^p.$$

35. Prove the analogs of Propositions 2.40, 2.41, and 2.42 when the operators W_a are replaced by the operators U_a.

36. Suppose ω_1 and ω_2 are strictly positive and Lebesgue measurable weight functions on the complex plane. If $1 \le p < \infty$ and $1/p + 1/q = 1$, then

$$[L^p(\mathbb{C}, \omega_1 dA)]^* = L^q(\mathbb{C}, \omega_2 dA),$$

with equal norms, where the duality pairing is given by the integral

$$\langle f, g\rangle_\omega = \int_\mathbb{C} f(z)\overline{g(z)}\,\omega(z)\,dA(z),$$

and

$$\omega(z) = \omega_1(z)^{\frac{1}{p}}\omega_2(z)^{\frac{1}{q}}$$

is a geometric mean of $\omega_1(z)$ and $\omega_2(z)$.

37. Suppose $1 \le p < \infty$ and $1/p + 1/q = 1$. For any positive parameters α and β, show that $(L_\alpha^p)^* = L_\beta^q$ under the integral pairing

$$\langle f, g\rangle_\gamma = \frac{\gamma}{\pi}\int_\mathbb{C} f(z)\overline{g(z)}e^{-\gamma|z|^2}\,dA(z),$$

where $\gamma = (\alpha + \beta)/2$ is the arithmetic mean of α and β.

38. If $0 < \beta < \alpha$, show that $F_\beta^p \subset F_\alpha^q$ for all $0 < p \le \infty$ and $0 < q \le \infty$.

39. If F is a bounded linear functional on F_α^p or f_α^∞, show that the function

$$g(w) = \overline{F_z(e^{\gamma z \overline{w}})}$$

is entire.

40. Suppose $1 \le p \le \infty$. Show that F_α^p is a complemented subspace of L_α^p, that is, there exists a closed subspace X_α^p of L_α^p such that $L_\alpha^p = F_\alpha^p \oplus X_\alpha^p$. Study the case when $0 < p < 1$.

Chapter 3
The Berezin Transform and BMO

In this chapter, we study the Berezin transform on F_α^2 and certain spaces of functions of bounded mean oscillation (BMO) on the complex plane. We first consider the Berezin symbol of a bounded linear operator on F_α^2 and show that this is a Lipschitz function in the Euclidean metric. We then consider the Berezin transform of a function and show that there is a semigroup property with respect to the parameter α. We also consider the action of the Berezin transform on L^p spaces and the behavior of the Berezin transform when it is iterated.

For every exponent $p \in [1, \infty)$, we define a space BMO^p of functions of bounded mean oscillation, based on Euclidean disks of a *fixed* radius, and study the structure of these spaces. When $1 < p < \infty$, we will show that the Berezin transform of every function in BMO^p is Lipschitz in the Euclidean metric.

As is well known, the Berezin transform is closely related to the notion of Carleson measures. So we include the discussion of Fock–Carleson measures in this chapter as well.

K. Zhu, *Analysis on Fock Spaces*, Graduate Texts in Mathematics 263,
DOI 10.1007/978-1-4419-8801-0_3,
© Springer Science+Business Media New York 2012

3.1 The Berezin Transform of Operators

Recall that for each $z \in \mathbb{C}$, we use k_z to denote the normalized reproducing kernel at z, namely,

$$k_z(w) = K(w,z)/\sqrt{K(z,z)} = e^{\alpha w \bar{z} - \frac{\alpha}{2}|z|^2}.$$

These are unit vectors in F_α^2.

If T is any linear operator on F_α^2 whose domain contains all the normalized reproducing kernels, then we can define a function $\widetilde{T}$ on $\mathbb{C}$ as follows:

$$\widetilde{T}(z) = \langle Tk_z, k_z \rangle, \qquad z \in \mathbb{C}, \tag{3.1}$$

where $\langle \, , \, \rangle$ is the inner product in F_α^2. We are going to call $\widetilde{T}$ the Berezin transform (or sometimes the Berezin symbol) of T. In particular, if T is a bounded linear operator on F_α^2, then the Berezin transform $\widetilde{T}$ is well defined and is actually real analytic in $\mathbb{C}$.

Proposition 3.1. *Let $L(F_\alpha^2)$ be the Banach space of all bounded linear operators on F_α^2. Then $T \mapsto \widetilde{T}$ is a bounded linear mapping from $L(F_\alpha^2)$ into $L^\infty(\mathbb{C})$. Furthermore, the mapping is one-to-one and order preserving.*

Proof. Everything is obvious except the one-to-one part. To see this, assume that T is a bounded linear operator on F_α^2 and that $\langle Tk_z, k_z \rangle = 0$ for all $z \in \mathbb{C}$. Then $\langle TK_z, K_z \rangle = 0$ for all $z \in \mathbb{C}$, where $K_z(w) = K(w,z)$. The function $F(z,w) = \langle TK_z, K_w \rangle$ is real analytic on $\mathbb{C} \times \mathbb{C}$, holomorphic in w, and conjugate holomorphic in z. Also, F vanishes on the diagonal of $\mathbb{C} \times \mathbb{C}$. It follows from a well-known theorem in several complex variables (see [142] for example) that F is identically zero on $\mathbb{C} \times \mathbb{C}$. Consequently, $TK_z(w) = 0$ for all z and w, or $TK_z = 0$ for all $z \in \mathbb{C}$. Since the set of finite linear combinations of kernel functions is dense in F_α^2, we conclude that $T = 0$. $\qquad\square$

Note that the proof above concerning the one-to-one property of the Berezin transform works for certain unbounded operators as well. More specifically, if T is an unbounded linear operator on F_α^2 such that its domain contains all finite linear combinations of kernel functions and $\langle TK_z, K_w \rangle$ is real analytic, then $\widetilde{T} = 0$ implies that $T = 0$.

Proposition 3.2. *If T is compact on F_α^2, then $\widetilde{T}(z) \to 0$ as $z \to \infty$.*

Proof. It is easy to see that $k_z \to 0$ weakly in F_α^2 as $z \to \infty$. This gives the desired result. $\qquad\square$

It is a classical result in functional analysis that if T is positive and compact on a Hilbert space H, then there exists an orthonormal set $\{e_n\}$ in H and a nonincreasing sequence $\{s_n\}$ of positive numbers such that

$$T(x) = \sum_n s_n \langle x, e_n \rangle e_n, \qquad x \in H.$$

The numbers s_n are uniquely determined by T and are called the singular values of T.

Let T be a positive and compact operator with singular values $\{s_n\}$, and let $0 < p < \infty$. We say that the operator T belongs to the Schatten class S_p if the sequence $\{s_n\}$ belongs to l^p. For a more general operator T, we say that it belongs to the Schatten class S_p if $|T| = (T^*T)^{1/2}$ belongs to S_p. If $\{s_n\}$ is the sequence of singular values for $|T|$, we write

$$\|T\|_{S_p} = \left[\sum_n s_n^p\right]^{1/p}.$$

Two special cases are worth mentioning: S_1 is called the trace class, and S_2 is called the Hilbert–Schmidt class. We refer the reader to [250] for more information about the Schatten classes.

Proposition 3.3. *If S is a trace-class operator or a positive operator, then*

$$\mathrm{tr}\,(S) = \frac{\alpha}{\pi}\int_{\mathbb{C}} \widetilde{S}(z)\,\mathrm{d}A(z). \tag{3.2}$$

Furthermore, a positive operator S belongs to the trace class if and only if the integral in (3.2) converges.

Proof. First, assume that S is positive, say $S = T^2$ for some $T \geq 0$. Then for any orthonormal basis $\{e_n\}$, it follows from Fubini's theorem that

$$\mathrm{tr}\,(S) = \sum_{n=1}^{\infty} \langle Se_n, e_n\rangle_\alpha = \sum_{n=1}^{\infty} \|Te_n\|_{2,\alpha}^2 = \sum_{n=1}^{\infty} \int_{\mathbb{C}} |Te_n(z)|^2\,\mathrm{d}\lambda_\alpha(z)$$

$$= \int_{\mathbb{C}} \left[\sum_{n=1}^{\infty} |Te_n(z)|^2\right]\mathrm{d}\lambda_\alpha(z) = \int_{\mathbb{C}} \left[\sum_{n=1}^{\infty} \langle Te_n, K_z\rangle_\alpha^2\right]\mathrm{d}\lambda_\alpha(z)$$

$$= \int_{\mathbb{C}} \left[\sum_{n=1}^{\infty} \langle e_n, TK_z\rangle_\alpha^2\right]\mathrm{d}\lambda_\alpha(z) = \int_{\mathbb{C}} \|TK_z\|_{2,\alpha}^2\,\mathrm{d}\lambda_\alpha(z)$$

$$= \int_{\mathbb{C}} \langle SK_z, K_z\rangle_\alpha\,\mathrm{d}\lambda_\alpha(z) = \int_{\mathbb{C}} \widetilde{S}(z)K(z,z)\,\mathrm{d}\lambda_\alpha(z)$$

$$= \frac{\alpha}{\pi}\int_{\mathbb{C}} \widetilde{S}(z)\,\mathrm{d}A(z).$$

Next, assume that S is self-adjoint and belongs to the trace class. Then we can write

$$S = \frac{|S|+S}{2} - \frac{|S|-S}{2},$$

where each of the two quotients above is a positive operator in the trace class. The desired trace formula then follows from the corresponding ones for positive trace-class operators.

Finally, an arbitrary trace-class operator S can be written as

$$S = \frac{S+S^*}{2} + i\frac{S-S^*}{2i},$$

where each of the two quotients above is a self-adjoint operator in the trace class. The desired trace formula for S follows from the corresponding ones for self-adjoint trace-class operators. $\square$

Lemma 3.4. *Suppose T is a positive operator on a Hilbert space H and x is a unit vector in H. Then $\langle T^p x, x \rangle \geq \langle Tx, x \rangle^p$ for $p \geq 1$ and $\langle T^p x, x \rangle \leq \langle Tx, x \rangle^p$ for all $0 < p \leq 1$.*

Proof. See Proposition 1.31 of [250]. $\square$

Proposition 3.5. *If $p \geq 1$ and T is in the Schatten class S_p, then $\widetilde{T}$ belongs to $L^p(\mathbb{C}, dA)$.*

Proof. If T is in the trace class, then we can write

$$T = T_1 - T_2 + i(T_3 - T_4),$$

where each T_k is a positive trace-class operator. By Proposition 3.3 above, the function

$$\widetilde{T} = \widetilde{T_1} - \widetilde{T_2} + i\widetilde{T_3} - i\widetilde{T_4}$$

is in $L^1(\mathbb{C}, dA)$.

If T is a bounded linear operator on F_α^2, the function $\widetilde{T}$ is in $L^\infty(\mathbb{C}, dA)$. It follows from complex interpolation that if T is any operator in the Schatten class S_p, $1 < p < \infty$, then the function $\widetilde{T}$ is in $L^p(\mathbb{C}, dA)$.

Alternatively, if $1 \leq p < \infty$ and T is in the Schatten class S_p, then by the decomposition $T = T_1 - T_2 + i(T_3 - T_4)$, we may assume that T is positive. But when T is positive, it is in the Schatten class S_p if and only if T^p is in the trace class, so the desired result follows from Proposition 3.3 and Lemma 3.4. $\square$

Note that we did not need the positivity of T above, while this is necessary in the next proposition.

Proposition 3.6. *Suppose $0 < p \leq 1$ and T is a positive operator on F_α^2. If $\widetilde{T} \in L^p(\mathbb{C}, dA)$, then T belongs to the Schatten class S_p.*

Proof. Since T is positive, it belongs to the Schatten class S_p if and only if S^p is in the trace class. The desired result then follows from Proposition 3.3 and Lemma 3.4. $\square$

Theorem 3.7. *Let T be any bounded linear operator on F_α^2. We have*

$$|\widetilde{T}(z) - \widetilde{T}(w)| \le 2\|T\| \left[1 - |\langle k_z, k_w \rangle|^2\right]^{1/2}$$

for all z and w in $\mathbb{C}$.

Proof. For any $z \in \mathbb{C}$, let P_z denote the rank-one projection from F_α^2 onto the one-dimensional subspace spanned by k_z. More specifically,

$$P_z(f) = \langle f, k_z \rangle k_z, \qquad f \in F_\alpha^2.$$

It is clear that P_z is a positive operator with $\operatorname{tr}(P_z) = 1$.

Let $\{e_k\}$ be an orthonormal basis of F_α^2 with $e_1 = k_z$. Then

$$\operatorname{tr}(TP_z) = \sum_{n=1}^\infty \langle TP_z e_n, e_n \rangle = \langle TP_z k_z, k_z \rangle = \langle Tk_z, k_z \rangle = \widetilde{T}(z).$$

It follows that

$$|\widetilde{T}(z) - \widetilde{T}(w)| = |\operatorname{tr}(T(P_z - P_w))| \le \|T\| \|P_z - P_w\|_{S_1},$$

where S_1 denotes the trace class as a Banach space. Note that we have just used the well-known inequality

$$|\operatorname{tr}(TS)| \le \|T\| \|S\|_{S_1}$$

from operator theory.

For any two different complex numbers z and w, the operator $P_z - P_w$ is a rank-two self-adjoint operator with trace 0. So there is an orthonormal basis in which $P_z - P_w$ is diagonal with two nonzero eigenvalues λ and $-\lambda$, where $\lambda = \|P_z - P_w\| > 0$. Consequently, the positive rank-two operator $(P_z - P_w)^2$ has a single nonzero eigenvalue λ^2 of multiplicity 2, and its trace equals $2\lambda^2$. It follows that the positive operator $|P_z - P_w|$ has a single positive eigenvalue λ with multiplicity 2, and its trace is 2λ, which is also the value of $\|P_z - P_w\|_{S_1}$.

Since

$$\operatorname{tr}(P_z - P_w)^2 = \operatorname{tr}(P_z - P_z P_w - P_w P_z + P_w) = 2 - 2\operatorname{tr}(P_z P_w),$$

we can expand the unit vector k_w to an orthonormal basis of F_α^2 and calculate the trace of $P_z P_w$ with respect to this basis to obtain

$$\operatorname{tr}(P_z P_w) = \langle P_z P_w k_w, k_w \rangle = \langle P_z k_w, k_w \rangle.$$

But $P_z k_w = \langle k_w, k_z \rangle k_z$, we have

$$\operatorname{tr}(P_z - P_w)^2 = 2\left[1 - |\langle k_z, k_w \rangle|^2\right].$$

It follows that $\lambda^2 = 1 - |\langle k_z, k_w \rangle|^2$, which gives the desired result. □

Corollary 3.8 *Let T be any bounded linear operator on F_α^2. Then*

$$|\widetilde{T}(z) - \widetilde{T}(w)| \leq 2\sqrt{\alpha}\|T\||z - w|$$

for all z and w in $\mathbb{C}$.

Proof. It is easy to see that

$$1 - |\langle k_z, k_w \rangle|^2 = 1 - e^{-\alpha|z-w|^2} \leq \alpha|z - w|^2$$

for all z and w. The desired Lispchitz estimate is then obvious. $\square$

Every bounded linear operator on F_α^2 also induces a function on $\mathbb{C} \times \mathbb{C}$. More specifically, if S is a bounded linear operator on F_α^2 and $z \in \mathbb{C}$, then

$$Sf(z) = \langle Sf, K_z \rangle_\alpha = \langle f, S^* K_z \rangle_\alpha$$

for all $f \in F_\alpha^2$. We then define

$$K_S(w, z) = S^* K_z(w) = \langle S^* K_z, K_w \rangle_\alpha = \langle K_z, S K_w \rangle_\alpha \qquad (3.3)$$

for all z and w in $\mathbb{C}$. It is easy to see that the function $K_S(w, z)$ is uniquely determined by the following two properties:

(a) $Sf(z) = \int_\mathbb{C} f(w)\overline{K_S(w, z)}\, d\lambda_\alpha(w)$ for all $f \in F_\alpha^2$ and $z \in \mathbb{C}$.
(b) $K_S(\cdot, z) \in F_\alpha^2$ for all $z \in \mathbb{C}$.

We collect in the following proposition some of the elementary properties of the kernel function $K_S(w, z)$ induced by S.

Proposition 3.9. *The mapping $S \mapsto K_S$ has the following properties:*

(1) $K_{S+T} = K_S + K_T$, $K_{cS} = cK_S$.
(2) $K_S(\cdot, z) \in F_\alpha^2$.
(3) $K_{S^*}(w, z) = \overline{K_S(z, w)}$.
(4) $K_I(w, z) = K(w, z)$.
(5) $K_{S_n} \to K_S$ pointwise whenever $S_n \to S$ weakly.
(6) $|K_S(w, z)| \leq \|S\|\sqrt{K(w, w)K(z, z)}$.
(7) $K_{S_n} \to K_S$ uniformly on compacta whenever $S_n \to S$ in norm.
(8) $K_S(z, z) = K(z, z)\widetilde{S^*}(z)$.
(9) $K_S(w, w) \equiv 0$ if and only if $S = 0$.

Proof. Properties (1)–(5) and (8) are direct consequences of the definition of K_S in (3.3) and the definition of the Berezin transform. Property (6) follows from (3.3) and the Cauchy–Schwarz inequality, and it implies property (7). Since the Berezin transform $S \mapsto \widetilde{S}$ is one-to-one, we see that (9) follows from (8). $\square$

Proposition 3.10. *Let S and T be bounded operators on F_α^2. Then*

$$K_{ST}(w,z) = \int_{\mathbb{C}} K_S(u,z) K_T(w,u) \, d\lambda_\alpha(u)$$

for all w and z in $\mathbb{C}$.

Proof. It follows from (3.3) that

$$
\begin{aligned}
K_{ST}(w,z) &= \langle T^* S^* K_z, K_w \rangle_\alpha = \langle S^* K_z, T K_w \rangle_\alpha \\
&= \int_{\mathbb{C}} S^* K_z(u) \overline{T K_w(u)} \, d\lambda_\alpha(u) \\
&= \int_{\mathbb{C}} \langle S^* K_z, K_u \rangle_\alpha \langle T^* K_u, K_w \rangle_\alpha \, d\lambda_\alpha(u) \\
&= \int_{\mathbb{C}} K_S(u,z) K_T(w,u) \, d\lambda_\alpha(u)
\end{aligned}
$$

for all z and w in $\mathbb{C}$. □

Proposition 3.11. *If S is a positive or trace-class operator, then*

$$\mathrm{tr}(S) = \int_{\mathbb{C}} \overline{K_S(z,z)} \, d\lambda_\alpha(z).$$

Proof. This follows from Proposition 3.3 and property (8) in Proposition 3.9. □

Corollary 3.12. *Let S and T be bounded linear operators on F_α^2 such that ST is trace class. Then*

$$\mathrm{tr}(ST) = \int_{\mathbb{C}} d\lambda_\alpha(w) \int_{\mathbb{C}} \overline{K_S(z,w)} \, \overline{K_T(w,z)} \, d\lambda_\alpha(z).$$

Proof. This is a direct consequence of Propositions 3.10 and 3.11. □

3.2   The Berezin Transform of Functions

We say that a Lebesgue measurable function φ satisfies *condition (I_p)*, where $0 < p < \infty$, if $\varphi \circ t_a \in L^p(\mathbb{C}, d\lambda_\alpha)$ for every $a \in \mathbb{C}$. In particular, any function satisfying condition (I_p) must be in $L^p(\mathbb{C}, d\lambda_\alpha)$.

By a change of variables, we see that a Lebesgue measurable function φ on $\mathbb{C}$ satisfies condition (I_p) if and only if

$$\int_{\mathbb{C}} |K(z,a)|^2 |\varphi(z)|^p \, d\lambda_\alpha(z) < \infty \tag{3.4}$$

for all $a \in \mathbb{C}$. By the exponential form of the kernel function $K(w,z)$, the above condition is equivalent to

$$\int_{\mathbb{C}} |K(z,a)| |\varphi(z)|^p \, d\lambda_\alpha(z) < \infty, \qquad a \in \mathbb{C}. \tag{3.5}$$

We are mostly interested in two particular cases: $p = 1$ and $p = 2$. The case $p = 1$ is needed in this section, while the case $p = 2$ will be used in Chap. 6 when we study Toeplitz operators with unbounded symbols. It is clear that every function in $L^\infty(\mathbb{C})$ satisfies condition (I_p).

Suppose f satisfies condition (I_1). We can then define a function $\widetilde{f}$ on $\mathbb{C}$ as follows:

$$\widetilde{f}(z) = \langle f k_z, k_z \rangle = \int_{\mathbb{C}} |k_z(w)|^2 f(w) \, d\lambda_\alpha(w). \tag{3.6}$$

We will also call $\widetilde{f}$ the Berezin transform of f. It is clear that we can write

$$\widetilde{f}(z) = \frac{\alpha}{\pi} \int_{\mathbb{C}} f(w) e^{-\alpha|z-w|^2} \, dA(w) = \int_{\mathbb{C}} f(z \pm w) \, d\lambda_\alpha(w). \tag{3.7}$$

Sometimes, we will need to emphasize the dependence on α. In such situations, we will use the notation

$$B_\alpha f(z) = \frac{\alpha}{\pi} \int_{\mathbb{C}} f(w) e^{-\alpha|z-w|^2} \, dA(w), \qquad z \in \mathbb{C}. \tag{3.8}$$

Thus, $\widetilde{f} = B_\alpha f$ if no parameter is specified.

Theorem 3.13. *Let $H_t = B_{1/t}$ for any positive parameter t. Then we have the following semigroup property: $H_s H_t = H_{s+t}$ for all positive parameters s and t.*

Proof. We check the semigroup property on $L^\infty(\mathbb{C})$. For $f \in L^\infty(\mathbb{C})$, we have

$$H_t f(z) = \frac{1}{\pi t} \int_{\mathbb{C}} f(w) e^{-\frac{1}{t}|z-w|^2} \, dA(w) \tag{3.9}$$

for $z \in \mathbb{C}$ and

$$H_s H_t f(z) = \frac{1}{\pi^2 st} \int_{\mathbb{C}} e^{-\frac{1}{s}|z-w|^2} dA(w) \int_{\mathbb{C}} f(u) e^{-\frac{1}{t}|w-u|^2} dA(u)$$

for $z \in \mathbb{C}$. By Fubini's theorem,

$$H_s H_t f(z) = \int_{\mathbb{C}} f(u) I(z,u) \, dA(u), \qquad z \in \mathbb{C},$$

where

$$I(z,u) = \frac{1}{\pi^2 st} \int_{\mathbb{C}} e^{-\frac{1}{s}|z-w|^2 - \frac{1}{t}|w-u|^2} dA(w).$$

Since

$$-\frac{1}{s}|z-w|^2 - \frac{1}{t}|w-u|^2 = -\left(\frac{1}{s} + \frac{1}{t}\right)|w|^2 - \frac{1}{s}|z|^2 - \frac{1}{t}|u|^2$$
$$+ \left(\frac{z}{s} + \frac{u}{t}\right)\overline{w} + \left(\frac{\overline{z}}{s} + \frac{\overline{u}}{t}\right)w,$$

we have

$$I(z,u) = \frac{1}{\pi^2 st} e^{-\frac{1}{s}|z|^2 - \frac{1}{t}|u|^2} \int_{\mathbb{C}} \left| e^{\left(\frac{z}{s} + \frac{u}{t}\right)\overline{w}} \right|^2 e^{-\left(\frac{1}{s} + \frac{1}{t}\right)|w|^2} dA(w)$$

$$= \frac{e^{-\frac{1}{s}|z|^2 - \frac{1}{t}|u|^2}}{\pi(s+t)} \cdot \frac{\frac{1}{s} + \frac{1}{t}}{\pi} \int_{\mathbb{C}} \left| e^{\left(\frac{1}{s} + \frac{1}{t}\right)\frac{tz+su}{s+t}\overline{w}} \right|^2 e^{-\left(\frac{1}{s} + \frac{1}{t}\right)|w|^2} dA(w).$$

Applying the reproducing formula in $F_{\frac{1}{s}+\frac{1}{t}}^2$, we obtain

$$I(z,u) = \frac{1}{\pi(s+t)} e^{-\frac{1}{s}|z|^2 - \frac{1}{t}|u|^2 + \left(\frac{1}{s} + \frac{1}{t}\right)\left|\frac{tz+su}{s+t}\right|^2}.$$

Elementary calculations then show that

$$I(z,u) = \frac{1}{\pi(s+t)} e^{-\frac{1}{s+t}|z-u|^2}.$$

Therefore,

$$H_s H_t f(z) = \frac{1}{\pi(s+t)} \int_{\mathbb{C}} f(u) e^{-\frac{1}{s+t}|z-u|^2} dA(u) = H_{s+t} f(z).$$

This proves the desired result. $\square$

Because of the following result, the operator H_t is sometimes called the heat transform.

Theorem 3.14. *The function $u(x,y,t) = H_t f(z)$, where $z = x+iy$, satisfies the heat equation*

$$\frac{\partial^2 u}{\partial x^2} + \frac{\partial^2 u}{\partial y^2} = 4\frac{\partial u}{\partial t}. \tag{3.10}$$

Moreover, if f is bounded and continuous on $\mathbb{C}$, then u also satisfies the initial condition

$$\lim_{t\to 0^+} H_t f(z) = f(z), \qquad z \in \mathbb{C}. \tag{3.11}$$

Proof. With $z = x+iy$ and $w = u+iv$, we have

$$u(x,y,t) = \frac{1}{\pi t}\int_{\mathbb{R}^2} f(u,v)e^{-\frac{1}{t}[(x-u)^2+(y-v)^2]}\,du\,dv.$$

Differentiating under the integral sign, we obtain

$$\frac{\partial u}{\partial t} = -\frac{1}{\pi t^2}\int_{\mathbb{R}^2} f(u,v)e^{-\frac{1}{t}[(x-u)^2+(y-v)^2]}\,du\,dv$$

$$+\frac{1}{\pi t^3}\int_{\mathbb{R}^2}[(x-u)^2+(y-v)^2]f(u,v)e^{-\frac{1}{t}[(x-u)^2+(y-v)^2]}\,du\,dv.$$

Similarly,

$$\frac{\partial u}{\partial x} = -\frac{2}{\pi t^2}\int_{\mathbb{R}^2}(x-u)f(u,v)e^{-\frac{1}{t}[(x-u)^2+(y-v)^2]}\,du\,dv,$$

and

$$\frac{\partial^2 u}{\partial x^2} = -\frac{2}{\pi t^2}\int_{\mathbb{R}^2} f(u,v)e^{-\frac{1}{t}[(x-u)^2+(y-v)^2]}\,du\,dv$$

$$+\frac{4}{\pi t^3}\int_{\mathbb{R}^2}(x-u)^2 f(u,v)e^{-\frac{1}{t}[(x-u)^2+(y-v)^2]}\,du\,dv.$$

Combining this with a similar calculation for $\partial^2 u/\partial y^2$ gives

$$\Delta u = -\frac{4}{\pi t^2}\int_{\mathbb{R}^2} f(u,v)e^{-\frac{1}{t}[(x-u)^2+(y-v)^2]}\,du\,dv$$

$$+\frac{4}{\pi t^3}\int_{\mathbb{R}^2}[(x-u)^2+(y-v)^2]f(u,v)e^{-\frac{1}{t}[(x-u)^2+(y-v)^2]}\,du\,dv$$

$$= 4\frac{\partial u}{\partial t},$$

where

$$\Delta u = \frac{\partial^2 u}{\partial x^2} + \frac{\partial^2 u}{\partial y^2}$$

is the Laplacian of u. Thus, u satisfies the heat equation (3.10).

To show that u also satisfies the initial condition (3.11), assume that f is bounded and continuous on $\mathbb{C}$. Fix a point $z \in \mathbb{C}$ and write

$$H_t f(z) - f(z) = \frac{1}{\pi t} \int_{\mathbb{C}} (f(w) - f(z)) e^{-\frac{1}{t}|z-w|^2} dA(w)$$

$$= \frac{1}{\pi t} \int_{|w-z|<\delta} + \frac{1}{\pi t} \int_{|w-z|>\delta}$$

$$=: I_1 + I_2.$$

Given any positive ε, we can choose a positive δ such that

$$|f(w) - f(z)| < \varepsilon, \qquad w \in B(z, \delta).$$

It follows that

$$|I_1| \leq \frac{\varepsilon}{\pi t} \int_{|w-z|<\delta} e^{-\frac{1}{t}|z-w|^2} dA(z) < \frac{\varepsilon}{\pi t} \int_{\mathbb{C}} e^{-\frac{1}{t}|z-w|^2} dA(w) = \varepsilon.$$

On the other hand,

$$|I_2| \leq 2\|f\|_\infty \frac{1}{\pi t} \int_{|z-w|>\delta} e^{-\frac{1}{t}|z-w|^2} dA(w)$$

$$= 2\|f\|_\infty \frac{1}{\pi t} \int_{|w|>\delta} e^{-\frac{1}{t}|w|^2} dA(w)$$

$$= 2\|f\|_\infty e^{-\delta^2/t} \to 0$$

as $t \to 0^+$. It follows that

$$\limsup_{t \to 0^+} |H_t f(z) - f(z)| \leq \varepsilon.$$

Since ε is arbitrary, we must have

$$\lim_{t \to 0^+} H_t f(z) = f(z),$$

which completes the proof of the theorem. $\square$

Note that in the heat equation (3.10), the value $u(x,y,t)$ represents the temperature at the point $(x,y) \in \mathbb{C}$ at time t. Thus, the function $f(z)$ represents the initial temperature distribution in the complex plane at time $t = 0$. With this interpretation, the assumption that f be bounded and continuous is reasonable. However, the initial condition in (3.11) can be shown to hold for certain functions that are more general than bounded and continuous ones.

The following result is a direct consequence of Theorems 3.13 and 3.14.

Corollary 3.15. *For any positive α and β, we have the identities*

$$B_\alpha B_\beta = B_{\frac{\alpha\beta}{\alpha+\beta}} = B_\beta B_\alpha.$$

If f is bounded and continuous, then

$$\lim_{\alpha \to +\infty} B_\alpha f(z) = f(z)$$

for every $z \in \mathbb{C}$.

We need the following result from Fourier analysis to generalize Proposition 3.1 to the Berezin transform of functions.

Lemma 3.16 *Suppose that n is a positive integer and f is a function on $\mathbb{R}^n$ such that the function*

$$x \mapsto f(x)e^{|tx|}e^{-x^2}$$

is integrable on $\mathbb{R}^n$ with respect to Lebesgue measure dx for any $t \in \mathbb{R}^n$. Here,

$$x = (x_1,\ldots,x_n), \quad t = (t_1,\ldots,t_n), \quad tx = t_1 x_1 + \cdots + t_n x_n,$$

and

$$x^2 = x_1^2 + \cdots + x_n^2, \quad dx = dx_1 \cdots dx_n.$$

If

$$\int_{\mathbb{R}^n} f(x)P(x)e^{-x^2}\, dx = 0$$

for every polynomial P, then $f = 0$ almost everywhere on $\mathbb{R}^n$.

Proof. Since

$$e^{itx} = \sum_{k=0}^{\infty} \frac{(itx)^k}{k!}$$

and

$$\left| \sum_{k=0}^{N} \frac{(itx)^k}{k!} \right| \leq \sum_{k=0}^{\infty} \frac{|tx|^k}{k!} = e^{|tx|}$$

for all $N \geq 0$, we apply the dominated convergence theorem to partial sums to obtain

$$\int_{\mathbb{R}^n} e^{itx} f(x) e^{-x^2} \, dx = \sum_{k=0}^{\infty} \frac{i^k}{k!} \int_{\mathbb{R}^n} (tx)^k f(x) e^{-x^2} \, dx = 0$$

for all $t \in \mathbb{R}^n$. By the Fourier inversion theorem, we have $f(x)e^{-x^2} = 0$, and hence $f(x) = 0$ for almost every $x \in \mathbb{R}^n$. $\qquad\square$

Note that the integral condition (3.12) in the next proposition is slightly stronger than condition (I_1) which was necessary for the definition of $B_\alpha f$.

Proposition 3.17. *The Berezin transform B_α is linear and order preserving. Furthermore, if $B_\alpha f = 0$ and f satisfies the condition that*

$$\int_{\mathbb{C}} |f(z)| e^{|tz|} e^{-\alpha|z|^2} \, dA(z) < \infty \tag{3.12}$$

for all real t, then $f(z) = 0$ for almost every $z \in \mathbb{C}$.

Proof. It is clear that each B_α is linear and order preserving.

If $B_\alpha f = 0$ and f satisfies the integral condition (3.12), then differentiating under the integral sign gives

$$\frac{\partial^{n+m}}{\partial z^n \partial \overline{z}^m} B_\alpha f(0) = c_{m,n} \int_{\mathbb{C}} f(w) w^m \overline{w}^n e^{-\alpha|w|^2} \, dA(w),$$

where $c_{m,n}$ is a nonzero constant. It follows that

$$\int_{\mathbb{C}} f(w) w^m \overline{w}^n e^{-\alpha|w|^2} \, dA(w) = 0$$

for all nonnegative integers m and n. The result then follows from Lemma 3.16. $\quad\square$

In the next few results, we describe some of the mapping properties of the Berezin transform. In particular, we will compare $B_\alpha f$ and $B_\beta f$ in various situations.

Theorem 3.18. *Let $1 \leq p \leq \infty$. Suppose $\alpha, \beta,$ and γ are positive weight parameters. Then $B_\alpha L_\beta^p \subset L_\gamma^p$ if and only if $\gamma(2\alpha - \beta) \geq 2\alpha\beta$.*

Proof. First, assume that $\gamma(2\alpha - \beta) \geq 2\alpha\beta$. Then, in particular, $\alpha > \frac{\beta}{2}$. If $f \in L_\beta^\infty$, we write

$$B_\alpha f(z) = \frac{\alpha}{\pi} e^{-\alpha|z|^2} \int_{\mathbb{C}} f(w) e^{-\frac{\beta}{2}|w|^2} |e^{\alpha z \overline{w}}|^2 e^{-\left(\alpha - \frac{\beta}{2}\right)|w|^2} \, dA(w).$$

It follows that

$$|B_\alpha f(z)| \leq \frac{\alpha\|f\|_{\infty,\beta}}{\pi} e^{-\alpha|z|^2} \int_{\mathbb{C}} \left| e^{(\alpha-\frac{\beta}{2})\frac{\alpha z}{\alpha-\frac{\beta}{2}}\overline{w}} \right|^2 e^{-\left(\alpha-\frac{\beta}{2}\right)|w|^2} \, dA(w)$$

$$= \frac{\alpha}{\alpha-\frac{\beta}{2}} \|f\|_{\infty,\beta} e^{-\alpha|z|^2} \left| e^{(\alpha-\frac{\beta}{2})\frac{\alpha z}{\alpha-\frac{\beta}{2}}} \right|^2.$$

Therefore,

$$|B_\alpha f(z)| e^{-\frac{\gamma}{2}|z|^2} \leq \frac{2\alpha\|f\|_{\infty,\beta}}{2\alpha-\beta} e^{-(\alpha+\frac{\gamma}{2}-\frac{2\alpha^2}{2\alpha-\beta})|z|^2}.$$

It is elementary to check that the condition $\gamma(2\alpha - \beta) \geq 2\alpha\beta$ is equivalent to

$$\alpha + \frac{\gamma}{2} - \frac{2\alpha^2}{2\alpha-\beta} \geq 0.$$

Thus, B_α maps L_β^∞ into L_γ^∞.

If $f \in L_\beta^1$, the integral

$$I = \int_{\mathbb{C}} \left| B_\alpha f(z) e^{-\frac{\gamma}{2}|z|^2} \right| \, dA(z)$$

equals

$$\frac{\alpha}{\pi} \int_{\mathbb{C}} \left| e^{-(\alpha+\frac{\gamma}{2})|z|^2} \int_{\mathbb{C}} f(w) |e^{\alpha z\overline{w}}|^2 e^{-\alpha|w|^2} \, dA(w) \right| \, dA(z),$$

which by Fubini's theorem is less than or equal to

$$\frac{\alpha}{\pi} \int_{\mathbb{C}} |f(w)| e^{-\alpha|w|^2} \, dA(w) \int_{\mathbb{C}} |e^{\alpha z\overline{w}}|^2 e^{-(\alpha+\frac{\gamma}{2})|z|^2} \, dA(z).$$

With the help of Corollary 2.5, we obtain

$$I \leq \frac{2\alpha}{2\alpha+\gamma} \int_{\mathbb{C}} |f(w)| e^{-\left(\alpha-\frac{2\alpha^2}{2\alpha+\gamma}\right)|w|^2} \, dA(w).$$

Again, it is elementary to check that the condition $\gamma(2\alpha - \beta) \geq 2\alpha\beta$ is equivalent to

$$\alpha - \frac{2\alpha^2}{2\alpha+\gamma} \geq \frac{\beta}{2}.$$

Thus, B_α maps L_β^1 into L_γ^1.

By complex interpolation, the Berezin transform B_α maps L_β^p into L_γ^p for all $1 \leq p \leq \infty$ whenever $\gamma(2\alpha - \beta) \geq 2\alpha\beta$.

To prove the other direction, observe that

$$B_\alpha f(z) = e^{-\alpha|z|^2} Q_\alpha f(2z).$$

It follows from this and a change of variables that $B_\alpha f \in L_\gamma^p$ if and only if $Q_\alpha f \in L_{\frac{\gamma}{4}+\frac{\alpha}{2}}^p$. Therefore, $B_\alpha L_\beta^p \subset L_\gamma^p$ is equivalent to $Q_\alpha L_\beta^p \subset L_{\frac{\gamma}{4}+\frac{\alpha}{2}}^p$, which implies that $P_\alpha L_\beta^p \subset L_{\frac{\gamma}{4}+\frac{\alpha}{2}}^p$. Combining this with Theorem 2.31, we conclude that $B_\alpha L_\beta^p \subset L_\gamma^p$ implies that

$$\alpha^2 \leq (2\alpha - \beta)\left(\frac{\gamma}{4} + \frac{\alpha}{2}\right),$$

which is equivalent to $\gamma(2\alpha - \beta) \geq 2\alpha\beta$. This completes the proof of the theorem. $\qquad\square$

Corollary 3.19. *Let $\alpha > 0$ and $\beta > 0$. For $1 \leq p \leq \infty$, we have*

(a) $B_\alpha : L_\alpha^p \to L_\beta^p$ *if and only if $\beta \geq 2\alpha$.*
(b) $B_\alpha : L_\beta^p \to L_\alpha^p$ *if and only if $2\alpha \geq 3\beta$.*

Proposition 3.20. *Let $\alpha > 0$ and $1 \leq p < \infty$. Then*

(a) $B_\alpha : L^\infty(\mathbb{C}) \to L^\infty(\mathbb{C})$ *is a contraction.*
(b) $B_\alpha : C_0(\mathbb{C}) \to C_0(\mathbb{C})$ *is a contraction.*
(c) $B_\alpha : L^p(\mathbb{C}, dA) \to L^p(\mathbb{C}, dA)$ *is a contraction.*

Proof. Part (a) is obvious. If $f \in C_c(\mathbb{C})$, namely, if f is a continuous function on $\mathbb{C}$ with compact support, then it is easy to see that $B_\alpha f \in C_0(\mathbb{C})$. Thus, part (b) follows from (a) and the fact that $C_c(\mathbb{C})$ is dense in $C_0(\mathbb{C})$ in the supremum norm.

To prove (c), we first consider the case $p = 1$. In this case, it follows from Fubini's theorem that

$$\int_{\mathbb{C}} |B_\alpha f(z)| \, dA(z) \leq \frac{\alpha}{\pi} \int_{\mathbb{C}} |f(w)| \, dA(w) \int_{\mathbb{C}} e^{-\alpha|z-w|^2} \, dA(z)$$

$$= \int_{\mathbb{C}} |f(w)| \, dA(w).$$

The case $1 < p < \infty$ then follows from complex interpolation. $\qquad\square$

Proposition 3.21. *Let $0 < \beta < \alpha$ and $1 \leq p < \infty$. Then*

(a) $B_\alpha f \in L^\infty(\mathbb{C})$ *implies $B_\beta f \in L^\infty(\mathbb{C})$ with*

$$\|B_\beta f\|_\infty \leq \|B_\alpha f\|_\infty$$

for all f.

(b) $B_\alpha f \in C_0(\mathbb{C})$ implies that $B_\beta f \in C_0(\mathbb{C})$.
(c) $B_\alpha f \in L^p(\mathbb{C}, dA)$ implies that $B_\beta f \in L^p(\mathbb{C}, dA)$ with

$$\int_{\mathbb{C}} |B_\beta f(z)|^p \, dA(z) \leq \int_{\mathbb{C}} |B_\alpha f(z)|^p \, dA(z)$$

for all f.

Proof. Choose a positive γ such that $1/\gamma + 1/\alpha = 1/\beta$. By Corollary 3.15, we have $B_\beta = B_\gamma B_\alpha$. The desired result then follows from Proposition 3.20. $\square$

Proposition 3.22. *If* $0 < \beta < \alpha$, $0 < p < \infty$, *and* $f \geq 0$. *Then*

$$B_\alpha f(z) \leq \frac{\alpha}{\beta} B_\beta f(z), \qquad z \in \mathbb{C}.$$

Consequently:

(a) $B_\beta f \in L^\infty(\mathbb{C})$ implies that $B_\alpha f \in L^\infty(\mathbb{C})$.
(b) $B_\beta f \in C_0(\mathbb{C})$ implies that $B_\alpha f \in C_0(\mathbb{C})$.
(c) $B_\beta f \in L^p(\mathbb{C}, dA)$ implies that $B_\alpha f \in L^p(\mathbb{C}, dA)$.

Proof. Since $f \geq 0$ and $0 < \beta < \alpha$, we have

$$\begin{aligned}
B_\alpha f(z) &= \frac{\alpha}{\pi} \int_{\mathbb{C}} f(w) e^{-\alpha|z-w|^2} \, dA(w) \\
&\leq \frac{\alpha}{\pi} \int_{\mathbb{C}} f(w) e^{-\beta|z-w|^2} \, dA(w) \\
&= \frac{\alpha}{\beta} \cdot \frac{\beta}{\pi} \int_{\mathbb{C}} f(w) e^{-\beta|z-w|^2} \, dA(w) \\
&= \frac{\alpha}{\beta} B_\beta f(z).
\end{aligned}$$

This proves the desired results. $\square$

Theorem 3.23. *Suppose* α *and* β *are positive weight parameters and* $f \geq 0$ *on* $\mathbb{C}$. *For* $0 < p \leq \infty$, *we have*

(a) $B_\alpha f \in L^p(\mathbb{C}, dA)$ *if and only if* $B_\beta f \in L^p(\mathbb{C}, dA)$.
(b) $B_\alpha f \in C_0(\mathbb{C})$ *if and only if* $B_\beta f \in C_0(\mathbb{C})$.

Proof. Part (a) in the case $1 \leq p \leq \infty$ and part (b) follow from Propositions 3.21 and 3.22. Part (a) in the case $0 < p < 1$ will be proved in Chap. 6. $\square$

Recall that for any $a \in \mathbb{C}$, we have

$$t_a(z) = z + a, \quad \tau_a(z) = z - a, \quad \varphi_a(z) = a - z.$$

The following result shows that the Berezin transform commutes with each of these maps.

Proposition 3.24. *If f is a function such that the Berezin transform $B_\alpha f$ is well defined, then for any $a \in \mathbb{C}$, we have*

(i) $B_\alpha(f \circ t_a) = (B_\alpha f) \circ t_a$.
(ii) $B_\alpha(f \circ \tau_a) = (B_\alpha f) \circ \tau_a$.
(iii) $B_\alpha(f \circ \varphi_a) = (B_\alpha f) \circ \varphi_a$.

Proof. By (3.7), we have

$$
\widetilde{f \circ t_a}(z) = \int_\mathbb{C} f \circ t_a(z + w) \, d\lambda_\alpha(w)
$$
$$
= \int_\mathbb{C} f(a + z + w) \, d\lambda_\alpha(w)
$$
$$
= \tilde{f}(a + z) = \tilde{f} \circ t_a(z)
$$

for any $z \in \mathbb{C}$. This proves (i). Replacing a by $-a$ in (i) leads to (ii).

Similarly, it follows from (3.7) that

$$
\widetilde{f \circ \varphi_a}(z) = \int_\mathbb{C} f \circ \varphi_a(z + w) \, d\lambda_\alpha(w)
$$
$$
= \int_\mathbb{C} f(a - z - w) \, d\lambda_\alpha(w)
$$
$$
= \tilde{f}(a - z) = \tilde{f} \circ \varphi_a(z).
$$

This proves (iii).   □

For any positive integer n, we use $B_\alpha^n f$ to denote the n-th iterate of the Berezin transform of f, that is, we take the Berezin transform of f repeatedly n times to obtain $B_\alpha^n f$.

Theorem 3.25. *Suppose $f \in L^\infty(\mathbb{C})$ and n is a positive integer. Then*

$$|B_\alpha^n f(z) - B_\alpha^n f(w)| \leq \frac{C\|f\|_\infty}{\sqrt{n}} |z - w| \tag{3.13}$$

for all z and w in $\mathbb{C}$, where $C = 2\sqrt{\alpha/\pi}$.

Proof. Recall that the Berezin transform of f is

$$B_\alpha f(z) = \frac{\alpha}{\pi} \int_{\mathbb{C}} f(u) e^{-\alpha|z-u|^2} \, dA(u).$$

It follows that the difference

$$D = B_\alpha f(z) - B_\alpha f(w)$$

can be written as

$$\frac{\alpha}{\pi} \int_{\mathbb{C}} f\left(u + \frac{z+w}{2}\right) \left[e^{-\alpha|u-(z-w)/2|^2} - e^{-\alpha|u+(z-w)/2|^2} \right] dA(u).$$

Let $(z-w)/2 = re^{i\theta}$ with $r \geq 0$. By the rotation invariance of the area measure,

$$|D| \leq \frac{\alpha\|f\|_\infty}{\pi} \int_{\mathbb{C}} \left| e^{-\alpha|u-r|^2} - e^{-\alpha|u+r|^2} \right| dA(u)$$

$$= \frac{\alpha\|f\|_\infty}{\pi} \int_{\mathbb{C}} e^{-\alpha(|u|^2+r^2)} \left| e^{\alpha(u+\bar{u})r} - e^{-\alpha(u+\bar{u})r} \right| dA(u).$$

Write $u = x + iy$ and $dA(u) = dxdy$. We obtain

$$|D| \leq \frac{\alpha\|f\|_\infty}{\pi} \int_{-\infty}^{\infty} e^{-\alpha y^2} \, dy \int_{-\infty}^{\infty} e^{-\alpha(x^2+r^2)} \left| e^{-2r\alpha x} - e^{2r\alpha x} \right| dx$$

$$= \frac{2\sqrt{\alpha}\|f\|_\infty}{\pi} \int_{-\infty}^{\infty} e^{-y^2} \, dy \int_{0}^{\infty} e^{-\alpha(x^2+r^2)} \left(e^{2r\alpha x} - e^{-2r\alpha x} \right) dx$$

$$= \frac{2\sqrt{\alpha}\|f\|_\infty}{\sqrt{\pi}} \int_{0}^{\infty} \left(e^{-\alpha(x-r)^2} - e^{-\alpha(x+r)^2} \right) dx$$

$$= \frac{2\sqrt{\alpha}}{\sqrt{\pi}} \|f\|_\infty \left(\int_{-r}^{\infty} e^{-\alpha x^2} \, dx - \int_{r}^{\infty} e^{-\alpha x^2} \, dx \right)$$

$$= \frac{2\sqrt{\alpha}}{\sqrt{\pi}} \|f\|_\infty \int_{-r}^{r} e^{-\alpha x^2} \, dx$$

$$\leq \frac{4r\sqrt{\alpha}}{\sqrt{\pi}} \|f\|_\infty = \frac{2\sqrt{\alpha}}{\sqrt{\pi}} \|f\|_\infty |z-w|.$$

Thus, we have proved that

$$|B_\alpha f(z) - B_\alpha f(w)| \le \frac{2\sqrt{\alpha}}{\sqrt{\pi}} \|f\|_\infty |z - w| \qquad (3.14)$$

for all $f \in L^\infty(\mathbb{C})$ and all z and w in $\mathbb{C}$.

By Corollary 3.15, we have

$$B_\alpha^n f(z) = B_{\frac{\alpha}{n}} f(z) = \frac{\alpha}{\pi n} \int_{\mathbb{C}} f(w) e^{-\frac{\alpha}{n}|z-w|^2} \, dA(w).$$

This, along with a simple change of variables, shows that

$$B_\alpha^n f(z) = B_\alpha g(z/\sqrt{n}),$$

where $g(z) = f(\sqrt{n} z)$. Combining this with the estimate in (3.14), we obtain the desired Lipschitz estimate in (3.13). $\square$

3.3 Fixed Points of the Berezin Transform

In the theory of Bergman spaces, it follows from a theorem of Ahern, Flores, and Rudin that a function is fixed by the Berezin transform in that context if and only if the function is harmonic, as long as the Berezin transform of the function is well defined. No other assumption on the function is necessary. See [1].

Therefore, it is natural to ask if the fixed points of the Berezin transform in our context here are exactly the harmonic functions as well. It turns out that the answer is negative in general, but positive under certain conditions.

Proposition 3.26. *Suppose f is a harmonic function on $\mathbb{C}$ satisfying condition (I_1). Then $\widetilde{f} = f$.*

Proof. If f is harmonic, then $f \circ t_z$ is harmonic for every z. It follows from the mean value theorem for harmonic functions that

$$f \circ t_z(0) = \int_{\mathbb{C}} f \circ t_z(w) \, d\lambda_\alpha(w).$$

This shows that $f(z) = \widetilde{f}(z)$ for every $z \in \mathbb{C}$.   □

The following result gives a partial converse to the proposition above.

Proposition 3.27. *If $f \in L^\infty(\mathbb{C})$, then the following conditions are equivalent:*

(a) $\widetilde{f} = f$.
(b) f is harmonic.
(c) f is constant.

Proof. Since f is bounded, the equivalence of (b) and (c) follows from the well-known maximum modulus principle for harmonic functions. If f is constant, then clearly $\widetilde{f} = f$. If $\widetilde{f} = f$, then $\widetilde{f}^{(n)} = f$ for all positive integers n. By Theorem 3.25, there exists a positive constant C such that

$$|f(z) - f(w)| \leq \frac{C}{\sqrt{n}} |z - w|$$

for all z and w in $\mathbb{C}$ with $z \neq w$. Let $n \to \infty$. We see that f must be constant.   □

Finally, in this section, we show by an example that there are more functions than the harmonic ones that are fixed by the Berezin transform.

Lemma 3.28. *For any complex ζ, let*

$$I(\zeta) = \frac{1}{\sqrt{\pi}} \int_{-\infty}^{\infty} e^{\zeta t - t^2} \, dt.$$

We have $I(\zeta) = e^{\zeta^2/4}$.

Proof. It is clear that $I(\zeta)$ is an entire function of ζ. Differentiating under the integral sign, we obtain

$$I'(\zeta) = \frac{1}{\sqrt{\pi}} \int_{-\infty}^{\infty} t e^{\zeta t - t^2} \, dt$$

$$= \frac{1}{\sqrt{\pi}} \int_{-\infty}^{\infty} \left(t - \frac{\zeta}{2} \right) e^{\zeta t - t^2} \, dt + \frac{\zeta}{2} I(\zeta)$$

$$= \frac{\zeta}{2} I(\zeta).$$

It follows that $I(\zeta) = C e^{\zeta^2/4}$ for some constant C and all $\zeta \in \mathbb{C}$. It is well known that $I(0) = 1$. Thus, $I(\zeta) = e^{\zeta^2/4}$ for all $\zeta \in \mathbb{C}$. □

Now fix two complex constants a and b such that $a^2 + b^2 = 8\alpha\pi i$ and consider the function

$$f(z) = e^{ax+by}, \quad z = x + iy \in \mathbb{C},$$

which clearly satisfies condition (I_1). A direct calculation shows that

$$\Delta f = (a^2 + b^2) f = 8\alpha\pi i f,$$

so f is not harmonic. On the other hand,

$$\tilde{f}(z) = \int_{\mathbb{C}} f(w + z) \, d\lambda_{\alpha}(w)$$

$$= f(z) \int_{\mathbb{C}} e^{au+bv} \, d\lambda_{\alpha}(w),$$

where $w = u + iv$. Separating the variables, we obtain

$$\tilde{f}(z) = f(z) I(a, \alpha) I(b, \alpha),$$

where

$$I(\zeta, \alpha) = \sqrt{\frac{\alpha}{\pi}} \int_{-\infty}^{\infty} e^{\zeta t - \alpha t^2} \, dt.$$

A simple change of variables gives

$$\tilde{f}(z) = f(z) I(a/\sqrt{\alpha}) I(b/\sqrt{\alpha}),$$

where $I(\zeta)$ is the function considered in Lemma 3.28 above. An application of Lemma 3.28 then gives

$$\widetilde{f}(z) = f(z)e^{(a^2+b^2)/(4\alpha)} = f(z).$$

This shows that the function f is fixed by the Berezin transform, but it is not harmonic.

3.4 Fock–Carleson Measures

The main result of this section is the following:

Theorem 3.29. *Suppose μ is a positive Borel measure on $\mathbb{C}$, $0 < p < \infty$, and $0 < r < \infty$. Then the following conditions are equivalent:*

(a) There exists a positive constant C such that

$$\int_{\mathbb{C}} |f(w)e^{-\frac{\alpha}{2}|w|^2}|^p \, d\mu(w) \le C \int_{\mathbb{C}} |f(w)e^{-\frac{\alpha}{2}|w|^2}|^p \, dA(w)$$

for all entire functions f.

(b) There exists a positive constant C such that

$$\int_{\mathbb{C}} e^{-\frac{p\alpha}{2}|z-w|^2} \, d\mu(w) \le C$$

for all $z \in \mathbb{C}$.

(c) There exists a constant $C > 0$ such that $\mu(B(z,r)) \le C$ for all $z \in \mathbb{C}$.

Proof. Fix a positive radius r and consider the lattice $r\mathbb{Z}^2$ in $\mathbb{C}$. Let $\{z_n\}$ denote any fixed arrangement of this lattice into a sequence. For any entire function f, we set

$$I(f) = \int_{\mathbb{C}} |f(w)e^{-\frac{\alpha}{2}|w|^2}|^p \, d\mu(w).$$

Then

$$I(f) \le \sum_n \int_{B(z_n,r)} |f(w)e^{-\frac{\alpha}{2}|w|^2}|^p \, d\mu(w).$$

By Lemma 2.32 and the triangle inequality, there exists a constant $C_1 > 0$ such that

$$|f(w)e^{-\frac{\alpha}{2}|w|^2}|^p \le C_1 \int_{B(z_n,2r)} |f(u)e^{-\frac{\alpha}{2}|u|^2}|^p \, dA(u)$$

for all $w \in B(z_n, r)$. If condition (c) holds, then we can find a positive constant C_2 (independent of f) such that

$$I(f) \le C_2 \sum_n \int_{B(z_n,2r)} |f(u)e^{-\frac{\alpha}{2}|u|^2}|^p \, dA(u)$$

for all entire functions f. It is clear that there exists a positive integer N such that every point in the complex plane belongs to at most N of the disks $B(z_n, 2r)$. Therefore,

$$I(f) \leq C_2 N \int_{\mathbb{C}} |f(u)e^{-\frac{\alpha}{2}|u|^2}|^p \, dA(u).$$

This shows that condition (c) implies condition (a).

To show that condition (a) implies condition (b), simply take $f = k_z$ and apply Lemma 2.33.

Finally, if condition (b) holds, then

$$\int_{B(z,r)} e^{-\frac{p\alpha}{2}|z-w|^2} \, d\mu(w) \leq C$$

for all $z \in \mathbb{C}$. This clearly implies that

$$\mu(B(z,r)) \leq Ce^{\frac{p\alpha}{2}r^2}$$

for all $z \in \mathbb{C}$.   □

It is interesting to notice that condition (c) is independent of p and α. It follows that if condition (a) holds for some $p > 0$ and some α, then it holds for every p and every α (with the constant C dependent on p and α).

Similarly, condition (a) is independent of r. Therefore, if condition (c) holds for some $r > 0$, then it holds for every $r > 0$ (with the constant C dependent on r).

From now on, we will call any positive Borel measure μ that satisfies any of the equivalent conditions (a)–(c) above a Fock–Carleson measure. Similarly, we say that a positive Borel measure μ on $\mathbb{C}$ is a vanishing Fock–Carleson measure if

$$\lim_{n\to\infty} \int_{\mathbb{C}} |f_n(z)e^{-\frac{\alpha}{2}|z|^2}|^p \, d\mu(z) = 0,$$

whenever $\{f_n\}$ is a bounded sequence in F_α^p that converges to 0 uniformly on compact subsets. We proceed to show that being a vanishing Fock–Carleson measure is also independent of p and α.

Theorem 3.30. *Suppose $p > 0$, $\alpha > 0$, $r > 0$, and μ is a positive Borel measure on $\mathbb{C}$. Then the following conditions are equivalent:*

(i) μ is a vanishing Fock–Carleson measure.

(ii) $\int_{\mathbb{C}} e^{-\frac{p\alpha}{2}|z-w|^2} \, d\mu(w) \to 0$ as $z \to \infty$.

(iii) $\mu(B(z,r)) \to 0$ as $z \to \infty$.

Proof. By the proof of Theorem 3.29, there exists a positive constant C (independent of z) such that

$$\mu(B(z,r)) \leq C \int_{\mathbb{C}} e^{-\frac{p\alpha}{2}|z-w|^2} \, d\mu(w)$$

for all $z \in \mathbb{C}$. So condition (ii) implies (iii).

For any sequence $z_n \to \infty$, it is easy to see that the sequence of functions

$$f_n(w) = k_{z_n}(w) = \frac{e^{\alpha \bar{z}_n w}}{e^{\alpha |z_n|^2/2}}, \qquad w \in \mathbb{C},$$

satisfy $\|f_n\|_{p,\alpha} = 1$ and $f_n(w) \to 0$ uniformly on compact sets. Therefore, condition (i) implies (ii).

On the other hand, carefully examining the proof of Theorem 3.29, we see that there is a positive constant C (independent of f) such that

$$\int_{\mathbb{C}} \left| f(w) e^{-\alpha |w|^2/2} \right|^p \, d\mu(w) \tag{3.15}$$

$$\leq C \sum_k \mu(B(z_k,r)) \int_{B(z_k,2r)} \left| f(w) e^{-\alpha |w|^2/2} \right|^p \, dA(w),$$

where $\{z_k\}$ is a fixed arrangement into a sequence of the lattice $r\mathbb{Z}^2$. If condition (iii) holds, then $z \mapsto \mu(B(z,r))$ is a bounded function, and for any $\varepsilon > 0$, there exists a positive integer N such that $\mu(B(z_k,r)) < \varepsilon$ whenever $k > N$. Thus, for any bounded sequence $\{f_n\}$ in F_α^p that converges to 0 uniformly on compact sets, we can estimate the sequence

$$I_n = \int_{\mathbb{C}} \left| f_n(w) e^{-\alpha |w|^2/2} \right|^p \, d\mu(w)$$

according to (3.15) as follows:

$$I_n \leq C \sum_{k=1}^{N} \int_{B(z_k,2r)} \left| f_n(w) e^{-\alpha |w|^2/2} \right|^p \, dA(w) \tag{3.16}$$

$$+ C\varepsilon \sum_{k=N+1}^{\infty} \int_{B(z_k,2r)} \left| f_n(w) e^{-\alpha |w|^2/2} \right|^p \, dA(w),$$

where C is a positive constant independent of n. Since $f_n(w) \to 0$ uniformly on compact sets in $\mathbb{C}$, we have

$$\lim_{n \to \infty} \sum_{k=1}^{N} \int_{B(z_k,2r)} \left| f_n(w) e^{-\alpha |w|^2/2} \right|^p \, dA(w) = 0.$$

Let $n \to \infty$ in (3.16). We obtain

$$\limsup_{n\to\infty} \int_{\mathbb{C}} \left| f_n(w)e^{-\alpha|w|^2/2} \right|^p d\mu(w)$$

$$\leq C\varepsilon \sum_{k=N+1} \int_{B(z_k,2r)} \left| f_n(w)e^{-\alpha|w|^2/2} \right|^p dA(w).$$

There is a positive integer m (depending on r only) such that every point in the complex plane belongs to at most m of the disks $D(z_k, 2r)$. Therefore,

$$\sum_{k=N+1}^{\infty} \int_{B(z_k,2r)} \left| f_n(w)e^{-\alpha|w|^2/2} \right|^p dA(w) \leq m \int_{\mathbb{C}} \left| f_n(w)e^{-\alpha|w|^2/2} \right|^p dA(w) \leq C,$$

where C is another positive constant independent of n (since $\{f_n\}$ is a bounded sequence in F_α^p). Therefore, we can find yet another positive constant C (independent of n and ε) such that

$$\limsup_{n\to\infty} \int_{\mathbb{C}} \left| f_n(w)e^{-\alpha|w|^2/2} \right|^p d\mu(w) \leq C\varepsilon.$$

Since ε is arbitrary, we have

$$\lim_{n\to\infty} \int_{\mathbb{C}} \left| f_n(w)e^{-\alpha|w|^2/2} \right|^p d\mu(w) = 0.$$

This shows that condition (iii) implies condition (i). The proof of the theorem is complete.   □

Carefully examining the proof of Theorems 3.29 and 3.30 above, we obtain the following characterization of Fock–Carleson and vanishing Fock–Carleson measures.

Corollary 3.31. *Suppose μ is a positive Borel measure on $\mathbb{C}$, $r > 0$, and $\{z_n\}$ is any arrangement into a sequence of the lattice $r\mathbb{Z}^2$. Then*

(a) *μ is a Fock–Carleson measure if and only if $\{\mu(B(z_k, r))\}$ is in l^∞.*
(b) *μ is a vanishing Fock–Carleson measure if and only if the sequence $\{\mu(B(z_k, r))\}$ is in c_0.*

Here, l^∞ denotes the space of all bounded sequences, and c_0 is the space of all sequences tending to 0.

Let μ be a complex, regular Borel measure μ on the complex plane. Define

$$\tilde{\mu}(z) = \frac{\alpha}{\pi} \int_{\mathbb{C}} |k_z(w)|^2 e^{-\alpha|w|^2} d\mu(w) = \frac{\alpha}{\pi} \int_{\mathbb{C}} e^{-\alpha|z-w|^2} d\mu(w),$$

whenever these integrals converge. If $d\mu(z) = f(z)dA(z)$ and f satisfies condition (I_1), it is clear that $\widetilde{\mu} = \widetilde{f}$. Thus, we are going to call $\widetilde{\mu}$ the Berezin transform of the measure μ.

Taking $p = 2$ in Theorems 3.29 and 3.30, we see that a positive Borel measure μ on $\mathbb{C}$ is a Fock–Carleson measure if and only if $\widetilde{\mu} \in L^{\infty}(\mathbb{C})$, and μ is a vanishing Fock–Carleson measure if and only if $\widetilde{\mu} \in C_0(\mathbb{C})$.

We also note that when the radius r is fixed, the function $z \mapsto \mu(B(z,r))$ is a constant multiple of the averaging function

$$\widehat{\mu}_r(z) = \frac{\mu(B(z,r))}{\pi r^2}.$$

Thus, conditions on the function $z \mapsto \mu(B(z,r))$ can be replaced with the corresponding conditions on the averaging function $\widehat{\mu}_r$.

3.5   Functions of Bounded Mean Oscillation

For any positive radius r and every exponent $p \in [1, \infty)$, we define BMO_r^p to be the space of locally area-integrable functions f on $\mathbb{C}$ such that

$$\|f\|_{\mathrm{BMO}_r^p} = \sup_{z \in \mathbb{C}} MO_{p,r}(f)(z) < \infty,$$

where

$$MO_{p,r}(f)(z) = \left[\frac{1}{\pi r^2} \int_{B(z,r)} |f - \widehat{f}_r(z)|^p \, dA \right]^{\frac{1}{p}}.$$

Here,

$$\widehat{f}_r(z) = \frac{1}{\pi r^2} \int_{B(z,r)} f \, dA$$

is the mean (average) of f over the Euclidean disk $B(z,r)$. Clearly, BMO_r^p is a linear space.

When $p = 2$, it is easy to see that

$$MO_{2,r}^2(f)(z) = \frac{1}{2(\pi r^2)^2} \int_{B(z,r)} \int_{B(z,r)} |f(u) - f(v)|^2 \, dA(u) \, dA(v). \tag{3.17}$$

It is also easy to check that

$$MO_{2,r}^2(f)(z) = \widehat{|f|^2}_r(z) - |\widehat{f}_r(z)|^2. \tag{3.18}$$

Lemma 3.32. *Let $1 \le p < \infty$, $r > 0$, and f be a locally area-integrable function on $\mathbb{C}$. Then $f \in \mathrm{BMO}_r^p$ if and only if there exists some $C > 0$ such that for any $z \in \mathbb{C}$, there is a complex constant c_z with*

$$\frac{1}{\pi r^2} \int_{B(z,r)} |f(w) - c_z|^p \, dA(w) \le C. \tag{3.19}$$

Proof. If $f \in \mathrm{BMO}_r^p$, then (3.19) holds with $C = \|f\|_{\mathrm{BMO}_r^p}^p$ and $c_z = \widehat{f}_r(z)$.

On the other hand, if (3.19) holds, then by the triangle inequality for the L^p integral,

$$MO_{p,r}(f)(z) = \left[\frac{1}{\pi r^2} \int_{B(z,r)} |f - \widehat{f}_r(z)|^p \, dA \right]^{\frac{1}{p}}$$

$$\le \left[\frac{1}{\pi r^2} \int_{B(z,r)} |f - c_z|^p \, dA \right]^{\frac{1}{p}} + |\widehat{f}_r(z) - c_z|.$$

By Hölder's inequality,

$$\left|\widehat{f_r}(z) - c_z\right| = \left|\frac{1}{\pi r^2}\int_{B(z,r)}(f - c_z)\,dA\right| \le \left[\frac{1}{\pi r^2}\int_{B(z,r)}|f - c_z|^p\,dA\right]^{\frac{1}{p}}.$$

It follows that $MO_{p,r}(f)(z) \le 2C$ for all $z \in \mathbb{C}$, so that $f \in \mathrm{BMO}_r^p$. □

For any $r > 0$, we consider the space BO_r of continuous functions f on $\mathbb{C}$ such that the function

$$\omega_r(f)(z) = \sup\{|f(z) - f(w)| : w \in B(z,r)\}$$

is bounded on $\mathbb{C}$. We think of $\omega_r(f)(z)$ as the local oscillation of f at the point z.

Lemma 3.33. *The space BO_r is independent of r. Moreover, a continuous function f on the complex plane belongs to BO_r if and only if there exists a constant $C > 0$ such that*

$$|f(z) - f(w)| \le C(|z - w| + 1) \tag{3.20}$$

for all z and w in $\mathbb{C}$.

Proof. If f satisfies the condition in (3.20), then clearly $f \in \mathrm{BO}_r$.

To prove the other direction, assume that $f \in \mathrm{BO}_r$. Thus, there exists a positive constant M such that

$$|f(u) - f(v)| \le M, \tag{3.21}$$

whenever $|u - v| \le r$.

Let z and w be two arbitrary points in the complex plane. We are going to show that (3.20) holds for some positive constant C that is independent of z and w.

If $|z - w| \le r$, then (3.20) holds with $C = M$. If $|z - w| > r$, we place points $z_0, \ldots, z_n$ on the line segment from z to w in such a way that $z_0 = z$, $z_n = w$, $|z_k - z_{k+1}| = r$ for $0 \le k < n - 1$, and $|z_{n-1} - z_n| \le r$. By the triangle inequality and (3.21),

$$|f(z) - f(w)| \le \sum_{k=0}^{n-1}|f(z_k) - f(z_{k+1})| \le nM.$$

Since $(n - 1)r \le |z - w| \le nr$, we have

$$nr \le |z - w| + r \le \max(1, 1/r)(|z - w| + 1).$$

With $C = \max(M, 1, 1/r)$, we obtain the desired estimate in (3.20). □

Since BO_r is actually independent of the radius r, we will write BO for BO_r. The initials in BO stand for *bounded oscillation*. It is clear that

$$\|f\|_{\text{BO}} = \sup\{|f(z) - f(w)| : |z - w| \leq 1\}$$

defines a complete seminorm on BO.

We will make the connection between BMO_r^p and the weighted Gaussian measures $d\lambda_\alpha$ with the help of Fock–Carleson measures. More specifically, for any $1 \leq p < \infty$ and $r > 0$, we use BA_r^p to denote the space of Lebesgue measurable functions f on $\mathbb{C}$ such that $\widehat{|f|^p}_r(z)$ is bounded. By the characterization of Fock–Carleson measures in Sect. 3.4, the space BA_r^p is independent of r. Therefore, we will write BA^p for BA_r^p. More specifically, a Lebesgue measurable function f on $\mathbb{C}$ belongs to BA_r^p if and only if

$$\|f\|_{\text{BA}^p}^p = \sup_{z \in \mathbb{C}} \widetilde{|f|^p}(z) < \infty,$$

where $\widetilde{|f|^p}$ is the Berezin transform of $|f|^p$ with respect to the Gaussian measure $d\lambda_\alpha$. Although the weight parameter α appears in the definition of the norm above, the space BA^p is independent of α.

The space BA^p depends on p. In fact, if $1 \leq p < q < \infty$, then $\text{BA}^q \subset \text{BA}^p$ and the containment is strict.

We now describe the structure of BMO_r^p in terms of the relatively simple spaces BO and BA^p. Recall that $\varphi_z(w) = z - w$.

Theorem 3.34. *Let $\alpha > 0$, $r > 0$, and $1 \leq p < \infty$. Suppose f is a locally area-integrable function on $\mathbb{C}$. Then the following conditions are equivalent:*

(a) $f \in \text{BMO}_r^p$.
(b) $f \in \text{BO} + \text{BA}^p$.
(c) f satisfies condition (I_1), and there exists a positive constant C such that

$$\int_{\mathbb{C}} |f \circ \varphi_z(w) - \tilde{f}(z)|^p \, d\lambda_\alpha(w) \leq C \tag{3.22}$$

for all $z \in \mathbb{C}$.
(d) There exists a positive constant C such that for any $z \in \mathbb{C}$, there is some complex number c_z with

$$\int_{\mathbb{C}} |f \circ \varphi_z(w) - c_z|^p \, d\lambda_\alpha(w) \leq C. \tag{3.23}$$

Proof. Let $f \in \text{BMO}_{2r}^p$ and $|z - w| \leq r$. We have

$$|\widehat{f}_r(z) - \widehat{f}_r(w)| \leq |\widehat{f}_r(z) - \widehat{f}_{2r}(z)| + |\widehat{f}_{2r}(z) - \widehat{f}_r(w)|$$

$$\leq \frac{1}{\pi r^2} \int_{B(z,r)} |f(u) - \widehat{f}_{2r}(z)| \, dA(u)$$

$$+ \frac{1}{\pi r^2} \int_{B(w,r)} |f(u) - \widehat{f}_{2r}(z)| \, dA(u).$$

Since $B(z,r)$ and $B(w,r)$ are both contained in $B(z,2r)$, it follows from Hölder's inequality that the two integral summands above are both bounded by a constant that is independent of z and w. This proves that $\widehat{f}_r$ belongs to $\mathrm{BO}_r = \mathrm{BO}$.

On the other hand, we can show that the function $g = f - \widehat{f}_r$ belongs to BA^p whenever $f \in \mathrm{BMO}_{2r}^p$. In fact, it follows from (3.17) that $f \in \mathrm{BMO}_{2r}^p$ implies that $f \in \mathrm{BMO}_r^p$, and it follows from the triangle inequality for L^p integrals that

$$\left[\widehat{|g|^p}_r(z) \right]^{\frac{1}{p}} = \left[\frac{1}{\pi r^2} \int_{B(z,r)} |f(u) - \widehat{f}_r(u)|^p \, dA(u) \right]^{\frac{1}{p}}$$

$$\leq \left[\frac{1}{\pi r^2} \int_{B(z,r)} |f(u) - \widehat{f}_r(z)|^p \, dA(u) \right]^{\frac{1}{p}}$$

$$+ \left[\frac{1}{\pi r^2} \int_{B(z,r)} |\widehat{f}_r(u) - \widehat{f}_r(z)|^p \, dA(u) \right]^{\frac{1}{p}}$$

$$\leq \|f\|_{\mathrm{BMO}_r^p} + \omega_r(\widehat{f}_r)(z).$$

Since $\widehat{f}_r \in \mathrm{BO}_r$ and $f \in \mathrm{BMO}_r^p$, we have $g \in \mathrm{BA}^p$.

Thus, we have proved that $f \in \mathrm{BMO}_{2r}^p$ implies

$$f = \widehat{f}_r + (f - \widehat{f}_r) \in \mathrm{BO} + \mathrm{BA}^p.$$

Since r is arbitrary, we conclude that $\mathrm{BMO}_r^p \subset \mathrm{BO} + \mathrm{BA}^p$, which proves that condition (a) implies condition (b).

It is clear that every function in BO satisfies condition (I_p). Also, every function in BA^p satisfies condition (I_p). Therefore, condition (b) implies that f satisfies condition (I_p). Since $p \geq 1$, f also satisfies condition (I_1). In particular, condition (b) implies that the Berezin transform of f is well defined.

By the triangle inequality and Hölder's inequality,

$$\|f \circ \varphi_z - \widetilde{f}(z)\|_{L^p(d\lambda_\alpha)} \leq \|f \circ \varphi_z\|_{L^p(d\lambda_\alpha)} + |\widetilde{f}(z)| \leq 2\widetilde{|f|^p}(z).$$

We see that condition (3.22) holds whenever $f \in \mathrm{BA}^p$. On the other hand, it follows from Hölder's inequality that

$$\|f \circ \varphi_z - \widetilde{f}(z)\|_{L^p(d\lambda_\alpha)}^p = \int_{\mathbb{C}} |f(z-w) - \widetilde{f}(z)|^p \, d\lambda_\alpha(w)$$

$$\leq \int_{\mathbb{C}} \int_{\mathbb{C}} |f(z-w) - f(z-u)|^p \, d\lambda_\alpha(w) d\lambda_\alpha(u).$$

This together with Lemma 3.33 shows that for any $f \in \mathrm{BO}$,

$$\|f \circ \varphi_z - \widetilde{f}(z)\|^p_{L^p(\mathrm{d}\lambda_\alpha)} \le C^p \int_{\mathbb{C}} \int_{\mathbb{C}} [|u - w| + 1]^p \, \mathrm{d}\lambda_\alpha(w) \, \mathrm{d}\lambda_\alpha(u).$$

The integral on the right-hand side above converges. Thus, condition (3.22) holds for all $f \in \mathrm{BO}$ as well, and we have proved that condition (b) implies condition (c).

Mimicking the proof of Lemma 3.32, we easily obtain the equivalence of conditions (c) and (d).

Finally, if condition (3.22) holds, we can find a positive constant C such that

$$\frac{C}{\pi r^2} \int_{B(z,r)} |f(w) - \widetilde{f}(z)|^p \, \mathrm{d}A(w)$$

$$\le \int_{\mathbb{C}} |f(w) - \widetilde{f}(z)|^p |k_z(w)|^2 \, \mathrm{d}\lambda_\alpha(w)$$

$$= \int_{\mathbb{C}} |f \circ \varphi_z(w) - \widetilde{f}(z)|^p \, \mathrm{d}\lambda_\alpha(w).$$

This, along with Lemma 3.32, then shows that condition (c) implies condition (a).

$\square$

As a consequence of Theorem 3.34, we see that the space BMO^p_r is independent of r and the Berezin transform of every function in BMO^p_r is well defined. Thus, we will write BMO^p for BMO^p_r and define a complete seminorm on BMO^p by

$$\|f\|_{\mathrm{BMO}^p} = \sup_{z \in \mathbb{C}} \|f \circ \varphi_z - \widetilde{f}(z)\|_{L^p(\mathrm{d}\lambda_\alpha)} = \sup_{z \in \mathbb{C}} \|f \circ t_z - \widetilde{f}(z)\|_{L^p(\mathrm{d}\lambda_\alpha)}.$$

One of the nice features of this seminorm is that it is invariant under the actions of t_a, τ_a, and φ_a.

The proof of Theorem 3.34 also shows that every function in BMO^p satisfies condition (I_p). In particular, $\mathrm{BMO}^p \subset L^p(\mathbb{C}, \mathrm{d}\lambda_\alpha)$.

Theorem 3.35. *If* $1 < p < \infty$, *then there exists a positive constant* $C = C(p, \alpha)$ *such that*

$$|\widetilde{f}(z) - \widetilde{f}(w)| \le C\|f\|_{\mathrm{BMO}^p} |z - w|$$

for all z *and* w *in* $\mathbb{C}$ *and all* $f \in \mathrm{BMO}^p$.

Proof. Fix any $z \in \mathbb{C}$ and fix any directional parameter θ. Consider the curve $\gamma(t) = z + e^{i\theta} t$, which is traced out by a particle that starts at z, with unit speed, and in the θ-direction. Recall that

$$\widetilde{f}(\gamma(t)) = \frac{\alpha}{\pi} \int_{\mathbb{C}} f(u) e^{-\alpha|\gamma(t) - u|^2} \, \mathrm{d}A(u).$$

Differentiating under the integral sign gives

$$\frac{d}{dt}\widetilde{f}(\gamma(t)) = -\frac{2\alpha^2}{\pi} \int_{\mathbb{C}} f(u)e^{-\alpha|\gamma(t)-u|^2} \text{Re}\left[\gamma'(t)(\overline{\gamma(t)}-\overline{u})\right] dA(u).$$

For any fixed t, the function

$$h(u) = \text{Re}\left[\gamma'(t)(\overline{\gamma(t)}-u)\right]$$

is harmonic, so it is fixed by the Berezin transform. It follows that

$$\frac{\alpha}{\pi} \int_{\mathbb{C}} e^{-\alpha|\gamma(t)-u|^2} \text{Re}\left[\gamma'(t)(\overline{\gamma(t)}-\overline{u})\right] dA(u) = \widetilde{h}(\gamma(t)) = 0.$$

Therefore, $d\widetilde{f}(\gamma(t))/dt$ is equal to

$$-\frac{2\alpha^2}{\pi} \int_{\mathbb{C}} (f(u) - \widetilde{f}(\gamma(t)))e^{-\alpha|\gamma(t)-u|^2} \text{Re}\left[\gamma'(t)(\overline{\gamma(t)}-\overline{u})\right] dA(u).$$

Let q be the conjugate exponent, $1/p + 1/q = 1$. Then by Hölder's inequality, $|d\widetilde{f}(\gamma(t))/dt|$ is less than or equal to

$$\frac{2\alpha^2}{\pi}\left[\int_{\mathbb{C}} |f(u) - \widetilde{f}(\gamma(t))|^p e^{-\alpha|\gamma(t)-u|^2} dA(u)\right]^{\frac{1}{p}}$$

times

$$\left[\int_{\mathbb{C}} |\gamma(t) - u|^q e^{-\alpha|\gamma(t)-u|^2} dA(u)\right]^{\frac{1}{q}}. \tag{3.24}$$

The integral in (3.24) is, via a simple change of variables, equal to

$$\int_{\mathbb{C}} |u|^q e^{-\alpha|u|^2} dA(u),$$

which is clearly convergent. Therefore, there exists a positive constant $C = C(\alpha, p)$ such that

$$\left|\frac{d}{dt}\widetilde{f}(\gamma(t))\right| \leq CMO_p(f)(\gamma(t)) \leq C\|f\|_{\text{BMO}^p}$$

for all t, where

$$\|f\|_{\text{BMO}^p} = \sup_{z\in\mathbb{C}} MO_p(f)(z) = \sup_{z\in\mathbb{C}} \|f \circ \varphi_z - \widetilde{f}(z)\|_{L^p(d\lambda_\alpha)}.$$

Integrating with respect to t, we obtain

$$|\widetilde{f}(z) - \widetilde{f}(w)| \leq C\|f\|_{\mathrm{BMO}^p}|z - w|$$

for all z and w in $\mathbb{C}$.   □

The following result gives another way to split the space BMO^p into the sum of two simpler spaces: a space of "smooth" functions and a space of "small" functions.

Theorem 3.36. *Suppose $f \in \mathrm{BMO}^p$ and $1 \leq p < \infty$. Then $\widetilde{f} \in \mathrm{BO}$ and $f - \widetilde{f} \in \mathrm{BA}^p$.*

Proof. It is easy to see that there is a positive constant C such that

$$\begin{aligned}
|\widetilde{f}(z) - \widehat{f}_r(z)| &\leq \frac{1}{\pi r^2}\int_{B(z,r)}|f(w) - \widetilde{f}(z)|\,dA(w)\\
&\leq C\int_{B(z,r)}|f(w) - \widetilde{f}(z)||k_z(w)|^2\,d\lambda_\alpha(w)\\
&\leq C\int_{\mathbb{C}}|f \circ \varphi_z(w) - \widetilde{f}(z)|\,d\lambda_\alpha(w)\\
&\leq C\|f \circ \varphi_z - \widetilde{f}(z)\|_{L^p(d\lambda_\alpha)},
\end{aligned}$$

where the last step follows from Hölder's inequality. This shows that $\widetilde{f} - \widehat{f}_r$ is a bounded function. Since a bounded continuous function belongs to both BO and BA^p, we have $\widetilde{f} - \widehat{f}_r \in \mathrm{BO} \cap \mathrm{BA}^p$.

Write

$$f - \widetilde{f} = (f - \widehat{f}_r) - (\widetilde{f} - \widehat{f}_r),$$

and recall from Theorem 3.34 that $f - \widehat{f}_r$ is in BA^p. We conclude that $f - \widetilde{f}$ belongs to BA^p. Similarly, we can write

$$\widetilde{f} = \widehat{f}_r + (\widetilde{f} - \widehat{f}_r)$$

and infer that $\widetilde{f} \in \mathrm{BO}$.   □

Corollary 3.37. *If $1 < p < \infty$, then*

$$\mathrm{BMO}^p = \mathrm{LIP} + \mathrm{BA}^p,$$

where LIP is the space of all Lipschitz functions on $\mathbb{C}$. Moreover, a canonical decomposition is given by $f = \widetilde{f} + (f - \widetilde{f})$.

The next result characterizes entire functions in BMO^p.

Proposition 3.38. *Suppose* $1 \leq p < \infty$ *and* f *is an entire function. Then* $f \in \text{BMO}^p$ *if and only if* f *is a linear polynomial.*

Proof. When f is entire, we have $\widehat{f}_r = f$ because of the mean value theorem. It follows from Theorem 3.34 (and its proof) that $f = \widehat{f}_r \in \text{BO}$ whenever $f \in \text{BMO}^p$. Thus, there exists a positive constant C such that

$$|f(z) - f(w)| \leq C(|z - w| + 1)$$

for all z and w. Let $w = 0$ and use Cauchy's estimate. We conclude that f must be a linear polynomial.

Conversely, if f is a linear polynomial, then f is Lipschitz in the Euclidean metric. In particular, $f \in \text{BO}$, and so $f \in \text{BMO}^p$. $\square$

Let VMO_r^p denote the space of locally area-integrable functions f such that

$$\lim_{z \to \infty} MO_{p,r}(f)(z) = 0.$$

It is clear that VMO_r^p is a subspace of BMO_r^p. Just like BMO_r^p, the space VMO_r^p is also independent of r, and we will write VMO^p for VMO_r^p.

Similarly, we consider the space VO_r consisting of continuous functions f such that

$$\lim_{z \to \infty} \omega_r(f)(z) = 0.$$

It can be shown that VO_r is independent of r, and we will write VO for VO_r. The initials in VO stand for "vanishing oscillation."

We also consider the space VA_r^p consisting of functions such that

$$\lim_{z \to \infty} \frac{1}{\pi r^2} \int_{B(z,r)} |f(w)|^p \, dA(w) = 0.$$

According to the characterizations of vanishing Fock–Carleson measures in Sect. 3.4, the space VA_r^p is independent of r and consists of functions f such that $\widetilde{|f|^p}(z) \to 0$ as $z \to \infty$. We will write VA^p for VA_r^p. The initials in VA^p stand for "vanishing average." The following theorem describes the structure of VMO^p.

Theorem 3.39. *Suppose* $1 \leq p < \infty$, $r > 0$, *and* f *is locally area integrable. Then the following conditions are equivalent:*

(i) $f \in \text{VMO}^p = \text{VMO}_r^p$.
(ii) $MO_p(f)(z) \to 0$ *as* $z \to \infty$.
(iii) $f \in \text{VO} + \text{VA}^p$.

Moreover, there are two canonical decompositions for condition (iii) above:

$$f = \widetilde{f} + (f - \widetilde{f}), \quad f = \widehat{f_r} + (f - \widehat{f_r}).$$

We omit the proof.

Corollary 3.40. *Suppose f is an entire function. Then $f \in \mathrm{VMO}^p$ if and only if f is constant.*

3.6 Notes

The Berezin transform was introduced in [23] and then studied systematically in [23–27] for a number of reproducing Hilbert spaces. It has become an indispensable tool in the study of operators on function spaces, including Hankel operators, Toeplitz operators, and composition operators. See [250] for applications of the Berezin transform in the theory of Bergman spaces. In particular, the proofs of Propositions 3.3–3.6 were adapted from the corresponding ones in [250].

In the setting of Fock spaces and when parametrized appropriately, the Berezin transform is nothing but the heat transform. This connection with the heat equation makes the Berezin transform on Fock spaces particularly useful. The semigroup property of the heat transforms was first observed in [30].

The Lipschitz estimate for the Berezin transform of a bounded linear operator on the Fock space is due to Coburn. See [54, 55]. Propositions 3.9–3.11 and Corollary 3.12 are taken from [55], and these results will be needed in Chap. 6 when we study Toeplitz operators on the Fock space.

Theorem 3.25, the Lipschitz estimate for the Berezin transform of a bounded function, was first proved in [29]. Together with the semigroup property, this result shows that the Berezin transform is a rapidly smoothing operation on bounded functions, and consequently, a bounded function that is fixed by the Berezin transform must be constant. On the other hand, there exist unbounded functions fixed by the Berezin transform that are not harmonic. The example in Sect. 3.3 was taken from [84]. This example shows the sharp contrast with the Bergman space theory, where the fixed points of the Berezin transform are exactly the harmonic functions; see [1].

The characterization of Fock–Carleson measures is analogous to the characterization of Carleson measures for Bergman spaces. The material in Sect. 3.4 is taken from [132]. See [250] for the corresponding results in the Bergman space theory. Note that the notion of Carleson measures was initially introduced in the Hardy space setting, where a geometric characterization is much more difficult. See [76].

The notion of BMO and VMO using a fixed Euclidean radius was first introduced in [32, 257]. This idea was then generalized to the setting of bounded symmetric domains in [21] and to the case of strongly pseudoconvex domains in [149], with the Euclidean metric replaced by the Bergman metric. The resulting spaces are independent of the particular radius used, but the dependence on the exponent p was observed and studied in [248] in the context of Bergman spaces on the unit ball. The extension to the Fock space setting is straightforward.

The Lipschitz estimate for the Berezin transform of a function in BMO was first proved in [21] in the context of Bergman spaces on bounded symmetric domains. The extension to the Fock space, Theorem 3.35, was first carried out in [13].

3.7 Exercises

1. Show that the Lipschitz constant $2\sqrt{\alpha}$ in Corollary 3.8 is best possible.
2. Show that the spaces BMO^p and VMO^p are complete under the norm

$$\|f\| = \|f\|_{\text{BMO}^p} + |\widetilde{f}(0)|.$$

3. Characterize the multipliers of the spaces BMO^p and VMO^p.
4. Show that the function $|z|$ belongs to BMO^p but the function $|z|^2$ does not belong to BMO^p.
5. Show that the function $\sqrt{|z|}$ belongs to VMO^p.
6. Show that the function $e^{i\sqrt{|z|}}$ belongs to VMO^p.
7. Study the behavior of the Berezin transform of the function $\ln|z|$, which is harmonic everywhere except the origin.
8. If $f \in L^\infty(\mathbb{C})$, show that the sequence $\{\widetilde{f}^{(n)}\}$ converges to a constant function as $n \to \infty$. Moreover, the convergence is uniform on any compact subset of $\mathbb{C}$.
9. If f is locally L^p-integrable and

$$\lim_{z \to \infty} f(z) = L$$

   exists, then $f \in \text{VMO}^p$.
10. A function f is "eventually slowly varying" if, for any $\varepsilon > 0$, there exist positive numbers R and δ such that $|f(z) - f(w)| < \varepsilon$ whenever $|z| > R$, $|w| > R$, and $|z - w| < \delta$. Show that every eventually slowly varying function is in VMO^p.
11. Characterize harmonic functions in BMO^p.
12. Suppose α, β, and γ are positive parameters. Show that for $1 \le p \le \infty$, we have $Q_\alpha L_\beta^p \subset L_\gamma^p$ if and only if $\alpha^2/\gamma \le 2\alpha - \beta$.
13. Show that the Berezin transform B_α is never bounded on L_β^p, where α and β are positive weight parameters.
14. If $f \in \text{BMO}^1$, show that $B_\alpha(|f|) - |B_\alpha f|$ is bounded for $\alpha > 0$.
15. Does the boundedness of $B_\alpha(|f|) - |B_\alpha f|$ imply $f \in \text{BMO}^1$?
16. Consider the previous two problems for $1 < p < \infty$.
17. Show that $B_\alpha f_r(z) = B_{\alpha/r^2} f(rz)$, where $f_r(z) = f(rz)$.
18. Show that B_α is a bounded and self-adjoint operator on $L^2(\mathbb{C}, dA)$.
19. Show that $\text{BA}^q \subset \text{BA}^p$ whenever $1 \le p \le q < \infty$. Furthermore, the inclusion is strict if $p < q$.
20. If $f \in \text{BMO}^p$, then $|f| \in \text{BMO}^p$. Similarly, if $f \in \text{VMO}^p$, then $|f| \in \text{VMO}^p$.

Chapter 4
Interpolating and Sampling Sequences

In this chapter, we characterize interpolating and sampling sequences for the Fock spaces F_α^p. The characterizations are based on a certain notion of uniform density on the complex plane. So we will first spend some time discussing this geometric notion of density which also has applications in other areas of analysis and physics.

K. Zhu, *Analysis on Fock Spaces*, Graduate Texts in Mathematics 263,
DOI 10.1007/978-1-4419-8801-0_4,
© Springer Science+Business Media New York 2012

4.1 A Notion of Density

Let $Z = \{z_n\}$ be a sequence of distinct points in $\mathbb{C}$. For any set S in $\mathbb{C}$, we let $n(Z,S) = |Z \cap S|$ denote the number of points in $Z \cap S$. There are two families of sets we are going to use in this chapter: Euclidean disks and squares. More specifically, we will use

$$S = B(w,r) = \{z \in \mathbb{C} : |z - w| < r\},$$

and

$$S = S(w,r) = \{z \in \mathbb{C} : |\mathrm{Re}\, z - \mathrm{Re}\, w| < r/2, |\mathrm{Im}\, z - \mathrm{Im}\, w| < r/2\}.$$

The area of $B(w,r)$ is πr^2, while the area of $S(w,r)$ is r^2.

The lower and upper densities of Z are then defined as

$$D^-(Z) = \liminf_{r \to \infty} \inf_{w \in \mathbb{C}} \frac{n(Z, B(w,r))}{\pi r^2},$$

and

$$D^+(Z) = \limsup_{r \to \infty} \sup_{w \in \mathbb{C}} \frac{n(Z, B(w,r))}{\pi r^2},$$

respectively.

The following result gives an alternative description of these densities in terms of squares. Note that in the definition above and the proposition below, the quotients $n(Z, B(w,r))/(\pi r^2)$ and $n(Z, S(w,r))/r^2$ represent the average number of points from Z per square unit in the disk $B(w,r)$ and the square $S(w,r)$, respectively.

Proposition 4.1. *For any sequence Z of distinct points in $\mathbb{C}$, let*

$$\widetilde{D}^-(Z) = \liminf_{r \to \infty} \inf_{w \in \mathbb{C}} \frac{n(Z, S(w,r))}{r^2},$$

and

$$\widetilde{D}^+(Z) = \limsup_{r \to \infty} \sup_{w \in \mathbb{C}} \frac{n(Z, S(w,r))}{r^2}.$$

Then we have $D^-(Z) = \widetilde{D}^-(Z)$ and $D^+(Z) = \widetilde{D}^+(Z)$.

Proof. Fix any positive number ε. It is clear that there exist a finite number of disjoint open squares $S(w_j, r_j)$, $1 \le j \le N$, in $B(0,1)$ such that

$$0 < \pi - (r_1^2 + \cdots + r_N^2) < \varepsilon.$$

For any $w \in \mathbb{C}$ and $r > 0$, it is easy to see that $z \in S(w + rw_j, rr_j)$ if and only if $(z - w)/r \in S(w_j, r_j)$. It follows that the squares $S(w + rw_j, rr_j)$ are disjoint and contained in $B(w,r)$. Thus,

$$n(Z, B(w, r)) \geq n\left(Z, \cup_{j=1}^N S(w + rw_j, rr_j)\right)$$

$$= \sum_{j=1}^N n(Z, S(w + rw_j, rr_j))$$

$$= \sum_{j=1}^N \frac{n(Z, S(w + rw_j, rr_j))}{(rr_j)^2} \cdot (rr_j)^2.$$

It follows that

$$\frac{n(Z, B(w, r))}{\pi r^2} \geq \sum_{j=1}^N \frac{n(Z, S(w + rw_j, rr_j))}{(rr_j)^2} \cdot \frac{r_j^2}{\pi}$$

$$\geq \sum_{j=1}^n \inf_{\zeta \in \mathbb{C}} \frac{n(Z, S(\zeta, rr_j))}{(rr_j)^2} \cdot \frac{r_j^2}{\pi}.$$

Taking the infimum over w, we obtain

$$\inf_{w \in \mathbb{C}} \frac{n(Z, B(w, r))}{\pi r^2} \geq \sum_{j=1}^N \inf_{w \in \mathbb{C}} \frac{n(Z, S(w, rr_j))}{(rr_j)^2} \cdot \frac{r_j^2}{\pi}.$$

Letting $r \to \infty$ then leads to

$$D^-(Z) \geq \widetilde{D}^-(Z) \sum_{j=1}^N \frac{r_j^2}{\pi} \geq \frac{\pi - \varepsilon}{\pi} \widetilde{D}^-(Z).$$

Since ε is arbitrary, we must have $D^-(Z) \geq \widetilde{D}^-(Z)$.

On the other hand, there exist a finite number of squares $S(w_j, r_j)$, $1 \leq j \leq N$, that cover the unit disk $B(0, 1)$ and satisfy

$$0 < r_1^2 + \cdots + r_N^2 - \pi < \varepsilon.$$

For any $w \in \mathbb{C}$ and $r > 0$, we have

$$B(w, r) \subset \bigcup_{j=1}^N S(w + rw_j, rr_j)$$

so that

$$n(Z, B(w, r)) \leq n\left(Z, \cup_{j=1}^N S(w + rw_j, rr_j)\right)$$

$$\leq \sum_{j=1}^N n(Z, S(w + rw_j, rr_j))$$

$$= \sum_{j=1}^N \frac{n(Z, S(w + rw_j, rr_j))}{(rr_j)^2} \cdot (rr_j)^2.$$

It follows that

$$\frac{n(Z,B(w,r))}{\pi r^2} \leq \sum_{j=1}^{N} \frac{n(Z,S(w+rw_j))}{(rr_j)^2} \cdot \frac{r_j^2}{\pi}$$

$$\leq \sum_{j=1}^{N} \sup_{\zeta \in \mathbb{C}} \frac{n(Z,S(\zeta,rr_j))}{(rr_j)^2} \cdot \frac{r_j^2}{\pi}.$$

First, take the supremum over $w \in \mathbb{C}$ and then let $r \to \infty$. We obtain

$$D^+(Z) \leq \widetilde{D}^+(Z) \sum_{j=1}^{N} \frac{r_j^2}{\pi} \leq \frac{\pi + \varepsilon}{\pi} \widetilde{D}^+(Z).$$

Since ε is arbitrary, we must have $D^+(Z) \leq \widetilde{D}^+(Z)$.

In the previous two paragraphs, we tried to cover the unit disk by a finite number of squares whose total area is arbitrarily close to the area of the unit disk. If we now try to cover the unit square $S(0,1)$ by a finite number of disks whose total area is arbitrarily close to the area of the unit square, then the same arguments show that $\widetilde{D}^-(Z) \geq D^-(Z)$ and $\widetilde{D}^+(Z) \leq D^+(Z)$. This completes the proof of the proposition. □

The following result shows that the upper and lower densities can also be defined in terms of arbitrary sets of Lebesgue measure 1. Note that the Euclidean disk $B(w,r)$ is just a translation of a dilation of the unit disk $|z| < 1$.

Theorem 4.2. *Let I be any subset of $\mathbb{C}$ of Lebesgue measure 1 whose boundary has Lebesgue measure 0. Then we have*

$$D^-(Z) = \liminf_{r \to \infty} \inf_{w \in \mathbb{C}} \frac{n(Z,w+rI)}{r^2},$$

and

$$D^+(Z) = \limsup_{r \to \infty} \sup_{w \in \mathbb{C}} \frac{n(Z,w+rI)}{r^2}.$$

Proof. The proof is similar to that of Proposition 4.1. We will not need the full strength of the theorem and will omit its proof here. We refer the interested reader to [36] for details. □

We conclude the section with an example for which we can explicitly compute the uniform densities.

Proposition 4.3. *For any lattice*

$$\Lambda = \{\omega + m\omega_1 + n\omega_2 : m \in \mathbb{Z}, n \in \mathbb{Z}\},$$

we have

$$D^+(\Lambda) = D^-(\Lambda) = \frac{1}{|\mathrm{Im}\,(\omega_1\,\overline{\omega}_2)|}.$$

Proof. The fundamental region of the lattice Λ is congruent to the parallelogram spanned by $\omega_1 = a_1 + ia_2$ and $\omega_2 = b_1 + ib_2$, whose area is

$$\left| \det \begin{pmatrix} a_1 & a_2 \\ b_1 & b_2 \end{pmatrix} \right| = |a_1 b_2 - a_2 b_1| = |\mathrm{Im}\,(\omega_1\,\overline{\omega}_2)|.$$

When r is very large, the number of points in $\Lambda \cap B(w,r)$ is roughly the area of $B(w,r)$ divided by the area of the fundamental region of Λ. It follows that

$$D^+(\Lambda) = D^-(\Lambda) = \lim_{r\to\infty} \frac{(\pi r^2)/|\mathrm{Im}\,(\omega_1\,\overline{\omega}_2)|}{\pi r^2} = \frac{1}{|\mathrm{Im}\,(\omega_1\,\overline{\omega}_2)|}.$$

□

As a special case, if r is any positive number, then the uniform densities of the square lattice $r\mathbb{Z}^2$ are given by

$$D^+(r\mathbb{Z}^2) = D^-(r\mathbb{Z}^2) = 1/r^2.$$

In particular, if $r = \sqrt{\pi/\alpha}$, then the uniform densities of the lattice $\Lambda_\alpha = \sqrt{\pi/\alpha}\,\mathbb{Z}^2$ are given by

$$D^+(\Lambda_\alpha) = D^-(\Lambda_\alpha) = \alpha/\pi.$$

4.2 Separated Sequences

Let $Z = \{z_n\}$ be a sequence of distinct points in the complex plane. We say that Z is separated if

$$\delta(Z) = \inf\{|z_n - z_m| : n \neq m\} > 0.$$

When Z is separated, the number $\delta = \delta(Z)$ will be called the *separation constant* of Z.

The next result is a necessary condition that the values of a function in F_α^p taken on a separated sequence must satisfy.

Proposition 4.4. *Let $Z = \{z_n\}$ be a separated sequence and $0 < p < \infty$. Then there exists a positive constant C, independent of f, such that*

$$\sum_{n=1}^{\infty} \left| f(z_n) e^{-\alpha|z_n|^2/2} \right|^p \leq C\|f\|_{p,\alpha}^p$$

for all $f \in F_\alpha^p$.

Proof. Let $\delta = \delta(Z)$ be the separation constant of Z. By Lemma 2.32, there exists a positive constant C, independent of n and f, such that

$$|f(z_n)e^{-\alpha|z_n|^2/2}|^p \leq C \int_{B(z_n,r)} |f(z)e^{-\alpha|z|^2/2}|^p \, dA(z)$$

for all $f \in F_\alpha^p$ and all $n \geq 1$, where $r = \delta/2$. By the definition of the separation constant, the Euclidean disks $B(z_n, r)$ are all disjoint. Therefore,

$$\sum_{n=1}^{\infty} |f(z_n)e^{-\alpha|z_n|^2/2}|^p \leq C \sum_{n=1}^{\infty} \int_{B(z_n,r)} |f(z)e^{-\alpha|z|^2/2}|^p \, dA(z)$$

$$\leq C \int_{\mathbb{C}} |f(z)e^{-\alpha|z|^2/2}|^p \, dA(z)$$

$$= \frac{2\pi C}{p\alpha} \|f\|_{p,\alpha}^p.$$

This proves the proposition.   □

Based on the proposition above, we now make the definition of interpolating sequences for F_α^p.

Let $Z = \{z_n\}$ denote a sequence of distinct points in the complex plane. We say that Z is an interpolating sequence for F_α^p, $0 < p < \infty$, if for every sequence $\{v_n\}$ of values satisfying

$$\sum_{k=1}^{\infty} \left| v_k e^{-\alpha|z_k|^2/2} \right|^p < \infty, \tag{4.1}$$

there exists a function $f \in F_\alpha^p$ such that $f(z_k) = v_k$ for all $k \geq 1$.

Similarly, we say that a sequence $Z = \{z_n\}$ of distinct points in $\mathbb{C}$ is an interpolating sequence for F_α^∞ if for every sequence $\{v_n\}$ of values satisfying

$$\sup_{n\geq 1}|v_n|e^{-\alpha|z_n|^2/2} < \infty, \qquad (4.2)$$

there exists a function $f \in F_\alpha^\infty$ such that $f(z_n) = v_n$ for all $n \geq 1$.

Given any sequence $Z = \{z_n\}$ and any entire function f, we write

$$\|f|Z\|_{p,\alpha} = \left[\sum_{n=1}^\infty \left|f(z_n)e^{-\frac{\alpha}{2}|z_n|^2}\right|^p\right]^{1/p}$$

for $0 < p < \infty$ and

$$\|f|Z\|_{\infty,\alpha} = \sup_{n\geq 1}|f(z_n)|e^{-\frac{\alpha}{2}|z_n|^2}.$$

The following result shows that if Z is an interpolating sequence for F_α^p, then interpolation can be performed in a stable way.

Lemma 4.5. *Suppose $0 < p \leq \infty$ and $Z = \{z_n\}$ is an interpolating sequence for F_α^p. Then there exists a positive constant C with the following property: whenever $\{v_n\}$ is a sequence such that $\{v_n e^{-\alpha|z_n|^2/2}\} \in l^p$ there exists a function $f \in F_\alpha^p$ such that $f(z_n) = v_n$ for all n and*

$$\|f\|_{p,\alpha} \leq C\|f|Z\|_{p,\alpha}. \qquad (4.3)$$

Proof. Let X_p denote the Banach space of sequences $\{v_k\}$ such that $\{v_k e^{-\frac{\alpha}{2}|z_k|^2}\} \in l^p$. Let J_Z denote the space of all functions $f \in F_\alpha^p$ such that $f(z) = 0$ for all $z \in Z$. It is clear that J_Z is a closed subspace of F_α^p. For any sequence $v = \{v_k\} \in X_p$, there exists a function $f \in F_\alpha^p$ such that $f(z_k) = v_k$ for all $k \geq 1$. We define $T(v) = f + J_Z$. Then T is a well-defined linear mapping from X_p into the quotient space F_α^p/J_Z. It is easy to check that T has a closed graph in $X_p \times (F_\alpha^p/J_Z)$. Therefore, by the closed-graph theorem, the mapping T is continuous, which implies the desired estimate. $\square$

If Z is an interpolating sequence for F_α^p, we are going to use $N_p(Z) = N_p(Z,\alpha)$ to denote the smallest constant C satisfying the inequality in (4.3). We put $N_p(Z) = N_p(Z,\alpha) = \infty$ when Z is not an interpolating sequence for F_α^p. We also use the convention that $N_p(\emptyset) = 0$.

We say that a sequence $Z = \{z_n\}$ of distinct points in $\mathbb{C}$ is a sampling sequence for F_α^p, $0 < p < \infty$, if there exists a constant $C > 0$ such that

$$C^{-1}\|f\|_{p,\alpha}^p \leq \sum_{n=1}^\infty \left|f(z_n)e^{-\frac{\alpha}{2}|z_n|^2}\right|^p \leq C\|f\|_{p,\alpha}^p \qquad (4.4)$$

for all $f \in F_\alpha^p$.

Sampling for F_α^∞ requires a slightly different treatment. More specifically, we say that an arbitrary set Z in $\mathbb{C}$ is a sampling set for F_α^∞ if there exists a constant $C > 0$ such that

$$\|f\|_{\infty,\alpha} \le C \sup_{z \in Z} |f(z)| e^{-\frac{\alpha}{2}|z|^2} \tag{4.5}$$

for all $f \in F_\alpha^\infty$. When Z is a sequence, we use the term "sampling sequence" instead of "sampling set."

We use $M_p(Z) = M_p(Z, \alpha)$ to denote the smallest constant C such that

$$\|f\|_{p,\alpha} \le C \|f|Z\|_{p,\alpha}$$

for all $f \in F_\alpha^p$. Thus, Z is a sampling set for F_α^∞ if and only if $M_\infty(Z) < \infty$, and it is a sampling sequence for F_α^p, $0 < p < \infty$, if and only if $M_p(Z) < \infty$ and $\|f|Z\|_{p,\alpha} < \infty$ for all $f \in F_\alpha^p$.

We use the convention that the empty set is not a sampling set for F_α^p, which should be easy to conceive and accept. In particular, we are going to write $M_\infty(\emptyset) = \infty$.

Recall that for any complex number a, the Weyl unitary operator W_a is defined by

$$W_a f(z) = e^{\alpha \bar{a} z - \frac{\alpha}{2}|a|^2} f(z - a).$$

Each W_a is a surjective isometry on F_α^p. As a consequence of this translation invariance, we immediately obtain

$$N_p(Z + a) = N_p(Z), \qquad M_p(Z + a) = M_p(Z), \tag{4.6}$$

which allows us to translate our analysis around an arbitrary point to the origin 0.

Our next step is to show that every interpolating sequence for F_α^p must be separated, and every sampling sequence for F_α^p must contain a separated sequence that is still sampling. The following estimate will be needed for this purpose as well as several other results.

Lemma 4.6. *Suppose* $0 < p < \infty$, f *is entire, and*

$$S(z) = f(z) e^{-\alpha |z|^2 / 2}.$$

For any positive radius δ, *there exists a constant* $C = C(\alpha, p, \delta) > 0$ *such that*

$$\big| |S(\zeta + z)| - |S(\zeta)| \big|^p \le C |z|^p \int_{B(\zeta, 3\delta)} \left| e^{-\frac{\alpha}{2}|u|^2} f(u) \right|^p \, dA(u)$$

for all $\zeta \in \mathbb{C}$ *and all* z *with* $|z| \le \delta$.

Proof. For convenience, we write

$$f_\zeta(w) = W_{-\zeta} f(w) = e^{-\alpha \bar{\zeta} w - \frac{\alpha}{2}|\zeta|^2} f(\zeta + w).$$

It is easy to see that

$$|S(\zeta + z)| = e^{-\frac{\alpha}{2}|z|^2}|f_\zeta(z)|, \quad |S(\zeta)| = |f_\zeta(0)|.$$

It follows that

$$
\begin{aligned}
||S(\zeta + z)| - |S(\zeta)|| &= \left|e^{-\frac{\alpha}{2}|z|^2}|f_\zeta(z)| - |f_\zeta(0)|\right| \\
&= \left|\left(e^{-\frac{\alpha}{2}|z|^2} - 1\right)|f_\zeta(z)| + |f_\zeta(z)| - |f_\zeta(0)|\right| \\
&\leq \left(1 - e^{-\frac{\alpha}{2}|z|^2}\right)|f_\zeta(z)| + |f_\zeta(z) - f_\zeta(0)| \\
&= \left(e^{\frac{\alpha}{2}|z|^2} - 1\right)\left|e^{-\frac{\alpha}{2}|z|^2} f_\zeta(z)\right| + |f_\zeta(z) - f_\zeta(0)| \\
&= \left(e^{\frac{\alpha}{2}|z|^2} - 1\right)\left|e^{-\frac{\alpha}{2}|z|^2} f_\zeta(z)\right| + |z||f'_\zeta(w)|,
\end{aligned}
$$

where the last step follows from the mean value theorem with some w satisfying $|w| < |z|$.

By Lemma 2.32, there exists a constant $C_1 > 0$ such that

$$
\begin{aligned}
\left|e^{-\frac{\alpha}{2}|z|^2} f_\zeta(z)\right|^p &\leq C_1 \int_{B(z,\delta)} \left|e^{-\frac{\alpha}{2}|u|^2} f_\zeta(u)\right|^p \, dA(u) \\
&\leq C_1 \int_{B(0,2\delta)} \left|e^{-\frac{\alpha}{2}|u|^2} f_\zeta(u)\right|^p \, dA(u) \\
&= C_1 \int_{B(\zeta,2\delta)} \left|e^{-\frac{\alpha}{2}|u|^2} f(u)\right|^p \, dA(u).
\end{aligned}
$$

The second inequality above follows from the triangle inequality, and the last equality follows from a change of variables.

On the other hand, it follows from Cauchy's integral formula that

$$f'_\zeta(w) = \frac{1}{2\pi i} \int_{|u-w|=\delta} \frac{f_\zeta(u)\, du}{(u-w)^2}.$$

Consequently,

$$|f'_\zeta(w)| \leq \frac{1}{\delta} \sup_{|u-w|=\delta} |f_\zeta(u)| \leq C_2 \sup_{|u-w|=\delta} |f_\zeta(u)| e^{-\alpha|u|^2/2}.$$

Another application of Lemma 2.32, followed by the triangle inequality and a change of variables, gives

$$|f'_\zeta(w)|^p \leq C_3 \int_{B(\zeta,3\delta)} \left|e^{-\frac{\alpha}{2}|u|^2} f(u)\right|^p \, dA(u).$$

The desired result now follows from the triangle inequality

$$|u+v|^p \leq 2^p(|u|^p + |v|^p)$$

and the elementary inequality

$$0 < e^{\frac{\alpha}{2}|z|^2} - 1 \leq C_4|z|^2 \leq C_4\delta|z|, \qquad |z| \leq \delta.$$

This completes the proof of the lemma. □

Corollary 4.7. *For $0 < p \leq \infty$, there is a positive constant $C = C(\alpha, p)$ such that*

$$\left| |S(z_1)| - |S(z_2)| \right| \leq C|z_1 - z_2| \|f\|_{p,\alpha}$$

for all $f \in F_\alpha^p$ and all complex numbers z_1 and z_2.

Proof. The case $|z_1 - z_2| \leq 1$ follows from the lemma above (and its proof, which gives a version for $p = \infty$), while the case $|z_1 - z_2| > 1$ is obvious. □

Lemma 4.8. *Suppose $0 < p \leq \infty$ and $Z = \{z_n\}$ is an interpolating sequence for F_α^p. Then Z must be separated.*

Proof. Fix any two different positive integers n and m. If $|z_n - z_m| > 1$, we do not do anything.

If $|z_n - z_m| \leq 1$, we consider the sequence $\{a_k\}$, where $a_n = 1$ and $a_k = 0$ for $k \neq n$. Since Z is an interpolating sequence for F_α^p, there exists a function $f \in F_\alpha^p$ such that $f(z_k)e^{-\alpha|z_k|^2/2} = a_k$ for all $k \geq 1$ and

$$\|f\|_{p,\alpha} \leq N_p(Z)\|f|Z\|_{p,\alpha} = N_p(Z).$$

With the notation $S(z) = e^{-\alpha|z|^2/2}f(z)$ from Lemma 4.6 and Corollary 4.7, we have

$$1 = \left| |a_n| - |a_m| \right| = \left| |S(z_n)| - |S(z_m)| \right| \leq CN_p(Z)|z_n - z_m|,$$

where C is a positive constant that only depends on α and p. This shows that the sequence Z is separated. □

We now proceed to show that every sampling sequence for F_α^p must contain a separated subsequence that is also a sampling sequence for F_α^p. We break the proof into two cases: $0 < p < \infty$ and $p = \infty$.

Lemma 4.9. *Suppose $0 < p < \infty$ and $Z = \{z_n\}$ is any sequence of complex numbers. Then the following two conditions are equivalent:*

(a) There exists a positive constant C such that

$$\sum_{n=1}^{\infty} \left| f(z_n)e^{-\frac{\alpha}{2}|z_n|^2} \right|^p \leq C\|f\|_{p,\alpha}^p$$

for all $f \in F_\alpha^p$.

(b) The sequence Z is a union of finitely many separated sequences.

Proof. Condition (a) above simply says that the measure

$$\mu = \sum_{n=1}^{\infty} \delta_{z_n}$$

is a Fock–Carleson measure for F_{α}^p, where δ_z is the unit point mass at z. Therefore, according to (an obvious variant of) Theorem 3.29, condition (a) is equivalent to the existence of a positive integer N such that any square $S \subset \mathbb{C}$ of side length 1 contains at most N points from Z, which is clearly equivalent to the condition that Z is the union of finitely many separated sequences. $\qquad\square$

An obvious consequence of the above result is that every sampling sequence for F_{α}^p, where $0 < p < \infty$, contains a separated subsequence. The following result shows that this is true for $p = \infty$ as well and we can do more than that.

Lemma 4.10. *If $Z = \{z_n\}$ is a sampling sequence for F_{α}^{∞}, then Z contains a separated subsequence Z' that is also a sampling sequence for F_{α}^{∞}.*

Proof. Fix a sufficiently small positive number ε whose exact value will be specified later. Let $z_1' = z_1$, discard the terms in the sequence $\{z_n\}$ that are within ε of z_1, and denote the remaining terms by $\{z_{11}, z_{12}, \cdots\}$ with the original order. Let $z_2' = z_{11}$, discard the terms in the sequence $\{z_{1n}\}$ that are within ε of z_2', and denote by $\{z_{21}, z_{22}, \cdots\}$ the remaining terms in the original order. Continuing this process, infinitely many times if necessary, we obtain a subsequence $Z' = \{z_n'\}$ of Z which clearly satisfies the condition $|z_i' - z_j'| \geq \varepsilon$ whenever $i \neq j$. In particular, Z' is separated. Furthermore, for any z_k, either it was discarded during the process above, in which case it is within ε of some point in the sequence Z', or it eventually gets picked as a term in Z'. Either way, we have $d(z_k, Z') < \varepsilon$ so that

$$Z = \bigcup_{z' \in Z'} \left[Z \cap B(z', \varepsilon) \right]. \qquad (4.7)$$

Write $Z = Z' \cup Z''$ as a disjoint union. Clearly,

$$\|f|Z\|_{\infty,\alpha} = \max \left(\|f|Z'\|_{\infty,\alpha}, \|f|Z''\|_{\infty,\alpha} \right) \leq \|f|Z'\|_{\infty,\alpha} + \|f|Z''\|_{\infty,\alpha}.$$

Given any $w \in Z''$, it follows from (4.7) that there exists some $z \in Z'$ such that $|w - z| \leq \varepsilon$. By the triangle inequality and Corollary 4.7,

$$|S(w)| \leq |S(z)| + \big||S(z)| - |S(w)|\big|$$
$$\leq \|f|Z'\|_{\infty,\alpha} + C\varepsilon\|f\|_{\infty,\alpha},$$

where C is a positive constant independent of ε and f. Therefore,

$$\|f|Z\|_{\infty,\alpha} \leq 2\|f|Z'\|_{\infty,\alpha} + C\varepsilon\|f\|_{\infty,\alpha}$$

for all $f \in F_\alpha^\infty$. Since Z is a sampling sequence for F_α^∞, there exists a positive constant c such that $c\|f\|_{\infty,\alpha} \leq \|f|Z\|_{\infty,\alpha}$ for all $f \in F_\alpha^\infty$. Thus,

$$(c - C\varepsilon)\|f\|_{\infty,\alpha} \leq 2\|f|Z'\|_{\infty,\alpha}$$

for all $f \in F_\alpha^\infty$. If the value of ε was chosen such that $c - C\varepsilon > 0$, then there is another positive constant $C' > 0$ such that $C'\|f\|_{\infty,\alpha} \leq \|f|Z'\|_{\infty,\alpha}$ for all $f \in F_\alpha^\infty$, which means that Z' is sampling for F_α^∞. □

We want to show that the lemma above holds for $p < \infty$ as well. But the proof is more complicated.

Lemma 4.11. *Suppose $0 < p < \infty$ and $Z = \{z_k\}$ is a sampling sequence for F_α^p. Then Z contains a separated subsequence that is also a sampling sequence for F_α^p.*

Proof. By Lemma 4.9, we can write $Z = Z_1 \cup Z_2 \cup \cdots \cup Z_n$ as a disjoint union of separated sequences. We prove the result by induction on n. If $n = 1$, there is nothing to prove. Thus, we assume $n > 1$ and proceed to show that we can find a subsequence Z' of Z such that:

(a) Z' is sampling for F_α^p.
(b) Z' is the disjoint union of $n - 1$ separated sequences.

Let δ be the separation constant for Z_n (so that $|z - w| \geq \delta$ for all z and w in Z_n with $z \neq w$) and write $\widetilde{Z} = Z_1 \cup \cdots \cup Z_{n-1}$. Fix any positive constant $\varepsilon < \delta/8$ and split Z_n into two parts:

$$\Gamma = \left\{ z \in Z_n : d(z, \widetilde{Z}) < \varepsilon \right\}, \quad \Gamma' = \left\{ z \in Z_n : d(z, \widetilde{Z}) \geq \varepsilon \right\}.$$

Let $Z' = \widetilde{Z} \cup \Gamma'$. Putting Γ' together with Z_1, we have

$$Z' = (Z_1 \cup \Gamma') \cup \cdots \cup Z_{n-1},$$

and each of the $n - 1$ sequences above is separated. We will show that Z' is sampling for F_α^p when ε is sufficiently small.

Since $Z = Z' \cup \Gamma$, we will be done if Γ is empty. If Γ is not empty, we write $\Gamma = \{\zeta_k\}$. For each k, there exists a point $a_k \in \widetilde{Z}$ such that $|\zeta_k - a_k| < \varepsilon$. For $i \neq j$, we have

$$|a_i - a_j| = |(a_i - \zeta_i) - (a_j - \zeta_j) + (\zeta_j - \zeta_i)|$$
$$\geq |\zeta_i - \zeta_j| - |(a_i - \zeta_i) - (a_j - \zeta_j)|$$
$$\geq \delta - 2\varepsilon > \frac{3}{4}\delta.$$

In particular, the points in the sequence $\{a_k\}$ are distinct.

Since $Z = Z' \cup \Gamma$ is sampling for F_α^p, there is a positive constant c such that

$$c\|f\|_{p,\alpha}^p \leq \|f|Z\|_{p,\alpha}^p = \|f|Z'\|_{p,\alpha}^p + \|f|\Gamma\|_{p,\alpha}^p$$

for all $f \in F_\alpha^p$. Using the notation $S(z) = f(z)e^{-\alpha|z|^2/2}$ and the triangle inequality, we have

$$\|f|\Gamma\|_{p,\alpha}^p = \sum_k |S(\zeta_k)|^p$$
$$= \sum_k [|S(\zeta_k)| - |S(a_k)| + |S(a_k)|]^p$$
$$\leq 2^p \sum_k \left[||S(\zeta_k)| - |S(a_k)||^p + |S(a_k)|^p \right].$$

Since the a_k's are distinct points from $\tilde{Z} \subset Z'$, we have

$$\sum_k |S(a_k)|^p \leq \|f|Z'\|_{p,\alpha}^p, \quad f \in F_\alpha^p.$$

By Lemma 4.6, with $\delta/8$ in place of δ, we can find a constant $C > 0$ that is independent of ε and f such that

$$\sum_k ||S(\zeta_k)| - |S(a_k)||^p \leq C\varepsilon^p \sum_k \int_{B(a_k,\delta/2)} \left| f(z)e^{-\frac{\alpha}{2}|z|^2} \right|^p dA(z).$$

Since the sequence $\{a_k\}$ is separated with separation constant at least $3\delta/4$, there is another constant $C' > 0$, independent of ε and f, such that

$$\sum_k ||S(\zeta_k)| - |S(a_k)||^p \leq C'\varepsilon^p \|f\|_{p,\alpha}^p$$

for all $f \in F_\alpha^p$. It follows that

$$c\|f\|_{p,\alpha}^p \leq 2^p C' \varepsilon^p \|f\|_{p,\alpha}^p + (2^p + 1)\|f|Z'\|_{p,\alpha}^p$$

so that

$$\left(c - 2^p C' \varepsilon^p \right) \|f\|_{p,\alpha}^p \leq (2^p + 1)\|f|Z'\|_{p,\alpha}^p$$

for all $f \in F_\alpha^p$. If the value of ε was chosen such that $c - 2^p C' \varepsilon > 0$, then the sequence Z' is sampling for F_α^p. This completes the proof of the lemma. $\square$

4.3   Stability Under Weak Convergence

In this section, we consider a notion of weak convergence for relatively closed subsets in the complex plane and establish several results about sampling and interpolation that are preserved under weak convergence.

We say that a set in the complex plane is relatively closed if its intersection with any compact set is still compact. Given a nonempty and relatively closed subset A of $\mathbb{C}$, let

$$A_t = \{z \in \mathbb{C} : d(z,A) < t\}, \quad 0 < t < 1.$$

So A_t is the set of all points in $\mathbb{C}$ that are within distance t of the set A. If A and B are two nonempty and relatively closed subsets of the complex plane, we define

$$[A,B] = \inf\{t : A \subset B_t, B \subset A_t\}$$

and call it the Hausdorff distance between A and B. It can be verified that this is indeed a metric. The assumption that A and B are relatively closed ensures that $[A,B] = 0$ only when $A = B$.

Alternatively,

$$[A,B] = \max\left(d^*(A,B), d^*(B,A)\right),$$

where

$$d^*(A,B) = \sup_{z \in A} d(z,B) = \sup_{z \in A} \inf_{w \in B} |z - w|$$

is the asymmetric "distance" from A to B.

From the definition above, we see that $[A,B] < \varepsilon$ if and only if the following two conditions hold:

1. For any $a \in A$, there exists $b \in B$ such that $|a - b| < \varepsilon$.
2. For any $b \in B$, there exists $a \in A$ such that $|a - b| < \varepsilon$.

Suppose $\{A_n\}$ and A are all nonempty and relatively closed subsets of the complex plane. We say that $\{A_n\}$ converges strongly to A if $[A_n, A] \to 0$ as $n \to \infty$. We say that $\{A_n\}$ converges weakly to A if $\{A_n \cap F\}$ converges strongly to $A \cap F$ for every compact set F such that none of $A_n \cap F$ and $A \cap F$ is empty. Since $[A,B]$ is a distance, the limit of strong and weak convergence is unique.

To simplify notation and statements, we say that a sequence $\{A_n\}$ of sequences converges weakly to the empty set if we can write

$$A_n = \{a_{n1}, a_{n2}, \cdots, \}, \quad |a_{n1}| \le |a_{n2}| \le \cdots$$

for each $n \ge 1$ and $a_{nk} \to \infty$ as $n \to \infty$ for each $k \ge 1$.

In what follows, whenever we consider a sequence, we assume that it consists of distinct points and has no finite accumulation point. In particular, such a sequence is relatively closed in $\mathbb{C}$ and can be rearranged so that the modulus of its terms is nondecreasing. We use the notation $W(Z)$ to denote the collection of weak limits of all the translates $Z + z$ of Z. The set $W(Z)$ will play a crucial role in our analysis.

We first prove a certain compactness property for uniformly separated sequences in the complex plane.

Proposition 4.12. *For each $n \geq 1$, let Z_n be a separated sequence. If $\delta = \inf_n \delta(Z_n) > 0$, then there exists a subsequence $\{Z_{n_k}\}$ and a separated sequence Z (possibly empty) such that $\{Z_{n_k}\}$ converges weakly to Z.*

Proof. We write $Z_n = \{z_{n1}, z_{n2}, \cdots\}$ with $|z_{n1}| \leq |z_{n2}| \leq \cdots$. If $z_{n1} \to \infty$ as $n \to \infty$, then for every k, we have $z_{nk} \to \infty$ as $n \to \infty$. In this case, $\{Z_n\}$ converges weakly to the empty set.

If $z_{n1} \not\to \infty$ as $n \to \infty$, we can find a subsequence $\{Z_{n_j}\}$ such that $z_{n_j1} \to z_1$ as $j \to \infty$. Then either $z_{n_j2} \to \infty$ as $j \to \infty$, which implies that for every $k \geq 2$, we have $z_{n_jk} \to \infty$ as $j \to \infty$, or $\{Z_{n_j}\}$ has a subsequence whose second components converge to some $z_2 \in \mathbb{C}$. In the latter case, the process continues.

There are now two possibilities: either the process terminates after a finite number, say N, of iterations, which produces a subsequence of $\{Z_n\}$ that converges weakly to a finite sequence $Z = \{z_1, \cdots, z_N\}$, or the process never stops, which via a diagonalization argument produces a subsequence of $\{Z_n\}$ that converges weakly to an infinite sequence $Z = \{z_1, z_2, \cdots\}$. The condition $\inf_n \delta(Z_n) > 0$ ensures that the limit sequence Z is separated as well. This proves the desired result. $\quad\square$

The following result gives an alternative description of weak convergence for separated sequences.

Proposition 4.13. *Suppose each Z_n is a separated sequence with $\delta = \inf_n \delta(Z_n) > 0$. Write $Z_n = \{z_{n1}, z_{n2}, \cdots\}$ with $|z_{n1}| \leq |z_{n2}| \leq \cdots$. Then $\{Z_n\}$ converges weakly to Z if and only if one of the following is true:*

(a) $Z = \emptyset$ is the empty set, and for every $k \geq 1$ we have $z_{nk} \to \infty$ as $n \to \infty$.
(b) $Z = \{z_1, \cdots, z_N\}$ is a finite set, $z_{nk} \to z_k$ for every $1 \leq k \leq N$, and $z_{nk} \to \infty$ for every $k > N$.
(c) $Z = \{z_1, z_2, \cdots\}$ is an infinite (separated) sequence and $z_{nk} \to z_k$ for every $k \geq 1$.

Proof. It is clear from the definition that any one of the above conditions implies that $\{Z_n\}$ converges weakly to Z. The other implication follows from Proposition 4.12 and its proof, if we start out with an arbitrary subsequence of $\{Z_n\}$. Here, we use the fact that $z_{nk} \to z_k$ (where z_k is either finite or infinite) if and only if each subsequence of $\{z_{1k}, z_{2k}, \cdots\}$ converges to z_k. $\quad\square$

We now prove that any weak limit of sampling sequences for F_α^∞ remains a sampling sequence for F_α^∞.

Proposition 4.14. *Suppose $\{Z_n\}$ converges weakly to Z. Then*

$$M_\infty(Z) \leq \liminf_{n \to \infty} M_\infty(Z_n),$$

where $M_\infty(Z)$ denotes the F_α^∞ sampling constant for Z.

Proof. If $Z = \emptyset$, we can write $Z_n = \{z_{nk}\}$ with $|z_{n1}| \le |z_{n2}| \le \cdots$ and have $z_{n1} \to \infty$ as $n \to \infty$, which implies that $z_{nk} \to \infty$ as $n \to \infty$ for every k. Choosing $f = 1$ in

$$\|f\|_{\infty,\alpha} \le M_\infty(Z_n) \sup_k e^{-\frac{\alpha}{2}|z_{nk}|^2} |f(z_{nk})|$$

shows that $M_\infty(Z_n) \to \infty$. The desired result is then obvious.

Next, assume that Z is nonempty. Since $M_\infty(Z)$ is the smallest M such that

$$\|f\|_{\infty,\alpha} \le M\|f|Z\|_{\infty,\alpha},$$

we can write

$$M_\infty(Z) = \sup_{f \in F_\alpha^\infty} \frac{\|f\|_{\infty,\alpha}}{\|f|Z\|_{\infty,\alpha}} = \sup_{\|f\|_{\infty,\alpha}=1} \frac{1}{\|f|Z\|_{\infty,\alpha}}.$$

It follows that the constant $M = M_\infty(Z)$ is given by

$$M^{-1} = \inf_{\|f\|_{\infty,\alpha}=1} \|f|Z\|_{\infty,\alpha}.$$

Thus, for any $\varepsilon \in (0,1)$, we can find a unit vector $f \in F_\alpha^\infty$ such that

$$\|f|Z\|_{\infty,\alpha} < M^{-1} + \varepsilon.$$

This is true even when $M = \infty$. Also, by translation invariance (namely, we can translate Z and Z_n simultaneously if necessary), we may assume that $|f(0)| > 1 - \varepsilon$.

By Corollary 4.7, there exists a positive number $\delta = C\varepsilon$, where $C > 0$ is independent of ε, such that

$$\left| e^{-\frac{\alpha}{2}|w|^2} |f(w)| - e^{-\frac{\alpha}{2}|z|^2} |f(z)| \right| < \varepsilon$$

whenever $|w - z| < \delta$. Since $\{Z_n\}$ converges weakly to Z, there exists a positive integer N such that

$$[Z_n \cap \bar{B}(0, \varepsilon^{-2}), Z \cap \bar{B}(0, \varepsilon^{-2})] < \delta/2$$

whenever $n > N$, where $\bar{B}(0, r)$ is the closed disk with center 0 and radius r. Here, we may assume that ε is small enough so that none of $Z_n \cap \bar{B}(0, \varepsilon^{-2})$ and $Z \cap \bar{B}(0, \varepsilon^{-2})$ is empty.

Let $a = 1 - (\delta\varepsilon^2/2)$ and assume that ε and δ are small enough so that $a \in (0,1)$. If $n > N$ and $w \in Z_n \cap \bar{B}(0, \varepsilon^{-2})$, there exists some $z \in Z \cap \bar{B}(0, \varepsilon^{-2})$ such that $|w - z| < \delta/2$. It follows from the triangle inequality that

$$|aw - z| \le a|z - w| + (1 - a)|z| < \frac{\delta}{2} + \frac{\delta}{2} = \delta.$$

Therefore,

$$\begin{aligned}
e^{-\frac{\alpha}{2}|w|^2}|f(aw)| &= e^{-\frac{\alpha}{2}|aw|^2}|f(aw)|e^{-\frac{\alpha}{2}(1-a^2)|w|^2} \le e^{-\frac{\alpha}{2}|aw|^2}|f(aw)| \\
&\le \left|e^{-\frac{\alpha}{2}|aw|^2}|f(aw)| - e^{-\frac{\alpha}{2}|z|^2}|f(z)|\right| + e^{-\frac{\alpha}{2}|z|^2}|f(z)| \\
&< \varepsilon + \|f|Z\|_{\infty,\alpha} < M^{-1} + 2\varepsilon.
\end{aligned}$$

On the other hand, if $|w| > \varepsilon^{-2}$, then

$$\begin{aligned}
e^{-\frac{\alpha}{2}|w|^2}|f(aw)| &= e^{-\frac{\alpha}{2}|aw|^2}|f(aw)|e^{-\frac{\alpha}{2}(1-a^2)|w|^2} \\
&\le \|f\|_{\infty,\alpha}e^{-\frac{\alpha}{2}(1-a)|w|^2} \\
&\le e^{-\frac{\alpha}{2}(1-a)\varepsilon^{-4}} = e^{-\frac{C\alpha}{4\varepsilon}}.
\end{aligned}$$

We may assume that ε is small enough so that

$$e^{-\frac{\alpha}{2}|w|^2}|f(aw)| \le e^{-(C\alpha)/(4\varepsilon)} < M^{-1} + 2\varepsilon$$

for all $|w| > \varepsilon^{-2}$. Combining this with the last estimate in the previous paragraph, we conclude that the function $g(z) = f(az)$ satisfies

$$\|g|Z_n\|_{\infty,\alpha} < M^{-1} + 2\varepsilon, \quad n > N.$$

Since $|f(0)| > 1 - \varepsilon$, we have

$$\|g\|_{\infty,\alpha} \ge |g(0)| = |f(0)| > 1 - \varepsilon.$$

It follows that

$$M_\infty(Z_n) \ge \frac{\|g\|_{\infty,\alpha}}{\|g|Z_n\|_{\infty,\alpha}} \ge \frac{1-\varepsilon}{M^{-1}+2\varepsilon}$$

for all $n > N$. Thus,

$$\liminf_{n\to\infty} M_\infty(Z_n) \ge \frac{1-\varepsilon}{M^{-1}+2\varepsilon}.$$

The desired result now follows by letting $\varepsilon \to 0$. □

As a consequence of the proposition above, we see that small perturbations of a sampling sequence for F_α^∞ remain sampling sequences for F_α^∞. More specifically, we have the following.

Corollary 4.15. *Suppose $Z = \{z_n\}$ is a sampling sequence for F_α^∞. There exists a positive number δ such that any sequence $W = \{w_n\}$ satisfying $|z_n - w_n| < \delta$, $n \ge 1$, is still a sampling sequence for F_α^∞.*

The discussion above was about the behavior of the sampling constant $M_\infty(Z)$ under weak convergence. The following result concerns the sampling constant $M_p(Z)$ when $p < \infty$. Recall that for any separated sequence Z, we use

$$\delta(Z) = \inf\{|z - w| : z \in Z, w \in Z, z \neq w\}$$

to denote the separation constant of Z.

Proposition 4.16. *For each n, let Z_n be a separated sequence in $\mathbb{C}$. If $\inf_n \delta(Z_n) > 0$, then*

$$M_p(Z, \alpha) \leq \liminf_{n \to \infty} M_p(Z_n, \alpha), \quad 0 < p < \infty,$$

whenever Z_n converges weakly to Z.

Proof. When $Z = \emptyset$, the desired result is proved just as in the case $p = \infty$. See the proof of Lemma 4.14. So we assume $Z \neq \emptyset$ in the rest of the proof.

Let $\delta = \inf_n \delta(Z_n)$. It follows from Proposition 4.13 that Z is separated and $\delta(Z) \geq \delta$.

Given any $\varepsilon > 0$, we follow the same argument at the beginning of the proof of Proposition 4.14 to find a unit vector f in F_α^p such that

$$\|f|Z\|_{p,\alpha} \leq M^{-1} + \varepsilon,$$

where $M = M_p(Z, \alpha)$ (which may be infinite).

For any fixed and large enough radius R, we can find a positive integer N such that

$$[Z_n \cap \bar{B}(0,R), Z \cap \bar{B}(0,R)] < \min(\delta/6, \varepsilon), \qquad n > N.$$

Thus, for any $n > N$ and $z \in Z_n \cap \bar{B}(0,R)$, we can find some $w \in Z$ such that

$$|z - w| < \frac{\delta}{6}, \quad |z - w| < \varepsilon.$$

Since Z is separated with separation constant at least δ, we see that different z correspond to different w. By Lemma 4.6, there exists a positive constant $C = C(\alpha, p, \delta)$ such that

$$\left| |f(z)|e^{-\frac{\alpha}{2}|z|^2} - |f(w)|e^{-\frac{\alpha}{2}|w|^2} \right|^p \leq C\varepsilon^p \int_{B(w,\delta/2)} |f(u)e^{-\alpha|u|^2/2}|^p \, dA(u).$$

If $0 < p \leq 1$, it follows from the triangle inequality that

$$\left| f(z)e^{-\frac{\alpha}{2}|z|^2} \right|^p \leq \left| f(w)e^{-\frac{\alpha}{2}|w|^2} \right|^p + \left| |f(z)|e^{-\frac{\alpha}{2}|z|^2} - |f(w)|e^{-\frac{\alpha}{2}|w|^2} \right|^p$$

$$\leq \left| f(w)e^{-\frac{\alpha}{2}|w|^2} \right|^p + C\varepsilon^p \int_{B(w,\delta/2)} \left| f(u)e^{-\frac{\alpha}{2}|u|^2} \right|^p \, dA(u).$$

Sum over all $z \in Z_n \cap \bar{B}(0,R)$, observe that different z correspond to different w, and use the facts that f is a unit vector in F_α^p and $\delta(Z) \geq \delta$. We obtain

$$\|f|Z_n \cap \bar{B}(0,R)\|_{p,\alpha}^p \leq \|f|Z\|_{p,\alpha}^p + C\varepsilon^p.$$

Since C is independent of R, letting $R \to \infty$ gives

$$\|f|Z_n\|_{p,\alpha}^p \leq \|f|Z\|_{p,\alpha}^p + C\varepsilon^p < (M^{-1}+\varepsilon)^p + C\varepsilon^p$$

for all $n > N$. It follows that

$$M_p(Z_n, \alpha) \geq \left[\frac{1}{(M^{-1}+\varepsilon)^p + C\varepsilon^p} \right]^{\frac{1}{p}}$$

for all $n > N$, and so

$$\liminf_{n \to \infty} M_p(Z_n, \alpha) \geq \left[\frac{1}{(M^{-1}+\varepsilon)^p + C\varepsilon^p} \right]^{\frac{1}{p}}.$$

Since ε is arbitrary, we must have

$$\liminf_{n \to \infty} M_p(Z_n, \alpha) \geq M = M_p(Z, \alpha).$$

If $1 \leq p < \infty$, we apply the version of the triangle inequality for $p > 1$ to get

$$\|f|Z_n \cap \bar{B}(0,R)\|_{p,\alpha} = \left[\sum_{z \in Z_n \cap \bar{B}(0,R)} \left| f(z) e^{-\frac{\alpha}{2}|z|^2} \right|^p \right]^{\frac{1}{p}}$$

$$\leq \left[\sum_{w \in Z \cap \bar{B}(0,R)} \left| f(w) e^{-\frac{\alpha}{2}|w|^2} \right|^p \right]^{\frac{1}{p}}$$

$$+ \left[C\varepsilon^p \sum_{w \in Z} \int_{B(w,\delta/2)} \left| f(u) e^{-\frac{\alpha}{2}|u|^2} \right|^p dA(u) \right]^{\frac{1}{p}}$$

$$\leq \|f|Z\|_{p,\alpha} + C^{1/p}\varepsilon \leq M^{-1} + (1 + C^{1/p})\varepsilon.$$

Since C is independent of R, letting $R \to \infty$ gives us

$$\|f|Z_n\|_{p,\alpha} \leq M^{-1} + (1 + C^{1/p})\varepsilon$$

for all $n > N$. It follows that

$$M_p(Z_n) \geq \left[M^{-1} + (1 + C^{1/p})\varepsilon \right]^{-1}, \qquad n > N,$$

so that

$$\liminf_{n\to\infty} M_p(Z_n,\alpha) \geq \left[M^{-1} + (1+C^{1/p})\varepsilon\right]^{-1}.$$

But ε is arbitrary and C is independent of ε, so we must have

$$\liminf_{n\to\infty} M_p(Z_n,\alpha) \geq M = M_p(Z,\alpha).$$

This completes the proof of the proposition. □

Corollary 4.17. *Suppose* $0 < p < \infty$ *and* Z *is a separated sequence with separation constant* δ. *If* Z *is sampling for* F_α^p *and* Z' *is another sequence such that* $[Z,Z']$ *is sufficiently small, then* Z' *is also a sampling sequence for* F_α^p.

Proof. This follows from Proposition 4.16. □

Carefully examining the proof of Lemmas 4.14 and 4.16, we see that more can be done. More specifically, if Z is separated, then there exists a constant $C > 0$ such that

$$M_p(Z',\alpha) \leq C[Z,Z']M_p(Z)$$

for sequences Z' that are sufficiently close to Z. Here, the constant C only depends on p and α.

This concludes the discussion about the stability of sampling sequences under weak convergence. Next, we consider the stability of interpolating sequences under weak convergence.

Proposition 4.18. *Suppose* $\{Z_n\}$ *converges to* Z *weakly. Then*

$$N_p(Z,\alpha) \leq \liminf_{n\to\infty} N_p(Z_n,\alpha)$$

for all $0 < p \leq \infty$.

Proof. The case $Z = \emptyset$ is obvious. Also, by working with a subsequence if necessary, we may assume that

$$\liminf_{n\to\infty} N_p(Z_n) = \lim_{n\to\infty} N_p(Z_n) < \infty.$$

In particular, we may assume that

$$S = \sup_n N_p(Z_n,\alpha) < \infty.$$

By the proof of Lemma 4.8, we have $\delta = \inf_n \delta(Z_n) > 0$. Then it follows easily from Proposition 4.13 that the sequence Z is also separated and its separation constant is at least δ.

With the help of Proposition 4.13, we may also assume that

$$Z_n = \{z_{n1}, z_{n2}, \cdots\}, \qquad Z = \{z_1, z_2, \cdots\},$$

with $z_{nk} \to z_k$, as $n \to \infty$, for every appropriate k (depending on whether Z is finite or infinite).

Fix a positive number ε and a sequence $v = \{v_k\} \in l^p$. If Z is a finite sequence of length m, we assume that $v_k = 0$ for $k > m$. For each n, there exists some function $f_n \in F_\alpha^p$ such that

$$f_n(z_{nk})e^{-\frac{\alpha}{2}|z_{nk}|^2} = v_k, \qquad k \geq 1,$$

and

$$\|f_n\|_{p,\alpha} \leq N_p(Z_n)\|f_n|Z_n\|_{p,\alpha} \leq S\|v\|_{l^p}.$$

By a normal family argument, we may assume that

$$\lim_{n\to\infty} f_n(z) = f(z)$$

uniformly on compact subsets of the complex plane. By Fatou's lemma, we have $f \in F_\alpha^p$ with

$$\|f\|_{p,\alpha} \leq \liminf_{n\to\infty}\|f_n\|_{p,\alpha} \leq \|v\|_{l^p}\liminf_{n\to\infty}N_p(Z_n).$$

Furthermore, for any fixed $z_k \in Z$, we have

$$f(z_k)e^{-\frac{\alpha}{2}|z_k|^2} = \lim_{n\to\infty} f_n(z_{nk})e^{-\frac{\alpha}{2}|z_{nk}|^2} = v_k.$$

It follows that $\|f|Z\|_{p,\alpha} = \|v\|_{l^p}$ so that

$$\|f\|_{p,\alpha} \leq \|f|Z\|_{p,\alpha}\liminf_{n\to\infty}N_p(Z_n).$$

This shows that

$$N_p(Z) \leq \liminf_{n\to\infty}N_p(Z_n)$$

and completes the proof of the proposition. □

Corollary 4.19. *Suppose $0 < p \leq \infty$ and Z is a separated sequence. If Z is an interpolating sequence for F_α^p, then there exists a positive constant σ such that Z' is interpolating for F_α^p whenever $[Z',Z] < \sigma$.*

Proof. This follows from Proposition 4.18. □

4.4 A Modified Weierstrass σ-Function

A key tool in our proof of the sufficiency of the sampling and interpolating conditions is a special, modified Weierstrass σ-function. Thus, we let $\Lambda_\alpha = \{\omega_{mn}\}$ denote the square lattice in $\mathbb{C}$ that is defined by

$$\omega_{mn} = \sqrt{\pi/\alpha}(m+in),$$

where m and n run over all integers. Recall that the Weierstrass σ-function associated to Λ_α is defined by

$$\sigma_\alpha(z) = z\prod_{m,n}' \left(1 - \frac{z}{\omega_{mn}}\right) \exp\left(\frac{z}{\omega_{mn}} + \frac{1}{2}\frac{z^2}{\omega_{mn}^2}\right),$$

where the prime denotes the omission of the factor corresponding to $m = n = 0$. By Proposition 1.19, $\sigma_\alpha(z)$ is an entire function with Λ_α as its zero set.

Also, recall that for any $a \in \mathbb{C}$, the Weyl unitary operator W_a is defined by

$$W_a f(z) = e^{\alpha \bar{a} z - \frac{\alpha}{2}|a|^2} f(z - a).$$

Proposition 4.20. *The function σ_α is quasiperiodic in the sense that*

$$W_{\omega_{mn}} \sigma_\alpha(z) = (-1)^{m+n+mn} \sigma_\alpha(z)$$

for all z and ω_{mn}. Consequently, if

$$R_\alpha = \left\{z = x + iy : |x| \leq \frac{1}{2}\sqrt{\pi/\alpha}, |y| \leq \frac{1}{2}\sqrt{\pi/\alpha}\right\}$$

is the fundamental region for Λ_α, then for any $z \in \mathbb{C}$, there exists some $w \in R_\alpha$ such that

$$|\sigma_\alpha(z)|e^{-\frac{\alpha}{2}|z|^2} = |\sigma_\alpha(w)|e^{-\frac{\alpha}{2}|w|^2}.$$

Furthermore, there exists a positive constant c such that

$$|\sigma_\alpha(z)|e^{-\frac{\alpha}{2}|z|^2} \geq cd(z, \Lambda_\alpha)$$

for all $z \in \mathbb{C}$, where

$$d(z, \Lambda_\alpha) = \min\{|z - w| : w \in \Lambda\}$$

is the Euclidean distance from z to Λ_α.

Proof. See Proposition 1.20 and Corollary 1.21. $\qquad\qquad\qquad\qquad\qquad$ $\square$

The reciprocal density parameter α in Λ_α is critical for the Fock spaces F_α^p. More precisely, we will see that Λ_β is interpolating for F_α^p if and only if $\beta < \alpha$; and Λ_β is sampling for F_α^p if and only if $\beta > \alpha$. When $\beta = \alpha$, Λ_β is neither interpolating nor sampling for F_α^p, but is a set of uniqueness for F_α^p; see Lemma 5.7.

We will need to perturb the zeros of the Weierstrass σ-function $\sigma_\alpha(z)$. Let $Z = \{z_{mn}\}$ be a sequence of distinct points in $\mathbb{C}$. If there exists a constant $Q > 0$ (not necessarily small!) such that $|\omega_{mn} - z_{mn}| \leq Q$ for all $\omega_{mn} \in \Lambda_\alpha$, then we say that Z is uniformly close to Λ_α. For any sequence $Z = \{z_{mn}\}$ that is uniformly close to Λ_α, we define an associated function as follows:

$$g(z) = g_Z(z) = (z - z_{00}) \prod_{m,n}{}' \left(1 - \frac{z}{z_{mn}}\right) \exp\left(\frac{z}{z_{mn}} + \frac{1}{2}\frac{z^2}{\omega_{mn}^2}\right). \tag{4.8}$$

Here, we assume that z_{00} is the point of Z closest to 0. Note that both z_{mn} and ω_{mn} appear in the formula above; it was not a misprint.

Lemma 4.21. *Let Z be uniformly close to $\Lambda = \Lambda_\alpha$ and let g be its associated function defined above. Then g is an entire function and the zero set of g is exactly Z. Moreover, there exist positive constants C_1, C_2, and c such that*

$$|g(z)|e^{-\frac{\alpha}{2}|z|^2} \geq C_1 e^{-c|z|\log|z|} d(z,Z) \tag{4.9}$$

and

$$|g(z)|e^{-\frac{\alpha}{2}|z|^2} \leq C_2 e^{c|z|\log|z|} \tag{4.10}$$

for all $z \in \mathbb{C}$. Moreover,

$$|g'(z_{mn})|e^{-\frac{\alpha}{2}|z_{mn}|^2} \geq C_1 e^{-c|z_{mn}|\log|z_{mn}|} \tag{4.11}$$

for all m and n.

Proof. The convergence of the infinite product defining g and the determination of the zero set of g are similar to the corresponding problems for the Weierstrass product in Chap. 1. We leave the routine details to the reader.

We may write

$$e^{-\frac{\alpha}{2}|z|^2} g(z) = \frac{e^{-\frac{\alpha}{2}|z|^2}\sigma_\alpha(z)}{d(z,\Lambda)} d(z,Z)h(z),$$

where the factor $e^{-\alpha|z|^2/2}\sigma_\alpha(z)/d(z,\Lambda)$ is bounded below (see Proposition 4.20) and

$$h(z) = \frac{g(z)d(z,\Lambda)}{\sigma_\alpha(z)d(z,Z)}.$$

It is easy to see that h is continuous and nonvanishing on the complex plane. So $|h(z)|$ is bounded below on $|z| \leq 2Q$. Here, Q is the constant that satisfies $|z_{mn} - \omega_{mn}| \leq Q$ for all (m,n). To show that $h(z)$ is bounded below for $|z| > 2Q$, we rewrite

$$h(z) = h_1(z)h_2(z)h_3(z),$$

where

$$h_1(z) = \exp\left[z \sum_{|z_{mn}| \le 2|z|}' \left(\frac{1}{z_{mn}} - \frac{1}{\omega_{mn}} \right) \right],$$

$$h_2(z) = \frac{d(z,\Lambda)}{d(z,Z)} \frac{z - z_{00}}{z} \prod_{|z_{mn}| \le 2|z|}' \frac{1 - z/z_{mn}}{1 - z/\omega_{mn}},$$

and

$$h_3(z) = \prod_{|z_{mn}| > 2|z|} \frac{(1 - z/z_{mn}) \exp(z/z_{mn})}{(1 - z/\omega_{mn}) \exp(z/\omega_{mn})}.$$

Since Z is uniformly close to Λ, we have

$$\left| \frac{1}{z_{mn}} - \frac{1}{\omega_{mn}} \right| \le \frac{C}{|\omega_{mn}|^2}$$

for some constant $C > 0$ and all $(m,n) \ne (0,0)$. Using this and the elementary estimates

$$|e^w| \ge e^{-|w|}, \qquad \sum_{n=1}^{N} \sum_{m=1}^{N} \frac{1}{n^2 + m^2} \sim \log N, \tag{4.12}$$

we can find positive constants C and c such that

$$|h_1(z)| \ge Ce^{-c|z|\log|z|}, \qquad z \in \mathbb{C}. \tag{4.13}$$

Rewrite $h_2(z)$ as

$$h_2(z) = \varphi(z) \frac{\prod'' [1 - (\omega_{mn} - z_{mn})/(\omega_{mn} - z)]}{\prod'' [1 - (\omega_{mn} - z_{mn})/\omega_{mn}]},$$

where

$$\varphi(z) = \frac{d(z,\Lambda)}{d(z,Z)} \frac{z - z_{00}}{z} \frac{1 - z/z_{kl}}{1 - z/\omega_{kl}},$$

ω_{kl} is the point in Λ that is closest to z, and the finite product $\prod''$ is taken over all (m,n) such that

$$(m,n) \ne (0,0), \quad (m,n) \ne (k,l), \quad |z_{mn}| \le 2|z|.$$

It is clear that $\varphi(z)$ is bounded below for $|z| \ge 2Q$.

Since Q satisfies $|z_{mn} - \omega_{mn}| \leq Q$ for all m and n, the condition $|z_{mn}| \leq 2|z|$ implies that

$$|\omega_{mn}| \leq 2|z| + Q, \qquad |\omega_{mn} - z| \leq 3|z| + Q.$$

It follows that

$$\left| \prod{}'' \left[1 - \frac{\omega_{mn} - z_{mn}}{\omega_{mn}} \right] \right| \leq \prod{}'' \left[1 + \frac{Q}{|\omega_{mn}|} \right]$$

$$\leq \prod \left\{ 1 + \frac{Q}{|\omega_{mn}|} : 0 < |\omega_{mn}| \leq 2|z| + Q \right\}.$$

To estimate the other product $\prod''$ in $h_2(z)$ above, we move a few additional factors into $\varphi(z)$ and further assume that $|z - \omega_{mn}| > Q$. Therefore, we can find a positive constant C, independent of z, such that

$$|h_2(z)| \geq C \frac{\prod\{(1 - Q/|\omega_{mn} - z|) : Q < |\omega_{mn} - z| \leq 3|z| + Q\}}{\prod\{(1 + Q/|\omega_{mn}|) : 0 < |\omega_{mn}| \leq 2|z| + Q\}}$$

for all $z \in \mathbb{C}$. If we write $z = w + \omega_{kl}$, where $|w|$ is a bounded function of z, then by the translation invariance of Λ, we have

$$|h_2(z)| \geq C \frac{\prod\{(1 - Q/|\omega_{mn} - w|) : Q < |\omega_{mn} - w| \leq 3|z| + Q\}}{\prod\{(1 + Q/|\omega_{mn}|) : 0 < |\omega_{mn}| \leq 2|z| + Q\}}$$

for all $z \in \mathbb{C}$. Take the logarithm of the above inequality, use the fact that $\log(1+x) \sim x$ when x is small, and observe that

$$\sum \left[\frac{1}{|\omega_{mn} - w|} : \delta < |\omega_{mn} - w| < R \right] \sim R$$

as $R \to \infty$ (which is easily obtained with the help of polar coordinates), we see that there are positive constants c and C such that

$$|h_2(z)| \geq C e^{-c|z|}, \qquad z \in \mathbb{C}. \tag{4.14}$$

To estimate $h_3(z)$, observe that $|z_{mn}| > 2|z|$ implies

$$1 - \frac{(1 - z/z_{mn}) \exp(z/z_{mn})}{(1 - z/\omega_{mn}) \exp(z/\omega_{mn})}$$

$$\sim (1 - z/z_{mn}) \exp(z/z_{mn}) - (1 - z/\omega_{mn}) \exp(z/\omega_{mn})$$

$$\sim \frac{z^2}{z_{mn}^2} - \frac{z^2}{\omega_{mn}^2} = O\left(\frac{z^2}{\omega_{mn}^3} \right).$$

It follows that

$$\log |h_3(z)| \geq -C_1 |z|^2 \sum_{|z_{mn}|>2|z|} \frac{1}{|\omega_{mn}|^3} \geq -C_2 |z|$$

so that

$$|h_3(z)| \geq Ce^{-c|z|}, \qquad z \in \mathbb{C}, \tag{4.15}$$

for some positive constants c and C.

Inserting the estimates (4.13), (4.14), and (4.15) into $h = h_1 h_2 h_3$ and then into the function $e^{-\alpha |z|^2/2} g(z)$, we have proved the inequality in (4.9), which in turn gives

$$\frac{|g(z) - g(z_{mn})|}{|z - z_{mn}|} e^{-\frac{\alpha}{2}|z|^2} \geq C_1 e^{-c|z|\log|z|} \frac{d(z,Z)}{|z - z_{mn}|}$$

for all $z \neq z_{mn}$. Fix z_{mn}, let $z \to z_{mn}$, and observe that $d(z,Z) = |z - z_{mn}|$ when z is sufficiently close to z_{mn}. We then obtain (4.11).

To prove (4.10), we write $g = \sigma_\alpha H$, or

$$e^{-\frac{\alpha}{2}|z|^2} g(z) = e^{-\frac{\alpha}{2}|z|^2} \sigma(z) H(z).$$

The quasiperiodicity of σ_α implies that the factor $e^{-\frac{\alpha}{2}|z|^2} \sigma_\alpha(z)$ is bounded. Rewrite $H = H_1 H_2 H_3$, where $H_1 = h_1$, $H_3 = h_3$, and

$$H_2(z) = \frac{z - z_{00}}{z} \prod_{|z_{mn}|\leq 2|z|}' \frac{1 - z/z_{mn}}{1 - z/\omega_{mn}},$$

and estimate the functions H_k the same way we did h_k, the result is (4.10). This completes the proof of the lemma.   □

Lemma 4.22. *Let g be the function associated to $Z = \{z_{mn}\}$. For any positive radius R, there exists a positive constant C such that*

$$\left| \frac{g(z)}{z - z_{mn}} \right| \leq C$$

for all (m,n) and all $|z| \leq R$.

Proof. It is clear that

$$\left| \frac{g(z)}{z - z_{mn}} \right| = \frac{|g(z)|}{d(z,Z)} \frac{d(z,Z)}{|z - z_{mn}|} \leq \frac{|g(z)|}{d(z,Z)}.$$

The desired result then follows from the fact that the function $g(z)/d(z,Z)$ is continuous on the whole complex plane.   □

Lemma 4.23. *Suppose Z is a sequence that is uniformly close to Λ_α. Then, $D^+(Z) = D^-(Z) = \alpha/\pi$.*

Proof. Suppose $Z = \{z_{mn}\}$, $\Lambda_\alpha = \{\omega_{mn}\}$, and $|z_{mn} - \omega_{mn}| \leq Q$ for all m and n, where Q is a positive constant. When r is much larger than Q, the number of points in $Z \cap B(w,r)$ is roughly the same as the number of points in $\Lambda_\alpha \cap B(w,r)$. More precisely, it is easy to see that

$$\lim_{r \to \infty} \frac{n(Z, B(w,r))}{n(\Lambda_\alpha, B(w,r))} = 1$$

and the convergence is uniform in $w \in \mathbb{C}$. This clearly gives the desired result.   □

The following result is usually referred to as a Lagrange-type interpolation formula.

Proposition 4.24. *Let $Z = \{z_{mn}\}$ be a separated sequence in $\mathbb{C}$ that is uniformly close to Λ_β and let g be the function associated to Z by (4.8). If $\alpha < \beta$, then every function $f \in F_\alpha^\infty$ can be written as*

$$f(z) = \sum_{m,n} \frac{f(z_{mn})}{g'(z_{mn})} \frac{g(z)}{z - z_{mn}},$$

where the series converges uniformly on compact subsets of $\mathbb{C}$.

Proof. Since $|f(z_{mn})| \leq Ce^{\alpha|z_{mn}|^2/2}$, it follows from (4.11) that

$$\left| \frac{f(z_{mn})}{g'(z_{mn})} \right| \leq C\exp\left(-\frac{1}{2}(\beta - \alpha)|z_{mn}|^2 + c|z_{mn}|\log|z_{mn}| \right)$$

for all m and n. This, along with Lemma 4.22, shows that the series converges uniformly on compact subsets of $\mathbb{C}$.

To show that the series actually converges to $f(z)$, we argue as follows. For each sufficiently large r, it is easy to see that we can find a simple closed pass $S = S_r$ such that

$$d(S, Z) \geq \delta(Z)/2, \quad d(S, 0) > r, \quad |S| \leq 8\pi r, \qquad (4.16)$$

where $\delta(Z)$ is the separation constant of Z. Let U be the region bounded by S. For any $z \in U - Z$, we have by the calculus of residues that

$$\frac{1}{2\pi i} \int_S \frac{f(\zeta)\,d\zeta}{(\zeta - z)g(\zeta)} = \frac{f(z)}{g(z)} - \sum_{z_{mn} \in U} \frac{f(z_{mn})}{g'(z_{mn})} \frac{1}{z - z_{mn}}.$$

By (4.9), with α replaced by β, (4.16), and the fact that

$$|f(\zeta)|e^{-\frac{\alpha}{2}|\zeta|^2} \leq \|f\|_{\infty,\alpha}, \qquad \zeta \in \mathbb{C},$$

we see that the integral on the left-hand side above tends to 0 as $r \to \infty$. This proves the desired expansion for f.   □

4.5 Sampling Sequences

We say that a set Z in $\mathbb{C}$ is a *set of uniqueness* for F_α^p if every function in F_α^p that vanishes on Z must be identically zero. Recall that a sequence Z is a zero set for F_α^p if there exists a function $f \in F_\alpha^p$ whose zero set is exactly Z. Thus, a zero sequence is not a set of uniqueness. But we cannot say that Z is a set of uniqueness if and only if Z is not a zero set for F_α^p. It is obvious that each sampling sequence for F_α^p is a set of uniqueness for F_α^p. We use the convention that the empty set is not a set of uniqueness for F_α^p, which is again easy to conceive and accept.

Recall that $W(Z)$ is the collection of weak limits of all the translates $Z + z$ of Z.

Lemma 4.25. *A separated sequence Z is sampling for F_α^∞ if and only if every $A \in W(Z)$ is a set of uniqueness (and hence nonempty) for F_α^∞.*

Proof. First assume that Z is a sampling sequence. Let $A \in W(Z)$ be the weak limit of some sequence $A_n = Z + \zeta_n$, $\zeta_n \in \mathbb{C}$. Although the set A may not be a sequence, it follows from the proof of Proposition 4.14 and the translation invariance of $M_\infty(Z)$ that

$$M_\infty(A) \le \liminf_{n\to\infty} M_\infty(A_n) = M_\infty(Z) < \infty,$$

where $M_\infty(A)$, just as in the case of sequences, is the smallest M such that

$$\|f\|_{\infty,\alpha} \le M \sup\left\{ |f(z)| e^{-\frac{\alpha}{2}|z|^2} : z \in A \right\}$$

for all $f \in F_\alpha^\infty$. So A is a sampling set for F_α^∞. In particular, A is a set of uniqueness for F_α^∞.

Next, assume that Z is not sampling for F_α^∞. Then there exists a sequence $\{f_n\}$ of unit vectors in F_α^∞ such that $\|f_n|Z\|_{\infty,\alpha} \to 0$ as $n \to \infty$. For each n, we use continuity to find some $z_n \in \mathbb{C}$ such that

$$|f_n(z_n)| e^{-\alpha|z_n|^2/2} = \frac{1}{2}.$$

Let

$$g_n(z) = f_n(z+z_n) e^{-\alpha \bar{z}_n z - \frac{\alpha}{2}|z_n|^2}.$$

Then for each n we have

$$\|g_n\|_{\infty,\alpha} = \|f_n\|_{\infty,\alpha} = 1, \qquad |g_n(0)| = 1/2.$$

Also,

$$\lim_{n\to\infty} \|g_n|A_n\|_{\infty,\alpha} = \lim_{n\to\infty} \|f_n|Z\|_{\infty,\alpha} = 0.$$

By a normal family argument, we may assume that $g_n(z) \to g(z)$ uniformly on compact subsets of $\mathbb{C}$. Clearly, $g \in F_\alpha^\infty$, $\|g\|_{\infty,\alpha} \le 1$, and $g(0) \ne 0$. Let A be a weak limit of the F_α^∞ sampling sets $A_n = Z - z_n$, possibly empty. The existence of such an A follows from Proposition 4.12.

If A is empty, it is certainly not a set of uniqueness for F_α^∞. If A is not empty, we fix any point $a \in A$. For any integer k, we can find a point ζ_k in some A_{n_k} such that $|a - \zeta_k| < 1/k$. By Corollary 4.7, there exists a positive constant C such that

$$\left| e^{-\frac{\alpha}{2}|a|^2} |g_{n_k}(a)| - e^{-\frac{\alpha}{2}|\zeta_k|^2} |g_{n_k}(\zeta_k)| \right| \le C|a - \zeta_k|$$

for all k. Let $k \to \infty$ and use the inequality

$$e^{-\frac{\alpha}{2}|\zeta_k|^2} |g_{n_k}(\zeta_k)| \le \|g_{n_k}|A_{n_k}\|_{\infty,\alpha}.$$

We obtain $g(a) = 0$. So g vanishes on A but $g(0) \ne 0$. Thus, A is not a set of uniqueness for F_α^∞. This completes the proof of the lemma. $\qquad\square$

Lemma 4.26. *If* $M_\infty(Z, \alpha) < \infty$, *then* $M_\infty(Z, \alpha + \varepsilon) < \infty$ *for all sufficiently small* $\varepsilon > 0$.

Proof. By Lemma 4.10, Z contains a separated subsequence which is also sampling for F_α^∞. By working with such a subsequence if necessary, we may assume that Z is already separated.

Suppose $M_\infty(Z, \alpha) < \infty$, but for a decreasing sequence of positive numbers ε_n approaching 0, we have $M_\infty(Z, \alpha + \varepsilon_n) = \infty$. We will obtain a contradiction.

For each n, we can find a unit vector f_n in $F_{\alpha+\varepsilon_n}^\infty$ such that

$$\|f_n|Z\|_{\infty,\alpha+\varepsilon_n} < \varepsilon_n.$$

Using the intermediate value theorem for continuous functions, we can also find a point $\zeta_n \in \mathbb{C}$ such that

$$|f_n(\zeta_n)| e^{-\frac{\alpha+\varepsilon_n}{2}|\zeta_n|^2} = \frac{1}{2}.$$

Let

$$g_n(z) = f_n(z + \zeta_n) e^{-(\alpha+\varepsilon_n)\overline{\zeta}_n z - \frac{\alpha+\varepsilon_n}{2}|\zeta_n|^2}, \qquad n \ge 1.$$

Then

$$\|g_n\|_{\infty,\alpha+\varepsilon_n} = \|f_n\|_{\infty,\alpha+\varepsilon_n} = 1, \quad |g_n(0)| = \frac{1}{2}.$$

Note that

$$\|g_n\|_{\infty,\alpha+\varepsilon_1} \le \|g_n\|_{\infty,\alpha+\varepsilon_n} = 1$$

for all n. With the help of a normal family argument and passing to a subsequence of $\{g_n\}$ if necessary, we may assume that $g_n(z) \to g(z)$ uniformly on compact subsets. The limit function g is entire, and $|g(0)| = 1/2$. For any $z \in \mathbb{C}$, we have

$$e^{-\frac{\alpha}{2}|z|^2} |g(z)| = \lim_{n\to\infty} e^{-\frac{\alpha+\varepsilon_n}{2}|z|^2} |g_n(z)| \le \lim_{n\to\infty} \|g_n\|_{\infty,\alpha+\varepsilon_n} = 1.$$

Thus $g \in F_\alpha^\infty$ with $\|g\|_{\infty,\alpha} \leq 1$.

Let $Z_n = Z - \zeta_n$ for every n. Then

$$\|g_n|Z_n\|_{\infty,\alpha+\varepsilon_n} = \|f_n|Z\|_{\infty,\alpha+\varepsilon_n} < \varepsilon_n.$$

Since Z is separated, we have $\inf_n \delta(Z_n) > 0$. By Proposition 4.12, $\{Z_n\}$ contains a weakly convergent subsequence. Let A be the weak limit of some sequence $\{Z_{n_k}\}$. Then $A \in W(Z)$.

If A is empty, it cannot be a set of uniqueness. Assume $A \neq \emptyset$ and fix some point $a \in A$. For any positive integer j, there exists some point $w_j \in Z_{n_{k_j}}$ such that $|a - w_j| < 1/j$. By Corollary 4.7, there exists a positive constant C such that

$$\left| e^{-\frac{\alpha+\varepsilon_{n_{k_j}}}{2}|a|^2} |g_{n_{k_j}}(a)| - e^{-\frac{\alpha+\varepsilon_{n_{k_j}}}{2}|w_j|^2} |g_{n_{k_j}}(w_j)| \right| < C|a - w_j|$$

for all j. Letting $j \to \infty$ leads to $g(a) = 0$. This shows that $g \in F_\alpha^\infty$, $g(0) \neq 0$, but g vanishes on A. So A is not a set of uniqueness for F_α^∞. This contradicts Lemma 4.25 as we are assuming that Z is a sampling sequence for F_α^∞. □

Lemma 4.27. *For any fixed positive number r, the sequence $\{\sigma_k(r)\}$ defined by*

$$\sigma_k(r) = \frac{1}{k!} \int_0^{\alpha r^2} t^k e^{-t} \, dt$$

is decreasing in k and tends to 0 as $k \to \infty$.

Proof. It is well known that the incomplete gamma function

$$\Gamma(a,z) = \int_z^\infty t^{a-1} e^{-t} \, dt$$

has the property that

$$\Gamma(k+1,z) = k!e^{-z} \sum_{j=0}^k \frac{z^j}{j!}.$$

It follows that

$$\sigma_k(r) = \frac{1}{k!} \left[\int_0^\infty t^k e^{-t} \, dt - \int_{\alpha r^2}^\infty t^k e^{-t} \, dt \right]$$

$$= \frac{1}{k!} \left[k! - \Gamma(k+1, \alpha r^2) \right]$$

$$= 1 - e^{-\alpha r^2} \sum_{j=0}^k \frac{(\alpha r^2)^j}{j!},$$

which is clearly decreasing in k and tends to 0 as $k \to \infty$. $\square$

Lemma 4.28. *If Z is a sampling sequence for F_α^∞, then $D^-(Z) > \alpha/\pi$.*

Proof. By Lemma 4.10, Z contains a separated subsequence which is also sampling for F_α^∞. Therefore, by working with such a subsequence if necessary, we may assume that Z is already separated.

In view of Lemma 4.26, we just need to show that $D^-(Z) \geq \alpha/\pi$. So let us assume the contrary and write $D^-(Z) = \alpha/\pi(1+2\varepsilon)$ for some positive number ε (the case $D^-(Z) = 0$ can be handled similarly). We will show that this leads to a contradiction.

Recall that

$$D^-(Z) = \liminf_{r \to \infty} \inf_{w \in \mathbb{C}} \frac{n(Z, B(w,r))}{\pi r^2},$$

where $n(Z, B(w,r))$ is the number of points in $Z \cap B(w,r)$. So the assumption $D^-(Z) = \alpha/\pi(1+2\varepsilon)$ implies that there exist sequences $\{r_n\}$ and $\{w_n\}$ such that $r_n \to \infty$ and

$$\frac{n(Z, B(w_n, r_n))}{r_n^2} < \frac{\alpha}{1+\varepsilon}, \qquad n \geq 1.$$

Let

$$R_n = r_n/\sqrt{1+\varepsilon} \qquad B_n = B(0, r_n) = B(0, \sqrt{1+\varepsilon}R_n),$$

and

$$N_n = n(Z, B(w_n, r_n)) = n(Z, B(w_n, \sqrt{1+\varepsilon}R_n)).$$

Then N_n is the number of points in $(Z - w_n) \cap B_n$ and

$$\frac{\alpha r_n^2}{1+2\varepsilon} \leq N_n < \alpha R_n^2.$$

In particular, $N_n \to \infty$ as $n \to \infty$.

To simplify the notation, we fix any n and write $B = B_n$, $R = R_n$, and $N = N_n$. Let $p = p_n$ be "the" (unique up to a unimodular constant multiple) polynomial with $(Z - w_n) \cap B_n$ as its zero set, normalized so that $\|p\|_{2,\alpha} = 1$.

We can write

$$p(z) = \sum_{k=0}^{N} a_k f_k(z), \qquad f_k(z) = \sqrt{\frac{\alpha^k}{k!}} z^k, \qquad \sum_{k=0}^{N} |a_k|^2 = 1.$$

It is easy to see that the functions $\{f_k\}$ are also orthogonal over the disk B:

$$\int_B f_k(z)\overline{f_m(z)} \, d\lambda_\alpha(z) = \sigma_k(\sqrt{1+\varepsilon}R)\delta_{k,m},$$

where the constants σ_k are from Lemma 4.27. It follows from this and Lemma 4.27 that

$$\int_B |p(z)|^2 \, d\lambda_\alpha(z) = \sum_{k=0}^{N} |a_k|^2 \int_B |f_k(z)|^2 \, d\lambda_\alpha(z)$$

$$= \sum_{k=0}^{N} |a_k|^2 \sigma_k(\sqrt{1+\varepsilon}R) \geq \sum_{k=0}^{N} |a_k|^2 \sigma_N(\sqrt{1+\varepsilon}R)$$

$$= \sigma_N(\sqrt{1+\varepsilon}R) = \frac{1}{N!} \int_0^{\alpha(1+\varepsilon)R^2} t^N e^{-t} \, dt$$

$$\geq \frac{1}{N!} \int_0^{(1+\varepsilon)N} t^N e^{-t} \, dt \geq \frac{1}{N!} \int_N^{(1+\varepsilon)N} t^N e^{-t} \, dt$$

$$\geq \frac{N^N}{N!} \int_N^{(1+\varepsilon)N} e^{-t} \, dt = \frac{N^N e^{-N}}{N!} (1 - e^{-\varepsilon N}).$$

This, together with Stirling's formula

$$N! \sim N^N e^{-N} \sqrt{N}, \qquad N \to \infty,$$

shows that there exists a constant $C = C(\alpha, \varepsilon) > 0$ (independent of N) such that

$$\int_B |p(z)|^2 \, d\lambda_\alpha(z) \geq \frac{C}{\sqrt{N}} \geq \frac{C}{\sqrt{\alpha}R}.$$

Since

$$\int_B |p(z)|^2 \, d\lambda_\alpha(z) \leq \frac{\alpha}{\pi} (1+\varepsilon) R^2 \sup_{z \in B} \left| p(z) e^{-\frac{\alpha}{2}|z|^2} \right|^2,$$

we can find another positive constant $C = C(\alpha, \varepsilon)$ (independent of R) such that

$$\|p\|_{\infty,\alpha} \geq \sup_{z \in B} \left| p(z) e^{-\frac{\alpha}{2}|z|^2} \right| \geq CR^{-\frac{3}{2}}.$$

On the other hand, for any z outside B and $0 \leq k \leq N$, we can write

$$|z|^2 = (1+t)R^2, \quad t \geq \varepsilon,$$

and deduce from

$$\frac{(\alpha R^2)^k}{k!} \leq \sum_{j=0}^{\infty} \frac{(\alpha R^2)^j}{j!} = e^{\alpha R^2}$$

that

$$|f_k(z)|^2 e^{-\alpha|z|^2} = \frac{\alpha^k}{k!}(1+t)^k R^{2k} e^{-\alpha(1+t)R^2}$$

$$= \frac{(\alpha R^2)^k e^{-\alpha R^2}}{k!} e^{-\alpha t R^2 + k \log(1+t)}$$

$$\leq e^{-\alpha t R^2 + k \log(1+t)}$$

$$\leq e^{-\alpha t R^2 + N \log(1+t)}$$

$$\leq e^{-\alpha t R^2 + \alpha R^2 \log(1+t)}.$$

Since $t \geq \varepsilon$, there exists another constant $c = c(\alpha, \varepsilon) > 0$ (independent of R) such that

$$|f_k(z)|^2 e^{-\alpha|z|^2} \leq e^{-2cR^2}$$

for all $0 \leq k \leq N$ and z outside B. By the Cauchy–Schwarz inequality and the fact that $\sum_{k=0}^{N} |a_k|^2 = 1$, we have

$$|p(z)|^2 e^{-\alpha|z|^2} = \left| \sum_{k=0}^{N} a_k f_k(z) \right|^2 e^{-\alpha|z|^2}$$

$$\leq \sum_{k=0}^{N} |a_k|^2 \sum_{k=0}^{N} |f_k(z)|^2 e^{-\alpha|z|^2}$$

$$\leq (N+1)e^{-2cR^2} \leq (\alpha R^2 + 1)e^{-2cR^2}.$$

for all z outside B. From this, we deduce that

$$\|p|Z_n\|_{\infty,\alpha} = \sup\left\{ |p(z)|e^{-\alpha|z|^2/2} : z \in Z_n \cap (\mathbb{C} - B) \right\}$$

$$\leq \sqrt{\alpha R^2 + 1} e^{-cR^2},$$

where $Z_n = Z - w_n$.

Finally, if we set

$$g_n(z) = e^{\alpha \bar{w}_n z - \frac{\alpha}{2}|w_n|^2} p_n(z - w_n),$$

then

$$\|g_n\|_{\infty,\alpha} = \|p_n\|_{\infty,\alpha} \geq CR_n^{-\frac{3}{2}}$$

and

$$\|g_n|Z\|_{\infty,\alpha} = \|p_n|Z_n\|_{\infty,\alpha} \leq \sqrt{\alpha R_n^2 + 1} e^{-c|R_n|^2}$$

so that

$$\frac{\|g_n|Z|\|_{\infty,\alpha}}{\|g_n\|_{\infty,\alpha}} \le C' R_n^{\frac{5}{2}} e^{-c|R_n|^2}$$

for all $n \ge 1$, where C' and c are positive constants independent of n. Since $R_n \to \infty$ as $n \to \infty$, we conclude that

$$\lim_{n\to+\infty} \frac{\|g_n|Z|\|_{\infty,\alpha}}{\|g_n\|_{\infty,\alpha}} = 0.$$

This contradicts with the assumption that Z is a sampling sequence for F_α^∞ and completes the proof of the lemma.   □

Lemma 4.29. *Suppose $0 < p \le \infty$ and Z is a sampling sequence for F_α^p. Then Z is a set of uniqueness for F_α^∞.*

Proof. By Lemmas 4.10 and 4.11, we may assume that Z is separated.

The case $p = \infty$ is obvious. Suppose $0 < p < \infty$, Z is sampling for F_α^p, but Z is not a set of uniqueness for F_α^∞. Then there exists a function $f \in F_\alpha^\infty$, not identically zero, such that f vanishes on Z. Let $g(z) = f(rz)$, where $0 < r < 1$. Then $g \in F_\alpha^p$, g is not identically zero, and g vanishes on Z/r. This is impossible because by Corollary 4.17, the sequence Z/r is sampling for F_α^p when r is sufficiently close to 1. Therefore, Z must be a set of uniqueness for F_α^∞.   □

Lemma 4.30. *Suppose $0 < p < \infty$ and Z is sampling for F_α^p. Then $D^-(Z) > \alpha/\pi$.*

Proof. Again, by working with a subsequence of Z if necessary, we may assume that Z is already separated.

Recall that $W(Z)$ consists of all weak limits of translates of Z. Since every translation of Z is also a sampling sequence for F_α^p with the same separation constant, it follows from Proposition 4.16 that every sequence in $W(Z)$ is sampling for F_α^p as well. Combining this with Lemmas 4.25 and 4.29, we conclude that Z is a sampling sequence for F_α^∞. Thus, $D^-(Z) > \alpha/\pi$ by Lemma 4.28.   □

This completes the proof for the necessity of the sampling condition $D^-(Z) > \alpha/\pi$ for F_α^p. We now proceed to prove the sufficiency. This will be accomplished with the help of the Weierstrass σ-function and its variant $g(z)$ discussed in the previous section. The first step is to show that every sequence contains a subsequence that is uniformly close to a square lattice Λ_γ and whose uniform lower density changes very little.

Lemma 4.31. *Suppose $0 < \alpha < \beta$ and Z is a sequence with $D^-(Z) = \beta/\pi$. There exists a subsequence Z' of Z such that Z' is uniformly close to Λ_γ for some $\alpha < \gamma < \beta$.*

Proof. Fix $\gamma \in (\alpha, \beta)$ and choose $\varepsilon > 0$ such that $\gamma + \varepsilon < \beta$. The condition $D^-(Z) = \beta/\pi$ implies that there exists a positive number r such that any square of side length r contains at least $(\gamma + \varepsilon)r^2/\pi$ points from Z.

We decompose $\mathbb{C}$ into the disjoint union of a sequence of squares (half open, half closed) of side length r: $\mathbb{C} = \cup\{S_k : k \geq 1\}$. Since the area of each S_k is r^2 and the area of the fundamental region of Λ_γ is π/γ, each S_k contains $r^2/(\pi/\gamma) = \gamma r^2/\pi$ points from Λ_γ (plus or minus a few points that can be neglected for our purpose). But S_k contains at least $(\gamma + \varepsilon)r^2/\pi$ points from Z. So for each k, we can choose $|\Lambda_\gamma \cap S_k|$ points from Z to match those in $\Lambda_\gamma \cap S_k$. We do this for each k, and the result is a subsequence of Z that is uniformly close to Λ_γ. More specifically, we have $|z_{mn} - \omega_{mn}| \leq \sqrt{2}r$ for all m and n, where $\sqrt{2}r$ is the length of the diagonal of each S_k. $\qquad\square$

We now prove that the condition $D^-(Z) > \alpha/\pi$ is sufficient for a separated sequence Z to be a sampling sequence of F_α^p. For clarity, we break the proof into three cases: $0 < p \leq 1$, $1 < p < \infty$, and $p = \infty$.

Lemma 4.32. *Suppose $1 < p < \infty$ and Z is a separated sequence in $\mathbb{C}$. If $D^-(Z) > \alpha/\pi$, then Z is a sampling sequence for F_α^p.*

Proof. Given a function $f \in F_\alpha^p \subset F_\alpha^\infty$ we need to estimate the integral

$$I = \int_{\mathbb{C}} \left| f(z) e^{-\frac{\alpha}{2}|z|^2} \right|^p dA(z)$$

from above. By Lemma 4.31, we may assume that Z is uniformly close to a square lattice Λ_β with $\beta > \alpha$. Let $\Omega = R_\alpha$ be the fundamental region for the square lattice $\Lambda_\alpha = \{\omega_{mn}\} = \{-\omega_{mn}\}$. Then by Lemma 1.13 and a change of variables, we have

$$I = \sum_{k,l} \int_{\Omega} \left| e^{-\frac{\alpha}{2}|z|^2} W_{\omega_{kl}} f(z) \right|^p dA(z),$$

where $W_{\omega_{kl}}$ are the Weyl unitary operators defined in Sect. 2.6.

To estimate each summand on the right-hand side above, first observe that $Z + \omega_{kl}$ is uniformly close to Λ_β as well, with a constant Q' that is independent of k and l. Thus, we can use Proposition 4.24 to write

$$W_{\omega_{kl}} f(z) = \sum_{m,n} \frac{W_{\omega_{kl}} f(z_{mn} + \omega_{kl})}{g'_{\omega_{kl}}(z_{mn} + \omega_{kl})} \frac{g_{\omega_{kl}}(z)}{z - z_{mn} - \omega_{kl}},$$

where $g_{\omega_{kl}}$ is the Weierstrass σ-type function associated to the sequences $Z + \omega_{kl}$ and Λ_β.

For $\varepsilon = (\beta - \alpha)/2$, we can write

$$\frac{|W_{\omega_{kl}} f(z_{mn} + \omega_{kl})|}{|g'_{\omega_{kl}}(z_{mn} + \omega_{kl})|} = \frac{e^{-\varepsilon|z_{mn} + \omega_{kl}|^2} e^{-\frac{\alpha}{2}|z_{mn}|^2} |f(z_{mn})|}{e^{-\frac{\beta}{2}|z_{mn} + \omega_{kl}|^2} |g'_{\omega_{kl}}(z_{mn} + \omega_{kl})|}. \qquad (4.17)$$

Let $q = p/(p-1)$ so that $1/p + 1/q = 1$. Then by Hölder's inequality and (4.11) in Lemma 4.21, we see that $|W_{\omega_{kl}} f(z)|^p$ is less than or equal to

$$Ch(z) \sum_{m,n} \left| e^{-\frac{\alpha}{2}|z_{mn}|^2} f(z_{mn}) \right|^p e^{-\varepsilon|z_{mn}+\omega_{kl}|^2 + c|z_{mn}+\omega_{kl}|\log|z_{mn}+\omega_{kl}|},$$

where

$$h(z) = h_{kl}(z) = \left[\sum_{m,n} e^{-\varepsilon|z_{mn}+\omega_{kl}|^2} \left| \frac{g_{\omega_{kl}}(z)}{z - z_{mn} - \omega_{kl}} \right|^q \right]^{\frac{p}{q}}.$$

By Lemmas 1.12 and 4.22, the positive function $h(z)$ is bounded on Ω with an upper bound that is independent of k and l. In particular, the integral

$$\int_{\Omega} h(z) e^{-\frac{p\alpha}{2}|z|^2} \, dA(z)$$

is dominated by a positive constant that is independent of k and l. Therefore, there exist positive constants C and C' such that

$$I \leq C \sum_{k,l} \sum_{m,n} \left| e^{-\frac{\alpha}{2}|z_{mn}|^2} f(z_{mn}) \right|^p e^{-\varepsilon|z_{mn}+\omega_{kl}|^2 + c|z_{mn}+\omega_{kl}|\log|z_{mn}+\omega_{kl}|}$$

$$= C \sum_{m,n} \left| e^{-\frac{\alpha}{2}|z_{mn}|^2} f(z_{mn}) \right|^p \sum_{k,l} e^{-\varepsilon|z_{mn}+\omega_{kl}|^2 + c|z_{mn}+\omega_{kl}|\log|z_{mn}+\omega_{kl}|}$$

$$\leq C' \|f|Z\|_{p,\alpha}^p,$$

which is the desired estimate. Note that the last estimate above follows from Lemma 1.12. $\qquad \square$

Lemma 4.33. *Suppose $0 < p \leq 1$ and Z is a separated sequence with $D^-(Z) > \alpha/\pi$. Then Z is sampling for F_α^p.*

Proof. With notation from the proof of the previous lemma, we use the assumption $0 < p \leq 1$ to get

$$|W_{\omega_{kl}} f(z)|^p \leq \sum_{m,n} \left| \frac{W_{\omega_{kl}} f(z_{mn} + \omega_{kl})}{g'_{\omega_{kl}}(z_{mn} + \omega_{kl})} \right|^p \left| \frac{g_{\omega_{kl}}(z)}{z - z_{mn} - \omega_{kl}} \right|^p.$$

Combining this with (4.11) and (4.17), we obtain positive constants C and c, both independent of k and l, such that

$$|W_{\omega_{kl}} f(z)|^p \leq C \sum_{m,n} \left| \frac{g_{\omega_{kl}}(z)}{z - z_{mn} - \omega_{kl}} \right|^p \left| e^{-\frac{\alpha}{2}|z_{mn}|^2} f(z_{mn}) \right|^p E(m,n,k,l),$$

where

$$E(m,n,k,l) = \mathrm{e}^{-p\varepsilon|z_{mn}+\omega_{kl}|^2+c|z_{mn}+\omega_{mn}|\log|z_{mn}+\omega_{kl}|}.$$

Integrate the above inequality over Ω with respect to $\mathrm{e}^{-p\alpha|z|^2/2}\mathrm{d}A(z)$ and notice that Lemma 4.22 implies

$$\int_{\Omega}\left|\frac{g_{\omega_{kl}}(z)}{z-z_{mn}-\omega_{kl}}\right|^p \mathrm{e}^{-\frac{p\alpha}{2}|z|^2}\,\mathrm{d}A(z) \leq C$$

for some constant $C > 0$ that is independent of k and l. We obtain another constant $C > 0$ such that

$$I \leq C\sum_{k,l}\sum_{m,n}\left|\mathrm{e}^{-\frac{\alpha}{2}|z_{mn}|^2}f(z_{mn})\right|^p E(m,n,k,l)$$

$$= C\sum_{m,n}\left|\mathrm{e}^{-\frac{\alpha}{2}|z_{mn}|^2}f(z_{mn})\right|^p\sum_{k,l}E(m,n,k,l)$$

$$\leq C'\sum_{m,n}\left|\mathrm{e}^{-\frac{\alpha}{2}|z_{mn}|^2}f(z_{mn})\right|^p$$

$$= C'\|f|Z\|_{p,\alpha}^p,$$

which is the desired estimate. $\qquad\square$

Lemma 4.34. *Any separated sequence Z with $D^-(Z) > \alpha/\pi$ is a sampling sequence for F_α^∞.*

Proof. With notation from the proof of the previous two lemmas, we have

$$\|f\|_{\infty,\alpha} = \sup_{k,l}S_{kl},$$

where

$$S_{kl} = \sup\left\{\mathrm{e}^{-\frac{\alpha}{2}|z|^2}|W_{\omega_{kl}}f(z)| : z \in \Omega\right\}.$$

To shorten the displays below, let

$$e(m,n,k,l) = \mathrm{e}^{-\varepsilon|z_{mn}+\omega_{kl}|^2+c|z_{mn}+\omega_{kl}|\log|z_{mn}+\omega_{kl}|}.$$

Then by (4.17), (4.11), and Lemmas 4.22 and 1.12, we have

$$S_{kl}\leq C\sup_{z\in\Omega}\sum_{m,n}\mathrm{e}^{-\frac{\alpha}{2}|z|^2}\left|\frac{g_{\omega_{kl}}(z)}{z-z_{mn}-\omega_{kl}}\right|\left|\mathrm{e}^{-\frac{\alpha}{2}|z_{mn}|^2}f(z_{mn})\right|e(m,n,k,l)$$

$$\leq C'\|f|Z\|_{\infty,\alpha}\sum_{m,n}e(m,n,k,l)$$

$$\leq C''\|f|Z\|_{\infty,\alpha},$$

which proves the desired result. $\qquad\square$

This completes the proof of the sufficiency of the condition $D^-(Z) > \alpha/\pi$ for Z to be a sampling sequence of F_α^p. We summarize the main results of this section as the following two theorems.

Theorem 4.35. *A set Z is sampling for F_α^∞ if and only if Z contains a separated sequence Z' such that $D^-(Z') > \alpha/\pi$.*

Theorem 4.36. *Let Z be a sequence in $\mathbb{C}$ and $0 < p < \infty$. Then, Z is sampling for F_α^p if and only if Z is the union of finitely many separated sequences and Z contains a separated subsequence Z' such that $D^-(Z') > \alpha/\pi$.*

Corollary 4.37. *If Z is separated and $0 < p \leq \infty$, then Z is sampling for F_α^p if and only if $D^-(Z) > \alpha/\pi$.*

4.6 Interpolating Sequences

In this section, we characterize interpolating sequences for F_α^p by the condition $D^+(Z) < \alpha/\pi$. We begin with the sufficiency, which is still based on the modified Weierstrass σ-function associated to a separated sequence that is uniformly close to a square lattice. The first step is to show that every separated sequence can be expanded to a sequence that is uniformly close to a square lattice and whose uniform upper density increases very little.

Lemma 4.38. *Let Z be a separated sequence in $\mathbb{C}$ with $D^+(Z) = \beta/\pi$ and $\beta < \alpha$. We can expand Z to a separated sequence Z' such that Z' is uniformly close to a square lattice Λ_γ with $\gamma \in (\beta, \alpha)$.*

Proof. Let $\gamma \in (\beta, \alpha)$ and choose $\varepsilon > 0$ such that $\beta < \gamma - \varepsilon$. The condition $D^+(Z) = \beta/\pi$ implies that there is some large r such that any square of side length r contains at most $(\gamma - \varepsilon)r^2/\pi$ points from Z.

Just as in the proof of Lemma 4.31, we decompose the complex plane into the disjoint union of squares (half open, half closed) of side length r: $\mathbb{C} = \cup S_k$. Each S_k contains at most $(\gamma - \varepsilon)r^2/\pi$ points from Z. On the other hand, each S_k contains $r^2/(\pi/\gamma) = \gamma r^2/\pi$ points from Λ_γ. Therefore, we can add a certain number of points in each S_k to Z to match the number of points in $\Lambda_\gamma \cap S_k$ so that the expanded sequence Z' will be uniformly close to Λ_γ. It is easy to see that we can also do the expansion in such a way that the new sequence Z' remains separated. $\square$

Lemma 4.39. *Suppose $0 < p \le \infty$ and Z is a separated sequence. If $D^+(Z) < \alpha/\pi$, then Z is interpolating for F_α^p.*

Proof. If we remove any number of points from an interpolating sequence for F_α^p, what remains is still an interpolating sequence for F_α^p: we just assign the value 0 to $f(z)$ for those removed z. So by Lemma 4.38, we may as well assume that Z is uniformly close to the square lattice $\Lambda_\beta = \{\omega_{mn}\}$ with $D^+(Z) = \beta/\pi$ and $\beta < \alpha$.

For any sequence $\{a_{kl}\}$ of values for which

$$\left\{ a_{kl} e^{-\frac{\alpha}{2}|z_{kl}|^2} \right\} \in l^p,$$

we claim that the interpolation problem $f(z_{kl}) = a_{kl}$ is solved explicitly by the function

$$f(z) = \sum_{m,n} a_{mn} e^{\alpha \bar{z}_{mn} z - \alpha |z_{mn}|^2} \frac{g_{mn}(z - z_{mn})}{z - z_{mn}}, \tag{4.18}$$

where g_{mn} denotes the generalized Weierstrass σ-function associated with the sequences $Z - z_{mn}$ and Λ_γ as given in (4.8) (it is easy to see that $Z - z_{mn}$ is uniformly close to Λ_γ). More specifically,

$$\frac{g_{mn}(z - z_{mn})}{z - z_{mn}}$$

is equal to

$$\prod_{(k,l)\neq(m,n)} \left(1 - \frac{z - z_{mn}}{z_{kl} - z_{mn}}\right) \exp\left(\frac{z - z_{mn}}{z_{kl} - z_{mn}} + \frac{1}{2}\frac{(z - z_{mn})^2}{\omega_{kl}^2}\right).$$

In particular,

$$g_{mn}(z_{mn} - z_{mn}) = g_{mn}(0) = 1, \qquad g_{mn}(z_{kl} - z_{mn}) = 0,$$

for $(k,l) \neq (m,n)$. Since

$$e^{-\frac{\alpha}{2}|z|^2}|f(z)| \leq \sum_{m,n}\left|e^{-\frac{\alpha}{2}|z_{mn}|^2}a_{mn}\right|e^{-\frac{\alpha}{2}|z-z_{mn}|^2}\left|\frac{g_{mn}(z - z_{mn})}{z - z_{mn}}\right|,$$

the series above can be written as

$$\sum_{m,n}\left|e^{-\frac{\alpha}{2}|z_{mn}|^2}a_{mn}\right|e^{-\frac{\alpha-\beta}{2}|z-z_{mn}|^2}e^{-\frac{\beta}{2}|z-z_{mn}|^2}\left|\frac{g_{mn}(z - z_{mn})}{z - z_{mn}}\right|.$$

By (4.10), there exist positive constants C, C', and c such that

$$e^{-\frac{\alpha}{2}|z|^2}|f(z)| \leq C\sum_{m,n}\left|e^{-\frac{\alpha}{2}|z_{mn}|^2}a_{mn}\right|e^{-\delta|z-z_{mn}|^2+c|z-z_{mn}|\log|z-z_{mn}|}$$

$$\leq C'\sum_{m,n}\left|e^{-\frac{\alpha}{2}|z_{mn}|^2}a_{mn}\right|e^{-\frac{\delta}{2}|z-z_{mn}|^2}$$

for all $z \in \mathbb{C}$, where $\delta = (\alpha - \beta)/2$. Since $Z = \{z_{mn}\}$ is uniformly close to the square lattice $\Lambda_\beta = \{\omega_{mn}\}$, we can find another positive constant C such that

$$e^{-\frac{\alpha}{2}|z|^2}|f(z)| \leq C\sum_{m,n}\left|e^{-\frac{\alpha}{2}|z_{mn}|^2}a_{mn}\right|e^{-\sigma|z-\omega_{mn}|^2} \tag{4.19}$$

for all $z \in \mathbb{C}$, where $\sigma = \delta/4$. Since the sequence $\{e^{-\frac{\alpha}{2}|z_{mn}|^2}a_{mn}\}$ is bounded, it follows from (4.19) and Lemma 1.12 that the series in (4.18) converges absolutely to an entire function f with $f(z_{kl}) = a_{kl}$ for all (k,l).

It remains for us to show that the function f defined in (4.18) belongs to F_α^p. Just as in the previous section, we break the proof into three cases: $0 < p \leq 1$, $1 < p < \infty$, and $p = \infty$.

The case $p = \infty$ is the easiest. In fact, if the sequence $e^{-\alpha|z_{mn}|^2/2}a_{mn}$ is bounded, then by (4.19), there is a positive constant C such that

$$e^{-\frac{\alpha}{2}|z|^2}|f(z)| \leq C\sum_{m,n}e^{-\sigma|z-\omega_{mn}|^2}.$$

This, along with Lemma 1.12, shows that $f \in F_\alpha^\infty$.

If $0 < p \le 1$, it follows from (4.19) and Hölder's inequality that

$$\left| f(z) e^{-\frac{\alpha}{2}|z|^2} \right|^p \le C \sum_{m,n} \left| a_{mn} e^{-\frac{\alpha}{2}|z_{mn}|^2} \right|^p e^{-p\sigma|z-\omega_{mn}|^2}.$$

Integrate term by term and use the translation invariance of the area measure. We see that

$$\int_{\mathbb{C}} \left| e^{-\frac{\alpha}{2}|z|^2} f(z) \right|^p \, dA(z) \le C \sum_{m,n} \left| a_{mn} e^{-\frac{\alpha}{2}|z_{mn}|^2} \right|^p \int_{\mathbb{C}} e^{-p\sigma|z|^2} \, dA(z).$$

This shows that $f \in F_\alpha^p$ whenever the series $\{a_{mn} e^{-\alpha|z_{mn}|^2/2}\}$ is in l^p.

The case $1 < p < \infty$ follows from complex interpolation. In fact, examining the arguments in the previous two paragraphs, we see that the linear operator

$$\{c_{mn}\} \mapsto \sum_{m,n} c_{mn} e^{-\sigma|z-\omega_{mn}|^2}$$

maps l^∞ to $L^\infty(\mathbb{C}, dA)$ and l^1 to $L^1(\mathbb{C}, dA)$. Therefore, this operator maps l^p to $L^p(\mathbb{C}, dA)$ for any $1 < p < \infty$. This, along with (4.19), shows that $f \in F_\alpha^p$ whenever the sequence $\{a_{mn} e^{-\alpha|z_{mn}|^2/2}\}$ belongs to l^p. $\quad\square$

The lemma above shows that the condition $D^+(Z) < \alpha/\pi$ is sufficient for a separated sequence Z to be interpolating for F_α^p. Next, we will prove that this density condition is also necessary.

Lemma 4.40. *Let $0 < p \le \infty$. There is no sequence in $\mathbb{C}$ that is both sampling for F_α^p and interpolating for F_α^p.*

Proof. Assume the contrary and let Z be a sequence that is both sampling and interpolating for F_α^p. Then Z is separated and sampling for $F_{\alpha+\varepsilon}^p$ for all sufficiently small ε, because we have characterized sampling sequences for F_α^p using the "open" condition $D^-(Z) > \alpha/\pi$.

Fix a point $\zeta \in Z$ and use the assumption that Z is interpolating for F_α^p to find a function $g \in F_\alpha^p$ such that $g(\zeta) = 1$ and $g(z) = 0$ for all $z \in Z - \{\zeta\}$. Then, the function $f(z) = (z - \zeta)g(z)$ is not identically zero, belongs to $F_{\alpha+\varepsilon}^p$, and vanishes on Z. Thus, Z cannot possibly be sampling for $F_{\alpha+\varepsilon}^p$. This contradiction shows that Z cannot be simultaneously sampling and interpolating for F_α^p. $\quad\square$

Lemma 4.41. *Suppose $0 < p \le \infty$ and Z is interpolating for F_α^p. If Z is a set of uniqueness for F_α^p, then it must be a sampling sequence for F_α^p.*

Proof. Since Z is interpolating for F_α^p, it must be separated by Lemma 4.8. Given any function $f \in F_\alpha^p$, the sequence $w_n = f(z_n)$ has the property that $\{w_n e^{-\alpha|z_n|^2/2}\} \in l^p$. By the definition of $N_p(Z)$, there exists some function $g \in F_\alpha^p$ such that $g(z_k) = w_k$ for all k and $\|g\|_{p,\alpha} \le N_p(Z)\|g|Z\|_{p,\alpha}$. Since Z is a set of uniqueness for F_α^p and $f(z_k) = w_k = g(z_k)$ for all k, we must have $g = f$, and so $\|f\|_{p,\alpha} \le N_p(Z)\|f|Z\|_{p,\alpha}$ for all $f \in F_\alpha^p$. This says that Z is sampling for F_α^p. $\quad\square$

As consequences of the two lemmas above, we obtain the following corollaries:

Corollary 4.42. *Let $0 < p \leq \infty$ and let Z be an interpolating sequence for F_α^p. Then, there exists a function $f \in F_\alpha^p$, not identically zero, such that f vanishes on Z.*

Note that the above corollary does NOT say that every interpolating sequence for F_α^p is an F_α^p-zero set because f may have additional zeros other than those in Z. In fact, there exist examples of F_α^p-interpolating sequences that are not F_α^p-zero sets. See Proposition 5.11.

Corollary 4.43. *Let $0 < p \leq \infty$ and let Z be a sampling sequence for F_α^p. For any $\zeta \in Z$, the sequence $Z - \{\zeta\}$ remains a sampling sequence for F_α^p.*

Proof. This is clear from the already-proved characterization of sampling sequences for F_α^p in terms of the lower density because deleting a single point from a sequence does not alter the density of the sequence.

We give another proof that only relies on the fact that if Z is sampling for F_α^p, then it is also sampling for $F_{\alpha+\varepsilon}^p$ for sufficiently small ε.

So suppose Z is sampling for F_α^p but $Z' = Z - \{\zeta\}$ is not, where $\zeta \in Z$. Without loss of generality, we may also assume that Z is separated. Then, there exists a sequence of unit vectors $\{f_n\}$ in F_α^p such that $\|f_n|Z'\|_{p,\alpha} \to 0$ as $n \to \infty$. By a normal family argument, we may as well assume that $f_n(z) \to f(z)$ uniformly on compact sets. By Fatou's lemma, we have $f \in F_\alpha^p$. From $\|f_n|Z'\|_{p,\alpha} \to 0$, we deduce that $f(z) = 0$ for all $z \in Z'$. Since

$$\|f_n|Z\|_{p,\alpha} \geq 1/M_p(Z) > 0$$

for all n, we see that $f(\zeta) \neq 0$. The function $(z - \zeta)f(z)$ is not identically zero, vanishes on Z, and belongs to $F_{\alpha+\varepsilon}^p$ for any $\varepsilon > 0$. This contradicts the fact that Z is a sampling sequence for $F_{\alpha+\varepsilon}^p$.   □

Thus, sampling sequences for F_α^p are stable under the following two operations: deleting a finite number of points or adding any number of separated points from outside the sequence.

Corollary 4.44. *Let $0 < p \leq \infty$. If $Z = \{z_n\}$ is an interpolating sequence for F_α^p, then so is $Z \cup \{\zeta\}$ for any $\zeta \notin Z$.*

Proof. By Corollary 4.42, there is a function $g \in F_\alpha^p$ that is not identically zero but vanishes on Z. By dividing out an appropriate power of $z - \zeta$ if necessary (which preserves membership in F_α^p), we may assume that $g(\zeta) \neq 0$. Multiplying g by a constant if necessary, we may further assume that $g(\zeta) = 1$.

Given a sequence $\{v\} \cup \{v_n\}$ of values with $\{v_n e^{-\alpha|z_n|^2/2}\} \in l^p$, we can find a function $f \in F_\alpha^p$ such that $f(z_n) = v_n$ for all n. The function

$$F(z) = f(z) + (v - f(\zeta))g(z)$$

belongs to F_α^p and satisfies

$$F(\zeta) = v, \qquad F(z_n) = v_n, \qquad n \geq 1.$$

This shows that $Z \cup \{\zeta\}$ is still an interpolating sequence for F_α^p. $\qquad\qquad\Box$

We see that interpolating sequences for F_α^p are stable under the following two operations: deleting any number of points from the sequence or adding a finite number of distinct points from outside the sequence.

A key tool for the rest of this section is the following quantity:

$$\rho_p(z, Z) = \sup_f |f(z)| e^{-\frac{\alpha}{2}|z|^2}, \qquad 0 < p \leq \infty,$$

where $Z = \{z_n\}$ and the supremum is taken over all unit vectors f in F_α^p such that $f(z_n) = 0$ for all n. We think of $\rho_p(z, Z)$ as some kind of distance from z to the sequence Z. A normal family argument shows that the supremum in the definition of $\rho_p(z, Z)$ is always attained.

By Corollary 2.8, we always have $0 \leq \rho_p(z, Z) \leq 1$. It is obvious that $\rho_p(z, Z) = 0$ when $z \in Z$. We are going to show that $\rho_p(z, Z) = 0$ only when $z \in Z$, provided that Z is an interpolating sequence for F_α^p.

Lemma 4.45. *If Z is interpolating for F_α^p, where $0 < p \leq \infty$, then $\rho_p(z, Z) > 0$ when $z \notin Z$.*

Proof. Actually, we only need to assume that Z is not a set of uniqueness (we already know that every interpolating sequence for F_α^p is not a set of uniqueness for F_α^p). In fact, if f is any function in F_α^p that is not identically zero and vanishes on Z, then f cannot possibly have a zero at z of infinite order. Therefore, by dividing out a finite and nonnegative power of $w - z$, which does not ruin membership in F_α^p, we arrive at a function in F_α^p that vanishes on Z but has a nonzero value at z. $\qquad\Box$

The following result is a quantitative version of Corollary 4.44.

Lemma 4.46. *Let $Z = \{z_1, z_2, \cdots\}$ and $z_0 \notin Z$. We have*

$$N_p(Z \cup \{z_0\}) \leq \frac{1 + 2N_p(Z)}{\rho_p(z_0, Z)}$$

for all $0 < p \leq \infty$.

Proof. We may assume that $N_p(Z) < \infty$, that is, Z is an interpolating sequence for F_α^p. Given a sequence of values $\{v_0, v_1, v_2, \cdots\}$ with the l^p norm of

$$\left\{ v_0 e^{-\frac{\alpha}{2}|z_0|^2}, v_1 e^{-\frac{\alpha}{2}|z_1|^2}, v_2 e^{-\frac{\alpha}{2}|z_2|^2}, \cdots \right\}$$

equal to 1, there is a function $f \in F_\alpha^p$ such that $f(z_n) = v_n$ for all $n \geq 1$ and

$$\|f\|_{p,\alpha} \leq N_p(Z) \|f|Z\|_{p,\alpha} \leq N_p(Z).$$

On the other hand, by Lemma 4.45, there exists a function $f_0 \in F_\alpha^p$ such that f_0 vanishes on Z, $\|f_0\|_{p,\alpha} \leq 1$, and

$$e^{-\frac{\alpha}{2}|z_0|^2} f_0(z_0) = \rho_p(z_0, Z).$$

Now the function

$$g(z) = f(z) + \frac{v_0 - f(z_0)}{\rho_p(z_0, Z)} f_0(z) e^{-\frac{\alpha}{2}|z_0|^2}$$

belongs to F_α^p, solves the interpolation problem $g(z_n) = v_n$ for all $n \geq 0$, and satisfies

$$\|g\|_{p,\alpha} \leq \|f\|_{p,\alpha} + \frac{|v_0 - f(z_0)|}{\rho_p(z_0, Z)} e^{-\frac{\alpha}{2}|z_0|^2}$$

$$\leq N_p(Z) + \frac{|v_0|e^{-\frac{\alpha}{2}|z_0|^2} + |f(z_0)|e^{-\frac{\alpha}{2}|z_0|^2}}{\rho_p(z_0, Z)}$$

$$\leq N_p(Z) + \frac{1 + \|f\|_{p,\alpha}}{\rho_p(z_0, Z)}$$

$$\leq N_p(Z) + \frac{1 + N_p(Z)}{\rho_p(z_0, Z)}$$

$$\leq \frac{1 + 2N_p(Z)}{\rho_p(z_0, Z)}$$

$$= \frac{1 + 2N_p(Z)}{\rho_p(z_0, Z)} \|g|(Z \cup \{z_0\})\|_{p,\alpha}.$$

This proves the desired estimate.   □

Lemma 4.47. *Given positive constants δ_0, l_0, and α, there exists a positive constant $C = C(\delta_0, l_0, \alpha)$ such that if $N_p(Z, \alpha) \leq l_0$ and $d(z, Z) \geq \delta_0$, then $\rho_p(z, Z) \geq C$. Here, $0 < p \leq \infty$.*

Proof. Let us assume the contrary, namely, there exists a sequence Z_n of interpolating sets for F_α^p and a sequence z_n of points in $\mathbb{C}$ such that

$$N_p(Z_n, \alpha) \leq l_0, \qquad d(z_n, Z_n) \geq \delta_0, \qquad n \geq 1,$$

and $\rho_p(z_n, Z_n) \to 0$ as $n \to \infty$.

By translation invariance, we may assume that each $z_n = 0$. Going down to a subsequence if necessary, we may also assume that Z_n converges weakly to Z', where Z' may be empty.

By Lemma 4.18, $N_p(Z', \alpha) \leq l_0$. Also, $d(0, Z_n) \geq \delta_0$ shows that 0 is not in Z'. By Lemma 4.45, there exists a function $f \in F_\alpha^p$ such that f vanishes on Z', $\|f\|_{p,\alpha} \leq 1$, and $f(0) = r > 0$. We may further assume that

$$\lim_{z \to \infty} f(z) e^{-\frac{\alpha}{2}|z|^2} = 0. \tag{4.20}$$

In fact, the above condition is automatically satisfied for $f \in F_\alpha^p$ when $0 < p < \infty$. If $p = \infty$, we modify the construction above as follows. Pick a complex number ζ such that $\zeta \notin Z'$ and $\zeta \neq 0$. Then $Z' \cup \{\zeta\}$ is still an interpolating sequence for F_α^p. Thus, there exists a function $g \in F_\alpha^p$ such that g vanishes on $Z' \cup \{\zeta\}$ and $g(0) \neq 0$. Then the function $f(z) = g(z)/(z - \zeta)$ belongs to F_α^p, vanishes on Z', satisfies the condition in (4.20), and $f(0) \neq 0$.

Since $\{Z_n\}$ converges weakly to Z', the sequence $\varepsilon_n = \|f|Z_n\|_{p,\alpha}$ converges to 0 as $n \to \infty$, which follows easily from (4.20). Now, choose $g_n \in F_\alpha^p$ with $g_n = f$ on Z_n and $\|g_n\|_{p,\alpha} \leq l_0 \varepsilon_n$ and define

$$f_n(z) = \frac{f(z) - g_n(z)}{\|f\|_{p,\alpha} + l_0 \varepsilon_n}.$$

For each n, it is clear that $\|f_n\|_{p,\alpha} \leq 1$ and $f_n = 0$ on Z_n. Since

$$|g_n(0)| \leq \|g_n\|_{p,\alpha} \leq l_0 \varepsilon_n \to 0$$

as $n \to \infty$, we also have

$$\rho_p(0, Z_n) \geq |f_n(0)| \to \frac{r}{\|f\|_{p,\alpha}} > 0,$$

which is a contradiction. □

Lemma 4.48. *Given positive constants l_0 and α, there is a constant $C = C(l_0, \alpha) > 0$ such that if $N_p(Z, \alpha) \leq l_0$, then*

$$\int_Q \log \rho_p(z, Z) \, dA(z) \geq -C|Q|^2$$

for every square Q with area $|Q| \geq 1$.

Proof. By the proof of Lemma 4.8, there exists a point $z_0 \in Q$ and a positive constant $\delta = \delta(\alpha, l_0)$ such that $d(z_0, Z) \geq \delta$. By translation invariance, we may assume that $z_0 = 0$. It then follows from Lemma 4.47 that there is a function f with $\|f\|_{p,\alpha} \leq 1$, $f|Z = 0$, and $|f(0)| \geq \sigma$, where $\sigma = \sigma(\alpha, l_0)$ is another positive constant. Since

$$\rho_p(z, Z) \geq e^{-\frac{\alpha}{2}|z|^2}|f(z)|, \qquad z \in \mathbb{C},$$

it follows from the subharmonicity of $\log|f(z)|$ that

$$\log|f(0)| \leq \frac{\alpha}{2} r^2 + \frac{1}{2\pi} \int_0^{2\pi} \log \rho_p(re^{i\theta}, Z) \, d\theta$$

for all $r \geq 0$. Multiply both sides by r, integrate with respect to r from 0 to $\sqrt{2|Q|}$, and observe that $Q \subset B(0, \sqrt{2|Q|})$ and $\rho_p \leq 1$. The desired result follows. □

We can now prove the necessity of the condition $D^+(Z) < \alpha/\pi$ for Z to be an interpolating sequence of F_α^p.

Lemma 4.49. *Suppose $0 < p \leq \infty$ and Z is an interpolating sequence for F_α^p. Then,* $D^+(Z) < \alpha/\pi$.

Proof. We consider an arbitrary large square Q of side length $R > 2$ and divide it into $N = [R] \times [R]$ squares Q_j, $1 \leq j \leq N$, each of side length $s = R/[R]$, where $[R]$ denotes the integer part of R. It is clear that $1 \leq s \leq 2$.

Since Z is interpolating for F_α^p, it is separated. Thus, for each j, we can find some point $z_j \in Q_j$ such that $d(z_j, Z) \geq \delta_0$, where δ_0 is a positive constant that only depends on $N_p(Z)$ and α. Let $Z_j = Z \cup \{z_j\}$ and use Lemmas 4.46 and 4.47 to find a positive constant l, independent of j, such that $N_p(Z_j) \leq l$ for all j. By Lemma 4.48, we can find a positive constant $C = C(l, \alpha)$ such that

$$\int_{Q_j} \log \rho_p(z, Z_j) \, dA(z) \geq -C, \qquad 1 \leq j \leq N.$$

For any $z \in Q_j$, we choose a function f such that f vanishes on $Z_j - z$, $\|f\|_{p,\alpha} \leq 1$, and $f(0) = \rho_p(0, Z_j - z) = \rho_p(z, Z_j)$. By Jensen's formula applied to the disk $|\zeta| < r$, where $2\sqrt{2} < r < R/2$ (Jensen's formula works for $f(0) \neq 0$, but the final estimate below clearly holds for $f(0) = 0$ as well),

$$\log \rho_p(z, Z_j) = \log|f(0)|$$

$$\leq \int_0^{2\pi} \log|f(re^{i\theta})| \frac{d\theta}{2\pi} - \sum_{\zeta \in Z, |z-\zeta| < r} \log \frac{r}{|z-\zeta|} - \log \frac{r}{|z-z_j|}$$

$$\leq \frac{\alpha r^2}{2} - \sum_{\zeta \in Z} \log^+ \frac{r}{|z-\zeta|} - \log \frac{r}{2\sqrt{2}}$$

$$\leq \frac{\alpha r^2}{2} - \sum_{\zeta \in Z \cap Q^-} \log^+ \frac{r}{|z-\zeta|} - \log \frac{r}{2\sqrt{2}},$$

where Q^- is the square of side length $R - 2r$ inside Q sharing the same center with Q and having sides parallel to the corresponding ones of Q. In other words, Q^- consists of those points whose distance to the complement of Q exceeds r. We integrate this inequality with respect to area measure over Q_j, use Lemma 4.48, and obtain

$$-C \leq \int_{Q_j} \log \rho_p(z, Z_j) \, dA(z)$$

$$\leq \frac{\alpha r^2}{2}|Q_j| - \sum_{\zeta \in Z \cap Q^-} \int_{Q_j} \log^+ \frac{r}{|z-\zeta|} \, dA(z) - |Q_j| \log \frac{r}{2\sqrt{2}}.$$

Summing over j, we obtain

$$-CN^2 \leq \frac{\alpha r^2}{2} R^2 - \sum_{\zeta \in Z \cap Q^-} \int_Q \log^+ \frac{r}{|z-\zeta|} \, dA(z) - R^2 \log \frac{r}{2\sqrt{2}}.$$

For any $\zeta \in Q^-$, the disk $|z - \zeta| < r$ is contained in Q so that

$$\int_Q \log^+ \frac{r}{|z - \zeta|} \, dA(z) = \int_{|z-\zeta|<r} \log \frac{r}{|z - \zeta|} \, dA(z)$$

$$= \int_{|z|<r} \log \frac{r}{|z|} \, dA(z) = \frac{\pi r^2}{2}.$$

Since $N^2 \le R^2$, it follows that

$$n(Z,Q^-)\frac{\pi r^2}{2} \le \left(\frac{\alpha r^2}{2} - \log \frac{r}{2\sqrt{2}} + C \right) R^2,$$

where $n(Z,Q^-)$ denotes the number of points from Z contained in Q^-. This can be rewritten as

$$\frac{n(Z,Q^-)}{(R-2r)^2} \le \left(\frac{\alpha}{\pi} - \frac{2}{\pi r^2} \log \frac{r}{2\sqrt{2}} + \frac{2C}{\pi r^2} \right) \frac{R^2}{(R-2r)^2}. \qquad (4.21)$$

Fix r and let $R \to \infty$. Then by Proposition 4.1,

$$D^+(Z) \le \frac{\alpha}{\pi} - \frac{2}{\pi r^2} \log \frac{r}{2\sqrt{2}} + \frac{2C}{\pi r^2}.$$

If r was chosen large enough so that

$$C - \log \frac{r}{2\sqrt{2}} < 0,$$

then $D^+(Z) < \alpha/\pi$. $\qquad \square$

We summarize the main result of this section as follows:

Theorem 4.50. *Suppose Z is a sequence in $\mathbb{C}$ and $0 < p \le \infty$. Then Z is an interpolating sequence for F_α^p if and only if Z is separated and $D^+(Z) < \alpha/\pi$.*

Corollary 4.51. *Suppose Z is a separated sequence in $\mathbb{C}$ and $0 < p \le \infty$. Then Z is interpolating for F_α^p if and only if $D^+(Z) < \alpha/\pi$.*

4.7 Notes

The main results of this chapter are due to Seip and Wallsten, and our presentation follows their papers [206] and [209] very closely. In turn, those two papers follow Beurling's 1977–1978 lectures on balayage and interpolation at the Mittag–Lefler Institute very closely. In particular, the density notion introduced in Sect. 4.1 can be found in Beurling's lectures [36].

We chose to follow the more classical and original arguments of estimating certain perturbations of the Weierstrass σ-function because this is more in line with the traditional approaches to entire functions. But we point out that there are now more modern and more powerful techniques for interpolation and sampling problems that work in much more general settings. For example, many ideas used in [203, 208] to characterize interpolating and sampling sequences for Bergman spaces can be adapted to work for Fock spaces as well.

See [205] for a complete description of interpolating and sampling sequences for Bergman spaces on the unit disk. The books [78, 119, 203] contain more details about the Bergman space results than Seip's original papers. The interested reader will find many additional papers in the bibliography about various interpolation and sampling problems.

4.8 Exercises

1. Suppose $Z = \{z_n\}$ is a sequence of interpolation for F_α^p and $\{v_n\}$ is a sequence of complex numbers such that $\{v_n e^{-\alpha|z_n|^2/2}\} \in l^p$. Show that the minimal interpolation problem

$$\inf\{\|f\|_{p,\alpha} : f(z_n) = v_n, n \geq 1\}$$

has a unique solution.

2. If $f \in F_\alpha^p$ for some $0 < p \leq \infty$ and $\alpha > 0$, then for any complex number a, the function $g(z) = (z-a)f(z)$ belongs to F_β^q for all $0 < q \leq \infty$ and $\beta > \alpha$.

3. Prove Theorem 4.2.

4. Show that there exist two interpolating sequences for F_α^p whose union is sampling for F_α^p.

5. If Z is not a set of uniqueness for F_α^p, then $\rho_p(z, Z) = 0$ if and only if $z \in Z$.

6. Show that for any $\varepsilon > 0$, there exists a positive constant $C = C(\varepsilon, \alpha, p)$ such that

$$\left| f(z) e^{-\frac{\alpha}{2}|z|^2} \right|^p \leq C \int_{\varepsilon < |w-z| < 2\varepsilon} \left| f(w) e^{-\frac{\alpha}{2}|w|^2} \right|^p \, dA(w)$$

for all $z \in \mathbb{C}$.

7. Show that the incomplete gamma function has the property that

$$\Gamma(k+1, z) = k! e^{-z} \sum_{j=0}^{k} \frac{z^j}{j!}$$

for all k and z.

8. Show that

$$\sum_{n=1}^{N} \sum_{m=1}^{N} \frac{1}{n^2 + m^2} \sim \log N$$

as $N \to \infty$.

9. Show that

$$\sum_{n^2 + m^2 > N^2} \frac{1}{(n^2 + m^2)^{3/2}} \sim \frac{1}{N}$$

as $N \to \infty$.

10. Let δ be a positive number. Show that for any $w \in \mathbb{C}$, we have

$$\sum \left[\frac{1}{|\omega_{mn} - w|} : \delta < |\omega_{mn} - w| < R \right] \sim R$$

as $R \to \infty$, where $\Lambda = \{\omega_{mn}\}$ is any lattice.

11. Suppose Z is uniformly close to Λ_α. Show that

$$D^+(Z) = D^-(Z) = \alpha/\pi.$$

12. Justify the last step in the proof of Lemmas 4.32 and 4.34.

13. If Z is an interpolating sequences for F_α^p, then any subset of Z is also an interpolating sequence for F_α^p.

14. Show that $|\sigma_\alpha'(\omega_{mn})|e^{-\frac{\alpha}{2}|\omega_{mn}|^2}$ is a positive constant independent of m and n, where σ_α' is the derivative of σ_α.

15. If Z is sampling for F_α^p, then adding any separated sequence to Z will create a sampling sequence for F_α^p again.

16. If $Z = \{z_n\}$ is a sequence in $\mathbb{C}$ such that

$$\inf\{|z_j - z_k| : j \neq k\} > \frac{2}{\sqrt{\alpha}},$$

then Z is an interpolating sequence for F_α^p. See Tung [225].

17. If $Z = \{z_n\}$ is a sequence in $\mathbb{C}$ and there is a positive number $\varepsilon < 1/\sqrt{\alpha}$ such that the disks $B(z_n, \varepsilon)$ cover the whole complex plane, then Z is a sampling sequence for F_α^p.

18. Suppose $Z = \{z_n\}$ is separated and T is the operator from F_α^p to l^p defined by

$$T(f) = \left\{ e^{-\frac{\alpha}{2}|z_n|^2} f(z_n) \right\}. \tag{4.22}$$

Show that:

(a) T is onto if and only if Z is interpolating for F_α^p.

(b) T is bounded below if and only if Z is sampling for F_α^p.

(c) T is one-to-one if and only if Z is a uniqueness set for F_α^p.

Prove or disprove that T has closed range if and only if Z is either interpolating or sampling for F_α^p.

19. Suppose $Z = \{z_n\}$ is separated, $1 \leq p < \infty$, and $1/p + 1/q = 1$. Then Z is an interpolating sequence for F_α^p if and only if there exists a positive constant c such that

$$\int_{\mathbb{C}} \left| \sum_{k=1}^\infty a_k e^{-\frac{\alpha}{2}|z-z_k|^2} \right|^q dA(z) \geq c \sum_{k=1}^\infty |a_k|^q$$

for every sequence $\{a_k\} \in l^q$.

20. Suppose $Z = \{z_n\}$ is separated, $1 \leq p < \infty$, and $1/p + 1/q = 1$. Then Z is a sampling sequence for F_α^p if and only if every function $f \in F_\alpha^q$ has the form

$$f(z) = \sum_{k=1}^\infty a_k e^{\alpha \bar{z}_k z - \frac{\alpha}{2}|z_k|^2}$$

for some $\{a_k\} \in l^q$.

21. Let μ and ν be two positive measures. If A_1 and A_2 are two sets that are measurable with respect to both μ and ν. Show that

$$\min\left(\frac{\nu(A_1)}{\mu(A_1)}, \frac{\nu(A_2)}{\mu(A_2)}\right) \leq \frac{\nu(A_1 \cup A_2)}{\mu(A_1 \cup A_2)} \leq \max\left(\frac{\nu(A_1)}{\mu(A_1)}, \frac{\nu(A_2)}{\mu(A_2)}\right).$$

22. Make precise the word "roughly" used in the proof of Proposition 4.3.
23. For a sequence $Z = \{z_n\}$ of distinct points in $\mathbb{C}$, show that the following conditions are equivalent:

(a) Z is sampling for F_α^2.
(b) Atomic decomposition holds on Z.
(c) The operator

$$Sf(z) = \sum_{n=1}^{\infty} f(z_n) e^{\alpha z \bar{z}_n - \alpha |z_n|^2} \tag{4.23}$$

is bounded and invertible on F_α^2.

24. Show that the operator S defined in (4.23) is bounded on F_α^2 if and only if Z is the union of finitely many separated sequences.
25. Handle the case $D^-(Z) = 0$ in the proof of Lemma 4.28.
26. Suppose $Z = \{z_{mn}\}$, $\Lambda_\alpha = \{\omega_{mn}\}$, and $\Lambda_\beta = \{\lambda_{mn}\}$. If

$$|z_{mn} - \lambda_{mn}| \leq Q$$

for all (m, n), then there exists a positive constant $Q' = Q'(\alpha, \beta, Q)$ such that for any (k, l), there exists some (k', l') with the property that

$$|(z_{mn} + \omega_{kl}) - (\lambda_{mn} + \lambda_{k'l'})| \leq Q'$$

for all (m, n).
27. Suppose $Z = \{z_{mn}\}$ is uniformly close to $\Lambda = \Lambda(\omega, \omega_1, \omega_2) = \{\omega_{mn}\}$ with $|z_{mn} - \omega_{mn}| \leq Q$ for all (m, n). Show that for any $\varepsilon > 0$, there exists some constant $C = C(\varepsilon, Q, \omega, \omega_1, \omega_2) > 0$ such that

$$\sum_{m,n} e^{-\varepsilon |z_{mn}|^2} \leq C.$$

Hint: write $|z|^2 = |\omega + (z - \omega)|^2 = |\omega|^2 |1 + (z - \omega)/\omega|^2$.

Chapter 5
Zero Sets for Fock Spaces

In this chapter, we study zero sets for the Fock spaces F_α^p. Throughout this book, we say that a sequence $Z = \{z_n\} \subset \Omega$ is a zero set for a space X of analytic functions in Ω if there exists a function $f \in X$, not identically zero, such that Z is *exactly* the zero sequence of f, counting multiplicities.

K. Zhu, *Analysis on Fock Spaces*, Graduate Texts in Mathematics 263,
DOI 10.1007/978-1-4419-8801-0_5,
© Springer Science+Business Media New York 2012

5.1   A Necessary Condition

Recall from Theorem 2.12 that every function $f \in F_\alpha^p$ is of order 2. Therefore, by Hadamard's factorization theorem, the zero sequence $\{z_n\}$ of f, with the origin removed, must satisfy

$$\sum_{n=1}^{\infty} \frac{1}{|z_n|^3} < \infty.$$

In this section, we improve upon this estimate and obtain the following necessary condition for a sequence $\{z_n\}$ to be a zero set for F_α^p.

Theorem 5.1. *Suppose $0 < p \le \infty$ and $\{z_n\}$ is the zero sequence of a function $f \in F_\alpha^p$ with $f(0) \neq 0$. Then there exist a positive constant c and a rearrangement of $\{z_n\}$ such that $|z_n| \ge c\sqrt{n}$ for all n.*

Proof. Without loss of generality, we may assume that $f(0) = 1$ and $p = \infty$. Let $\{z_n\}$ denote the zero sequence of f, repeated according to multiplicity and arranged so that $0 < |z_1| \le |z_2| \le |z_3| \le \cdots$.

Fix any positive radius r such that f has no zero on $|z| = r$ and let $n(r)$ denote the number of zeros of f in $|z| < r$. By Jensen's formula,

$$\sum_{k=1}^{n(r)} \log \frac{r}{|z_k|} = \frac{1}{2\pi} \int_0^{2\pi} \log|f(re^{i\theta})| \, d\theta.$$

Since $f \in F_\alpha^\infty$, we have

$$|f(re^{i\theta})| \le \|f\|_{\infty,\alpha} e^{\frac{\alpha}{2}r^2}, \qquad 0 \le \theta \le 2\pi, r > 0.$$

It follows that

$$\sum_{k=1}^{n(r)} \log \frac{r}{|z_k|} \le \frac{\alpha}{2} r^2 + C,$$

where $C = \log \|f\|_{\infty,\alpha}$. Rewrite the above inequality as

$$\prod_{k=1}^{n(r)} \frac{r}{|z_k|} \le \exp\left(\frac{\alpha}{2} r^2 + C\right)$$

and observe that

$$\prod_{k=1}^{n} \frac{r}{|z_k|} \le \prod_{k=1}^{n(r)} \frac{r}{|z_k|}.$$

for any positive integer n (independent of r). Then

$$\prod_{k=1}^{n} \frac{r}{|z_k|} \leq \exp\left(\frac{\alpha}{2}r^2 + C\right)$$

for all positive integers n and all $r > 0$ such that f has no zero on $|z| = r$. Since $\{|z_k|\}$ is nondecreasing, we have

$$\frac{r^n}{|z_n|^n} \leq \exp\left(\frac{\alpha}{2}r^2 + C\right),$$

or

$$\frac{1}{|z_n|} \leq \frac{1}{r}\exp\left(\frac{\alpha}{2n}r^2 + \frac{C}{n}\right), \tag{5.1}$$

where n is any positive integer and r is any radius such that f has no zero on $|z| = r$.

There are only a countable number of radius r such that f has zeros on $|z| = r$. Therefore, for any positive integer n, we can choose a sequence $\{r_k\}$ such that $r_k \to \sqrt{n}$ as $k \to \infty$ and f has no zero on each $|z| = r_n$. Combining this with (5.1), we conclude that

$$\frac{1}{|z_n|} \leq \frac{1}{\sqrt{n}}\exp\left(\frac{\alpha}{2} + \frac{\log\|f\|_{\infty,\alpha}}{n}\right), \qquad n \geq 1.$$

It is then clear that there is some positive constant c such that $|z_n| \geq c\sqrt{n}$ for all $n \geq 1$. □

Note that the assumption $f(0) \neq 0$ is not a critical one. In fact, if $f \in F_\alpha^p$ and it has a zero of order m at the origin, then the function g defined by $g(z) = f(z)/z^m$ is in F_α^p and does not vanish at the origin.

Corollary 5.2. *Suppose $0 < p \leq \infty$ and $\{z_n\}$ is the zero sequence of some $f \in F_\alpha^p$ with $f(0) \neq 0$. Then*

$$\sum_{n=1}^{\infty} \frac{1}{|z_n|^r} < \infty$$

for every $r > 2$.

The function

$$f(z) = \frac{\sin(\delta z^2)}{\delta z^2}$$

used in the proof of Theorem 5.4 shows that the estimate in Theorem 5.1 is best possible. More specifically, we can find a positive constant C in this case such that

$$C^{-1}\sqrt{n} \leq |z_n| \leq C\sqrt{n}$$

for all $n \geq 1$.

5.2   A Sufficient Condition

The purpose of this section is to prove the following sufficient condition for zero sequences of F_α^p.

Theorem 5.3. *Suppose that* $\{z_n\}$ *is a sequence of complex numbers such that*

$$\sum_{n=1}^{\infty} \frac{1}{|z_n|^2} < \infty. \tag{5.2}$$

Then $\{z_n\}$ *is a zero set for* F_α^p, *where* $0 < p \leq \infty$.

Proof. Suppose that $\{z_n\}$ satisfies condition (5.2). We may also assume that the sequence $\{z_n\}$ has been ordered in such a way that $\{|z_n|\}$ is nondecreasing. Consider the Weierstrass product

$$f(z) = \prod_{n=1}^{\infty} E_1\left(\frac{z}{z_n}\right),$$

where $E_1(z) = (1 - z)e^z$. By Theorem 1.6, f is entire, and $\{z_n\}$ is the zero sequence of f. We will show that this function f belongs to all the Fock spaces F_α^p, where $0 < p \leq \infty$ and $\alpha > 0$.

If $|z| < 1/2$, we have

$$\begin{aligned}
\log|E_1(z)| &= \mathrm{Re}\left[\log(1-z) + z\right] \\
&= \mathrm{Re}\left[-\frac{z^2}{2} - \frac{z^3}{3} - \frac{|z|^4}{4} - \cdots\right] \\
&\leq |z|^2\left[\frac{1}{2} + \frac{|z|}{3} + \frac{|z|^2}{4} + \cdots\right] \\
&\leq \frac{1}{2}|z|^2\left[1 + \frac{1}{2} + \frac{1}{2^2} + \cdots\right] \\
&= |z|^2.
\end{aligned}$$

On the other hand, we have

$$|E_1(z)| \leq (1 + |z|)e^{|z|}, \qquad \log|E_1(z)| \leq |z| + \log(1 + |z|), \tag{5.3}$$

for all z. It follows that for any positive A, there exists a positive number R such that

$$\log|E_1(z)| \leq A|z|^2, \qquad |z| > R.$$

On the annulus $1/2 \leq |z| \leq R$, the function $|z|^2 \log |E_1(z)|$ is continuous except at $z = 1$, where it tends to $-\infty$. Hence, there is a constant B such that

$$\log |E_1(z)| \leq B|z|^2, \qquad \frac{1}{2} \leq |z| \leq R.$$

Combining the estimates from the last three paragraphs, we conclude that

$$\log |E_1(z)| \leq M|z|^2, \qquad z \in \mathbb{C},$$

where $M = \max(1, A, B)$.

Given any positive ε, we can find a positive integer N such that

$$\sum_{n=N+1}^{\infty} \frac{1}{|z_n|^2} < \frac{\varepsilon}{2M}.$$

From this, we deduce that

$$\sum_{n=N+1}^{\infty} \log |E_1(z/z_n)| \leq M \sum_{n=N+1}^{\infty} \left| \frac{z}{z_n} \right|^2 \leq \frac{\varepsilon}{2} |z|^2$$

for all $z \in \mathbb{C}$. Using (5.3) again, we can find some $r_1 > 0$ such that

$$\log |E_1(z)| \leq \frac{\varepsilon}{2S} |z|^2, \qquad |z| > r_1,$$

where

$$S = \sum_{n=1}^{N} \frac{1}{|z_n|^2}.$$

Set $r_2 = r_1 |z_N|$. Then $|z| > r_2$ implies that $|z/z_n| > r_1$ for $1 \leq n \leq N$. It follows that

$$\sum_{n=1}^{N} \log |E_1(z/z_n)| \leq \frac{\varepsilon}{2} |z|^2, \qquad |z| > r_2.$$

Therefore,

$$\log |f(z)| = \sum_{n=1}^{\infty} \log |E_1(z/z_n)| < \varepsilon |z|^2$$

for all $|z| > r_2$, or $|f(z)| < e^{\varepsilon |z|^2}$ for all $|z| > r_2$. Since ε is arbitrary, we see that $f \in F_\alpha^p$ for all $\alpha > 0$ and $0 < p \leq \infty$.   □

Note that the proof above can easily be adapted to show that the function $P(z)f(z)$ belongs to F_α^p for any polynomial $P(z)$. Therefore, if $\{z_n\}$ satisfies (5.2), then $\{z_n\} \cup F$ is also a zero set for F_α^p, where F is any finite set. It is permitted to have the origin contained in F.

5.3 Pathological Properties

In this section, we present examples to show certain pathological properties of zero sequences of Fock spaces. More specifically, we will show that:

(i) The union of two zero sequences for F_α^p is not necessarily a zero sequence for F_α^p again.
(ii) A subsequence of a zero sequence for F_α^p is not necessarily a zero sequence for F_α^p again.
(iii) If $\alpha \neq \beta$, then the spaces F_α^p and F_β^q have different zero sequences.
(iv) An interpolating sequence for F_α^p is not necessarily a zero sequence for F_α^p.

Theorem 5.4. *Suppose $\alpha > 0$ and $0 < p \leq \infty$. There exist two zero sequences for F_α^p whose union is no longer a zero sequence for F_α^p.*

Proof. Fix $\delta \in (\pi\alpha/8, \alpha/2)$ and consider the sequence

$$Z = \left\{ e^{k\pi i/2}\sqrt{n\pi/\delta} : k = 0,1,2,3; n = 1,2,3,\cdots \right\}.$$

It is easy to see that Z is the zero sequence of the entire function

$$f(z) = \frac{\sin(\delta z^2)}{\delta z^2}.$$

Converting the sine function above to complex exponential functions and using the assumption that $\delta < \alpha/2$, we easily check that $f \in F_\alpha^p$. Therefore, Z is a zero sequence for F_α^p.

Let $Z' = \{e^{\pi i/4}z : z \in Z\}$ be a rotation of the sequence Z above. Then Z' is also an F_α^p zero sequence. Clearly, Z and Z' are disjoint. We now arrange $Z \cup Z'$ into a single sequence $\{z_n\}$ such that

$$|z_1| \leq |z_2| \leq |z_3| \leq \cdots .$$

If $\{z_n\}$ is a zero sequence for $F_\alpha^p \subset F_\alpha^\infty$, it follows from the proof of Theorem 5.1 that there exists a positive constant C such that

$$\prod_{k=1}^{n} \frac{r}{|z_k|} \leq Ce^{\frac{\alpha}{2}r^2}$$

for all $n \geq 1$ and $r > 0$. Square both sides, replace n by $8n$, and integrate from 0 to ∞ with respect to the measure $re^{-\beta r^2}$, where $\beta > \alpha$. We obtain another positive constant C such that

$$\frac{(8n)!}{\beta^{8n}} \prod_{k=1}^{8n} \frac{1}{|z_k|^2} \leq C$$

for all $n \geq 1$. It is easy to see that this reduces to

$$\left(\frac{\delta}{\pi\beta}\right)^{8n}\frac{(8n)!}{(n!)^8} \leq C, \qquad n \geq 1.$$

By Stirling's formula, there exists yet another positive constant C, independent of n, such that

$$\left(\frac{8\delta}{\pi\beta}\right)^{8n}\frac{\sqrt{n}}{n^4} \leq C$$

for all $n \geq 1$. This clearly implies that $8\delta \leq \pi\beta$. Since β can be arbitrarily close to α, we have $\delta \leq \pi\alpha/8$, which is a contradiction. This shows that $\{z_n\}$ is not an F_α^p zero set and completes the proof of the theorem. $\qquad\square$

Theorem 5.5. *Let $\alpha > 0$ and $0 < p \leq \infty$. There exists an F_α^p zero sequence $\{z_n\}$ and a subsequence $\{z_{n_k}\}$ which is not an F_α^p zero sequence.*

Proof. Fix a positive constant δ such that $\delta < \alpha/2$ and consider the following entire function:

$$f(z) = \frac{e^{i\delta z^2} - 1}{i\delta z^2}.$$

It is easy to check that $f \in F_\alpha^p$. Thus, its zero set

$$\left\{\pm\sqrt{\frac{2n\pi}{\delta}} : n = 1,2,3,\cdots\right\} \cup \left\{\pm i\sqrt{\frac{2n\pi}{\delta}} : n = 1,2,3,\cdots\right\}$$

is an F_α^p zero sequence. Let $\{z_n\}$ denote the subsequence consisting of real elements in the above set. We proceed to show that $\{z_n\}$ is not an F_α^p zero set.

Again, aiming to arrive at a contradiction later, we assume that g is a function in F_α^p that vanishes precisely on $\{z_n\}$. It is clear that $\rho_1(g) = m(g) = 2$; see Sect. 1.1 for definitions and properties of these numbers. By Theorem 1.10, we always have $\rho(g) \geq \rho_1(g)$, so g must be of order greater than or equal to 2. Combining this with Theorem 2.12, we conclude that g must be of order 2. By Lindelöf's theorem (see Theorem 1.11), the function g must be of maximum (infinite) type since the sums

$$S(r) = \sum_{|z_n| \leq r} \frac{1}{z_n^2} \sim \log r, \quad r > 1,$$

are clearly unbounded. By Theorem 2.12 again, the function g cannot possibly be in F_α^p. This contradiction shows that $\{z_n\}$ is not an F_α^p zero set. $\qquad\square$

We now consider zero sets for different Fock spaces. The Weierstrass σ-functions play a significant role here.

Recall that for any positive α,

$$\Lambda_\alpha = \left\{ \omega_{mn} = \sqrt{\frac{\pi}{\alpha}}(m+in) : m \in \mathbb{Z}, n \in \mathbb{Z} \right\}$$

is the square lattice in the complex plane with fundamental region

$$\Omega_\alpha = \left\{ z = x+iy : |x| < \frac{1}{2}\sqrt{\frac{\pi}{\alpha}}, |y| < \frac{1}{2}\sqrt{\frac{\pi}{\alpha}} \right\}.$$

The Weierstrass σ-function associated to Λ_α is the following infinite product:

$$\sigma_\alpha(z) = z \prod_{m,n}' \left(1 - \frac{z}{\omega_{mn}}\right) \exp\left(\frac{z}{\omega_{mn}} + \frac{1}{2}\frac{z^2}{\omega_{mn}^2}\right),$$

where the product is taken over all integers m and n with $\omega_{mn} \neq 0$.

Lemma 5.6. *Let $0 < \alpha_1 < \alpha < \alpha_2 < \infty$. We have:*

(a) $\sigma_\alpha \in F_{\alpha_2}^p$ *for all $0 < p \leq \infty$.*
(b) $\sigma_\alpha \notin F_{\alpha_1}^p$ *for any $0 < p \leq \infty$.*
(c) $\sigma_\alpha \in F_\alpha^\infty$.
(d) $\sigma_\alpha \notin f_\alpha^\infty$, *and so $\sigma_\alpha \notin F_\alpha^p$ for any $0 < p < \infty$.*

Proof. It follows from the quasiperiodicity of σ_α that if $z = \omega_{mn} + w$ and $w \in \Omega_\alpha$, then

$$|\sigma_\alpha(z)|e^{-\frac{\alpha}{2}|z|^2} = |\sigma_\alpha(w)|e^{-\frac{\alpha}{2}|w|^2}. \tag{5.4}$$

Since the function $|\sigma_\alpha(w)|e^{-\alpha|w|^2/2}$ is bounded on the relatively compact set Ω_α, there exists a positive constant C such that

$$|\sigma_\alpha(z)| \leq Ce^{\frac{\alpha}{2}|z|^2}, \qquad z \in \mathbb{C}.$$

This clearly implies that $\sigma_\alpha \in F_\alpha^\infty$ and $\sigma_\alpha \in F_{\alpha_2}^p$ for all $0 < p \leq \infty$.

If S is any compact set contained in the fundamental region of Λ_α, then there exists a positive constant δ such that

$$|\sigma_\alpha(w)|e^{-\frac{\alpha}{2}|w|^2} \geq \delta, \qquad w \in S.$$

This together with (5.4) shows that

$$|\sigma_\alpha(z)|e^{-\frac{\alpha}{2}|z|^2} \geq \delta, \qquad z \in S + \omega_{mn},$$

for all (m,n). This clearly shows that $\sigma_\alpha \notin f_\alpha^\infty$. Since $F_\alpha^p \subset f_\alpha^\infty$ for $0 < p < \infty$, we have $\sigma_\alpha \notin F_\alpha^p$ for any $0 < p < \infty$. Also, $F_{\alpha_1}^p \subset f_\alpha^\infty$ for all $0 < p \leq \infty$. So $\sigma_\alpha \notin F_{\alpha_1}^p$ for all $0 < p \leq \infty$. □

Lemma 5.7. *Suppose $0 < p < \infty$ and $f \in F_\alpha^p$. If $f(z) = 0$ for all $z \in \Lambda_\alpha$, then f is identically zero.*

Proof. By the Weierstrass factorization theorem, we can write $f = h\sigma_\alpha$, where h is an entire function. In view of the quasiperiodicity of σ_α, we have

$$\int_\mathbb{C} \left| f(z)e^{-\frac{\alpha}{2}|z|^2} \right|^p dA(z) = \sum_{m,n} \int_{\Omega_\alpha} |h(z+\omega_{mn})|^p \left| \sigma_\alpha(z)e^{-\frac{\alpha}{2}|z|^2} \right|^p dA(z),$$

where Ω_α is the fundamental region of σ_α. Let D be any small disk centered at 0 and contained in $\frac{1}{2}\Omega_\alpha$. Then by Corollary 1.21, there exists a positive constant C such that

$$\int_\mathbb{C} \left| f(z)e^{-\frac{\alpha}{2}|z|^2} \right|^p dA(z) \geq C \sum_{m,n} \int_{\Omega_\alpha - D} |h(z+\omega_{mn})|^p dA(z).$$

Since the function $z \mapsto |h(z+\omega_{mn})|^p$ is subharmonic, there exists a positive constant δ (independent of (m,n)) such that

$$\int_{\Omega_\alpha - D} |h(z+\omega_{mn})|^p dA(z) \geq \delta \int_{\Omega_\alpha} |h(z+\omega_{mn})|^p dA(z)$$

for all (m,n). It follows that there is another positive constant C such that

$$\int_\mathbb{C} \left| f(z)e^{-\frac{\alpha}{2}|z|^2} \right|^p dA(z) \geq C \int_\mathbb{C} |h(z)|^p dA(z).$$

This is impossible unless h is identically zero. □

Theorem 5.8. *Suppose $0 < p \leq \infty$, $0 < q \leq \infty$, and $\alpha_1 \neq \alpha_2$. Then $F_{\alpha_1}^p$ and $F_{\alpha_2}^q$ have different zero sets.*

Proof. Without loss of generality, let us assume that $\alpha_1 < \alpha < \alpha_2$. By Lemma 5.6, the Weierstrass function σ_α belongs to $F_{\alpha_2}^q$, so its zero sequence Λ_α is a zero set for $F_{\alpha_2}^q$. On the other hand, if $f \in F_{\alpha_1}^p \subset F_\alpha^2$ and f vanishes on Λ_α, then it follows from Lemma 5.7 that f is identically zero. Therefore, Λ_α cannot possibly be a zero set for $F_{\alpha_1}^p$. □

The remaining question for us now is this: do F_α^p and F_α^q have different zero sets whenever $p \neq q$? As of this writing, there is no complete answer, but it is easy to produce examples of such pairs that do not have the same zero sets. The simplest example is $Z = \Lambda_\alpha$, which is a zero set for F_α^∞, but not a zero set for any F_α^p when $0 < p < \infty$. This again follows from Lemmas 5.6 and 5.7.

Similarly, the sequence $Z = \Lambda_\alpha - \{0\}$ is a zero set for F_α^p when $p > 2$ because the function $f(z) = \sigma_\alpha(z)/z$ belongs to F_α^p if and only if $p > 2$. However, this sequence Z is not a zero set for F_α^2. To see this, suppose f is a function in F_α^2, not identically zero, such that f vanishes on Z. By Weierstrass factorization, we have $f(z) = [\sigma_\alpha(z)/z]g(z)$ for some entire function g that is not identically zero. Mimicking the proof of Lemma 5.7, we can show that

$$\int_{|z|>1} \left| \frac{g(z)}{z} \right|^2 \, dA(z) < \infty.$$

It follows from polar coordinates and the Taylor expansion of g that this is impossible unless g is identically zero. This actually shows that $Z = \Lambda_\alpha - \{0\}$ is a uniqueness set for F_α^2. In the above arguments, the point 0 can be replaced by any other point in Λ_α.

On the other hand, if Z is the resulting sequence when two points a and b are removed from Λ_α, then the function

$$f(z) = \frac{\sigma_\alpha(z)}{(z-a)(z-b)}$$

belongs to F_α^2 and has Z as its zero sequence. Therefore, Z is a zero set for F_α^2. Consequently, it is possible to go from a uniqueness set to a zero set by removing just one point. Equivalently, it is possible to add just a single point to a zero set of F_α^2 so that the resulting sequence becomes a uniqueness set for F_α^2. This shows how delicate the problem of characterizing zero sets for F_α^p is.

We can also show by an example that it is generally very difficult to distinguish between zero sets for F_α^p and F_α^q. More specifically, for any positive integer N with $Np > 2$, if Z is an F_α^q zero set and if N points $\{z_1, \cdots, z_N\}$ are removed from Z, then the remaining sequence Z' is an F_α^p zero set. To see this, let Z be the zero sequence of a function $f \in F_\alpha^q$, not identically zero, then Z' is the zero sequence of the function

$$g(z) = \frac{f(z)}{(z-z_1) \cdots (z-z_N)},$$

which is easily seen to be in F_α^p. Therefore, zero sets for F_α^p and F_α^q may be different, but they are not too much different.

Let Z be a zero sequence for F_α^p and let I_Z denote the set of functions f in F_α^p such that f vanishes on Z. In the classical theories of Hardy and Bergman spaces, the space I_Z is always infinite dimensional. This is no longer true for Fock spaces.

Theorem 5.9. *For any $0 < p \leq \infty$ and $k \in \{1, 2, \cdots\} \cup \{\infty\}$, there exists a zero set Z for F_α^p such that $\dim(I_Z) = k$.*

Proof. The case $k = \infty$ is trivial; any finite sequence Z will work. So we assume that k is a positive integer in the rest of the proof.

We first consider the case $p = \infty$ and $k > 1$. In this case, we consider $Z = \Lambda_\alpha - \{a_1, \cdots, a_{k-1}\}$, where $a_1, \cdots, a_{k-1}$ are (any) distinct points in Λ_α and

$$f(z) = \frac{\sigma_\alpha(z)}{(z-a_1)\cdots(z-a_{k-1})}.$$

It follows from Corollary 1.21 that $f \in F_\alpha^\infty$ and Z is exactly the zero sequence of f. Furthermore, if h is a polynomial of degree less than or equal to $k-1$, then the function $f(z)h(z)$ is still in F_α^∞.

On the other hand, if F is any function in F_α^∞ that vanishes on Z, then we can write

$$F(z) = f(z)g(z) = \frac{\sigma_\alpha(z)g(z)}{(z-a_1)\cdots(z-a_{k-1})},$$

where g is an entire function. For any positive integer n, let C_n be the boundary of the square centered at 0 with horizontal and vertical side length $(2n+1)\sqrt{\pi/\alpha}$. It is clear that

$$d(C_n, \Lambda_\alpha) \geq \sqrt{\pi/\alpha}/2, \qquad n \geq 1.$$

So there exists a positive constant C such that

$$|\sigma_\alpha(z)|e^{-\frac{\alpha}{2}|z|^2} \geq C, \qquad z \in C_n, n \geq 1.$$

This together with the assumption that $F \in F_\alpha^\infty$ implies that there exists another positive constant C such that

$$|g(z)| \leq C|z-a_1|\cdots|z-a_{k-1}| \qquad (5.5)$$

for all $z \in C_n$ and $n \geq 1$. By Cauchy's integral estimates, the function g must be a polynomial of degree at most $k-1$.

Therefore, when $p = \infty$, $k > 1$, and $Z = \Lambda_\alpha - \{a_1, \cdots, a_{k-1}\}$, we have shown that a function $F \in F_\alpha^\infty$ vanishes on Z if and only if

$$F(z) = \frac{\sigma_\alpha(z)h(z)}{(z-a_1)\cdots(z-a_{k-1})},$$

where h is a polynomial of degree less than or equal to $k-1$. This shows that $\dim(I_Z) = k$.

When $p = \infty$ and $k = 1$, we simply take $Z = \Lambda_\alpha$. The arguments above can be simplified to show that a function $F \in F_\alpha^\infty$ vanishes on Z if and only if $F = c\sigma_\alpha$ for some constant c.

Next, we assume that $0 < p < \infty$ and k is a positive integer. In this case, we let N denote the smallest positive integer such that $Np > 2$, or equivalently,

$$\int_{|z|>1} \left| \frac{\sigma_\alpha(z) e^{-\frac{\alpha}{2}|z|^2}}{z^N} \right|^p \, dA(z) < \infty. \tag{5.6}$$

Remove any $N+k-1$ points $\{a_1, \cdots, a_{N+k-1}\}$ from Λ_α and denote the remaining sequence by Z. Then Z is the zero sequence of the function

$$\frac{\sigma_\alpha(z)}{(z-a_1)\cdots(z-a_{N+k-1})},$$

which belongs to F_α^p in view of (5.6). In fact, if g is any polynomial of degree less than or equal to $k-1$, then it follows from (5.6) that g times the above function belongs to I_Z.

Conversely, if f is any function in F_α^p that vanishes on Z, then we can write

$$f(z) = \frac{\sigma_\alpha(z)g(z)}{(z-a_1)\cdots(z-a_{N+k-1})},$$

where g is an entire function. Since $F_\alpha^p \subset F_\alpha^\infty$, it follows from (5.5) and Cauchy's integral estimates that g is a polynomial with degree less than or equal to $N+k-1$. If the degree of g is $j > k-1$, then

$$\frac{g(z)}{(z-a_1)\cdots(z-a_{N+k-1})} \sim \frac{1}{z^{N+k-1-j}}, \qquad z \to \infty.$$

This together with $f \in F_\alpha^p$ shows that (5.6) still holds when N is replaced by $N+k-1-j$, which contradicts our minimality assumption on N. Thus, $j \le k-1$, which shows that I_Z is k dimensional. $\qquad\qquad\qquad\qquad\qquad\qquad\square$

The following result describes the structure of I_Z when it is finite dimensional.

Theorem 5.10. *Suppose Z is a zero set for F_α^p and $\dim(I_Z) = k$ is a positive integer. Then there exists a function $g \in I_Z$ such that $I_Z = gP_{k-1}$, where P_{k-1} is the set of all polynomials of degree less than or equal to $k-1$.*

Proof. First, observe that if $\dim(I_Z) = k < \infty$, then $Z' = Z \cup \{a_1, \cdots, a_k\}$ is a uniqueness set for F_α^p for all $\{a_1, \cdots, a_k\}$. Here, the union in Z' should be understood in the sense of zero sequences, where multiplicities are taken into account. In fact, if there exists a function $f \in F_\alpha^p$, not identically zero, such that f vanishes on Z', then the functions

$$f(z), \quad \frac{f(z)}{z-a_1}, \quad \cdots, \quad \frac{f(z)}{z-a_k}$$

all belong to F_α^p and vanish on Z. Here again, if zeros of higher multiplicity are involved, then some obvious adjustments should be made. It is clear that the functions listed above are linearly independent, so the dimension of I_Z is at least $k+1$, a contradiction.

Next, observe that if $\dim(I_Z) > m$, then $Z' = Z \cup \{a_1, \cdots, a_m\}$ is not a uniqueness set for F_α^p for any collection $\{a_1, \cdots, a_m\}$. To see this, pick any $m+1$ linearly independent functions $f_1, \cdots, f_{m+1}$ from I_Z, let

$$ f = c_1 f_1 + \cdots + c_{m+1} f_{m+1}, $$

and consider the system of linear equations

$$ c_1 f_1(a_j) + \cdots + c_{m+1} f_{m+1}(a_j) = 0, \quad 1 \le j \le m. $$

Once again, obvious adjustments should be made when there are zeros of higher multiplicity. The homogeneous system above has m equations but $m+1$ unknowns, so it always has nonzero solutions c_j, $1 \le j \le m+1$. With such a choice of c_j, the function f is not identically zero but vanishes on Z', so Z' is not a uniqueness set.

It follows that if $1 \le j < k$ and $Z' = Z \cup \{a_1, \cdots, a_j\}$, then Z' is not a uniqueness set for F_α^p. We can actually show that Z' is a zero set for F_α^p. In fact, if f is a function in F_α^p, not identically zero, such that f vanishes on Z' (but not necessarily exactly on Z'), then the conclusion of the previous paragraph shows that the number of zeros of f in addition to those in Z' cannot exceed $k-j$. If these additional zeros a are divided out of f by the appropriate powers of $z-a$, the resulting function is still in F_α^p and vanishes exactly on Z'. Thus, Z' is a zero set for F_α^p.

Fix a function $g \in I_Z$ that has exactly Z as its zero set. If f is any function in I_Z, not identically zero, then just as in the previous paragraph, we can show that the zeros of f must be of the form $Z' = Z \cup \{a_1, \cdots, a_j\}$, where $j \le k-1$. Thus, we can factor f as follows: $f = gPe^h$, where $P \in P_{k-1}$ and h is entire. It is clear that dividing a polynomial out of f, whenever the division is possible, always results in a function in F_α^p. Therefore, the function ge^h belongs to I_K as well. It follows that the function $ge^h - g = g(e^h - 1)$ belongs to I_Z. If h is not constant, then by Picard's theorem, $e^h - 1$ has infinitely many zeros, so $ge^h - g$ is a function in I_Z that has infinitely many zeros in addition to those in Z, a contradiction. This shows that h is constant and $I_Z \subset gP_{k-1}$. A count of dimension then gives $I_Z = gP_{k-1}$. $\qquad \Box$

In the classical theories of Hardy and Bergman spaces, every interpolating sequence is necessarily a zero sequence. We now show that this is not true for Fock spaces.

Proposition 5.11. *There exists an interpolating sequence for F_α^p that is not a zero set for F_α^p.*

Proof. Fix some $\delta > 2/\sqrt{\alpha}$. For any positive integer k, let Z_k denote the set of $k+1$ points evenly spaced in the first quadrant on the circle $|z| = k\delta$, including the end-points $k\delta$ and $k\delta i$. Let

$$Z = \bigcup_{k=1}^{\infty} Z_k = \{z_1, z_2, \cdots, z_n, \cdots\}.$$

Since the distance between any two neighboring points in Z_k is

$$2k\delta \sin\frac{\pi}{4k} > \delta,$$

the sequence Z is separated with a separation constant greater than $2/\sqrt{\alpha}$. This implies that Z is an interpolating sequence for F_α^p; see Exercise 16 in Chap. 4.

If Z is the zero sequence of some function $f \in F_\alpha^p$, then by Theorem 5.1, the order ρ of f is less than or equal to 2. On the other hand, for the sequence Z, we have $m = \rho_1 = 2$; see Sect. 1.1 for the definition of these constants. By Theorem 1.10, we have $\rho \geq m = 2$. Thus, $\rho = 2$, and Lindelöf's theorem (Theorem 1.11) applies.

For $r \in (m\delta, (m+1)\delta)$, we have

$$S(r) = \sum_{|z_k| < r} \frac{1}{z_k^2} = \sum_{k=1}^{m} \frac{1}{(k\delta)^2} \sum_{j=0}^{k} e^{-i\pi j/k}$$

$$= \sum_{k=1}^{m} \frac{1}{(k\delta)^2} \frac{1 + e^{-i\pi/k}}{1 - e^{-i\pi/k}} = \sum_{k=1}^{m} \frac{1}{(k\delta)^2} \frac{\cos(\pi/2k)}{\sin(\pi/2k)}$$

$$\sim -\frac{2i}{\pi\delta^2} \sum_{k=1}^{m} \frac{1}{k} \sim -\frac{2i}{\pi\delta^2} \log m \sim -\frac{2i}{\pi\delta^2} \log r$$

as $r \to \infty$. This shows that $S(r)$ is not bounded in r. By Lindelöf's theorem, f has infinite type. This contradicts with Theorem 2.12, which asserts that f must have type less than or equal to $\alpha/2$ when f is of order 2. Therefore, Z cannot be a zero sequence for F_α^p. $\qquad\square$

5.4 Notes

Theorem 5.1, the necessary condition for zero sets of Fock spaces, was obtained in [249]. Theorem 5.3, the sufficient condition for zero sets of Fock spaces, is classical and follows from the general theory of entire functions. The proof of Theorem 5.3 here is basically from [67].

The results in Sect. 5.3 were mostly from [249, 258]. The motivation for [249] was Horowitz's study of zero sets for Bergman spaces; see [127–129]. The most intriguing results concerning zero sequences for Fock spaces are probably Theorems 5.9 and 5.10, which were proved in [258]. One interesting problem that remains open is the following: if $p \neq q$, do F_α^p and F_α^q always have different zero sequences?

Lemma 5.7 shows that Λ_α is a set of uniqueness for F_α^p when $0 < p < \infty$. This result as well as its proof are from [209]. Proposition 5.11, which is a little surprising when compared to the corresponding questions in the Hardy and Bergman space settings, is from [225].

5.5 Exercises

1. We say that an entire function $f(z)$ belongs to the Nevanlinna–Fock class F_α^* if

$$\int_{\mathbb{C}} \log^+ |f(z)| \, d\lambda_\alpha(z) < \infty.$$

Show that the zero sequence $\{z_n\}$ of any function f in F_α^* with $f(0) \neq 0$ satisfies the following condition:

$$\sum_{n=1}^{\infty} \frac{e^{-\alpha|z_n|^2}}{|z_n|^2} < \infty.$$

2. Let a be a nonzero complex number. Solve the extremal problem

$$\sup\{\mathrm{Re}\, f(0) : \|f\|_{2,\alpha} \leq 1, f(a) = 0\}.$$

3. Suppose Z is a zero set for F_α^p and k is a positive integer. Show that the following conditions are equivalent:

   (a) $\dim(I_Z) \leq k$.
   (b) $Z \cup \{a_1, \cdots, a_k\}$ is a uniqueness set for F_α^p for *all* $\{a_1, \cdots, a_k\}$.
   (c) $Z \cup \{a_1, \cdots, a_k\}$ is a uniqueness set for F_α^p for *some* $\{a_1, \cdots, a_k\}$.

4. Suppose Z is a zero set for F_α^p and k is a positive integer. Show that the following conditions are equivalent:

   (a) $\dim(I_Z) = k$.
   (b) For *any* $\{a_1, \cdots, a_k\}$, the sequence $Z \cup \{a_1, \cdots, a_{k-1}\}$ is not a uniqueness set for F_α^p but $Z \cup \{a_1, \cdots, a_k\}$ is.
   (c) For *some* $\{a_1, \cdots, a_{k-1}\}$, the sequence $Z \cup \{a_1, \cdots, a_{k-1}\}$ is not a uniqueness set for F_α^p, but for *some* $\{b_1, \cdots, b_k\}$, the sequence $Z \cup \{b_1, \cdots, b_k\}$ is a uniqueness set for F_α^p.

5. If Z is a zero set for F_α^p, then the sequence remains a zero set for F_α^p after any finite number of points are removed from it.
6. Suppose $0 < p < \infty$ and Z is uniformly close to Λ_α. Show that Z is a uniqueness set for F_α^p.
7. If $Z = \{z_n\}$ is a zero set for F_α^p, then

$$\sum_{n=1}^{\infty} \frac{1}{|z_n|^2 \log^{1+\varepsilon} |z_n|} < \infty$$

for all $\varepsilon > 0$, provided that $|z_n| \neq 0, 1$. Show that this is false in general if $\varepsilon = 0$.

8. Suppose $\{z_n\}$ is the zero sequence of a function $f \in F_\alpha^p$, where $f(0) = 1, 0 < p < \infty$, and $\{|z_n|\}$ is nondecreasing. Show that

$$\prod_{k=1}^n \frac{1}{|z_n|^p} \le \frac{C}{\sqrt{n}} \left(\frac{\alpha e}{n}\right)^{\frac{np}{2}} \|f\|_{p,\alpha}^p$$

   for all $n \ge 1$, where C is a positive constant independent of n and f.
9. Suppose $0 < p < q \le \infty$, Z is a zero set for F_α^q, and N is a positive integer with $Np > 2$. Show that if any N points are removed from Z, the remaining sequence becomes a zero set for F_α^p.
10. Let Z be a zero set for F_α^2 with $0 \notin Z$. Show that there is no function $G_Z \in F_\alpha^2$ such that $G_Z(0) > 0$, $\|G_Z\|_{2,\alpha} = 1$, $Z(G_Z) = Z$, and $\|f/G_Z\|_{2,\alpha} \le \|f\|_{2,\alpha}$ for all $f \in F_\alpha^2$ with $f|Z = 0$. See [119] for information about the corresponding problem in the Bergman space setting.
11. Suppose $f \in F_\alpha^p$ has order 2 and type $\alpha/2$. Then f must have infinitely many zeros. See [22].
12. Show that the function

$$f(z) = \sum_{n=1}^\infty \frac{1}{n\sqrt{n!}} z^n$$

   belongs to F_1^2 but the function $zf(z)$ is no longer in F_1^2. See [22].
13. If Z is a zero set for F_α^p and $\dim(I_Z) < \infty$, then every function in I_Z has order 2 and type $\alpha/2$.
14. If Z is a zero set for F_α^p and $\dim(I_Z) < \infty$, then any two functions in I_Z whose zeros are exactly those in Z can only differ by a constant multiple. Thus, there is essentially just one function that vanishes exactly on Z.

Chapter 6
Toeplitz Operators

There is a rich history of Toeplitz operators, especially those on the Hardy space. In particular, Toeplitz operators on the Hardy space provide ample examples of shifts, isometries, and Fredholm operators. They also provide motivating examples in index theory and the theory of invariant subspaces.

In this chapter, we study Toeplitz operators on the Fock space F_α^2. Problems considered include boundedness, compactness, and membership in the Schatten classes. The approach here is more closely related to the theory of Toeplitz operators on the Bergman space that was developed over the past thirty years or so.

K. Zhu, *Analysis on Fock Spaces*, Graduate Texts in Mathematics 263,
DOI 10.1007/978-1-4419-8801-0_6,
© Springer Science+Business Media New York 2012

6.1 Trace Formulas

Recall that for any fixed weight parameter α, the orthogonal projection

$$P : L_\alpha^2 \to F_\alpha^2$$

is an integral operator,

$$Pf(z) = \int_{\mathbb{C}} K(z, w) f(w) \, d\lambda_\alpha(w),$$

where

$$K(z, w) = e^{\alpha z \bar{w}}$$

is the reproducing kernel of the Hilbert space F_α^2.

Given $\varphi \in L^\infty(\mathbb{C})$, we define a linear operator $T_\varphi : F_\alpha^2 \to F_\alpha^2$ by

$$T_\varphi(f) = P(\varphi f), \qquad f \in F_\alpha^2.$$

We call T_φ the Toeplitz operator on F_α^2 with symbol φ. It is clear that T_φ is bounded with $\|T_\varphi\| \le \|\varphi\|_\infty$.

Proposition 6.1. *For any complex numbers a and b, and for any bounded functions φ and ψ, we have:*

(i) $T_{a\varphi + b\psi} = aT_\varphi + bT_\psi$.
(ii) $T_{\bar{\varphi}} = T_\varphi^*$.
(iii) $T_\varphi \ge 0$ *if* $\varphi \ge 0$.

Proof. These follow easily from the definitions. We omit the routine details. □

One of the main differences between Toeplitz operators on the Fock space and those on Hardy and Bergman type spaces is the lack of bounded analytic and harmonic symbols in the Fock space setting. In fact, by the maximum modulus principle, if an analytic or harmonic function on $\mathbb{C}$ is bounded, it has to be a constant.

By the integral representation for the orthogonal projection P, we can write

$$T_\varphi(f)(z) = \int_{\mathbb{C}} K(z, w) f(w) \varphi(w) \, d\lambda_\alpha(w)$$

$$= \frac{\alpha}{\pi} \int_{\mathbb{C}} K(z, w) f(w) e^{-\alpha|w|^2} \varphi(w) \, dA(w).$$

This motivates us to define Toeplitz operators on F_α^2 with much more general symbols.

If μ is a Borel measure on $\mathbb{C}$, we define the Toeplitz operator T_μ as follows:

$$T_\mu(f)(z) = \frac{\alpha}{\pi} \int_{\mathbb{C}} K(z,w) f(w) e^{-\alpha|w|^2} \, d\mu(w), \qquad z \in \mathbb{C}.$$

Note that T_φ is very loosely defined here, because it is not clear when the integrals above will converge, even if the measure μ is finite, as the kernel function $K(z,w)$ is unbounded for any fixed $z \neq 0$.

To make things a little more precise, we say that a complex Borel measure μ on $\mathbb{C}$ satisfies *condition (M)* if

$$\int_{\mathbb{C}} |K(z,w)| e^{-\alpha|w|^2} \, d|\mu|(w) < \infty \tag{6.1}$$

for all $z \in \mathbb{C}$. Because of the exponential form of the reproducing kernel, it is clear that the above is equivalent to

$$\int_{\mathbb{C}} |K(z,w)|^2 e^{-\alpha|w|^2} \, d|\mu|(w) < \infty \tag{6.2}$$

for all $z \in \mathbb{C}$. When $d\mu(z) = \varphi(z) \, dA(z)$, the measure μ satisfies condition (M) if and only if the function φ satisfies condition (I_1). See Sect. 3.2 for the definition of condition (I_p).

If μ satisfies condition (M), then the Toeplitz operator T_μ above is well defined on a dense subset of F_α^2. In fact, if

$$f(w) = \sum_{k=1}^{N} c_k K(w, a_k)$$

is any finite linear combination of kernel functions in F_α^2, then it follows from condition (M) and the Cauchy–Schwarz inequality that $T_\mu(f)$ is well defined. Recall from Lemma 2.11 that the set of all finite linear combinations of kernel functions is dense in F_α^2.

If μ satisfies condition (M), the Berezin transform of μ (see Sect. 3.4) is well defined:

$$\widetilde{\mu}(z) = \frac{\alpha}{\pi} \int_{\mathbb{C}} |k_z(w)|^2 e^{-\alpha|w|^2} \, d\mu(w) = \frac{\alpha}{\pi} \int_{\mathbb{C}} e^{-\alpha|z-w|^2} \, d\mu(w),$$

where

$$k_z(w) = K(w,z)/\sqrt{K(z,z)} = e^{\alpha w \bar{z} - \frac{\alpha}{2}|z|^2}$$

are the normalized reproducing kernels of F_α^2.

If μ is positive or if the Toeplitz operator T_μ happens to be a bounded operator on F_α^2, then it is easy to see that

$$\widetilde{\mu}(z) = \langle T_\mu k_z, k_z \rangle_\alpha, \qquad z \in \mathbb{C}.$$

When $d\mu(z) = \varphi(z) \, dA(z)$, we get back to T_φ and $\widetilde{\varphi}$.

In the rest of this section, we focus on the case of trace-class Toeplitz operators. We will obtain several trace formulas related to Toeplitz operators. These trace formulas will then be used in the next two sections to study bounded and compact Toeplitz operators.

The *definition* of the Berezin transform $\widetilde{\varphi}$ and the Toeplitz operator T_φ requires that the function φ satisfy condition (I_1). But in the study of Toeplitz operators, we often need to require that φ satisfy condition (I_2), which is slightly stronger than condition (I_1).

If the Toeplitz operator T_φ is bounded on F_α^2, we have

$$T_\varphi f(z) = \int_{\mathbb{C}} f(w)\varphi(w)K(z,w)\,d\lambda_\alpha(w),$$

and it is easy to check that

$$K_{T_\varphi}(w,z) = \int_{\mathbb{C}} \overline{\varphi(u)}K(u,z)K(w,u)\,d\lambda_\alpha(u). \tag{6.3}$$

See Sect. 3.1 for the definition of $K_S(w,z)$ for any bounded linear operator S on F_α^2. If we further assume that φ satisfies condition (I_2), then it is also easy to check that the uniqueness of K_{T_φ} implies

$$\overline{\varphi}K(\,\cdot\,,z) - K_{T_\varphi}(\,\cdot\,,z) \perp F_\alpha^2 \tag{6.4}$$

for all $z \in \mathbb{C}$.

Theorem 6.2. *Suppose φ is Lebesgue measurable on $\mathbb{C}$ and S is a bounded linear operator on F_α^2. If*

(1) φ satisfies condition (I_2),
(2) T_φ is bounded on F_α^2,
(3) $T_\varphi S$ is trace class,
(4) $\int_{\mathbb{C}} \int_{\mathbb{C}} |\varphi(z)||K(w,z)||K_S(w,z)|\,d\lambda_\alpha(w)\,d\lambda_\alpha(z) < \infty$,

then we have

$$\operatorname{tr}(T_\varphi S) = \int_{\mathbb{C}} \varphi(z)\overline{K_S(z,z)}\,d\lambda_\alpha(z) = \frac{\alpha}{\pi}\int_{\mathbb{C}} \varphi(z)\widetilde{S}(z)\,dA(z). \tag{6.5}$$

Proof. By hypothesis (1), each function $\overline{\varphi}K(\,\cdot\,,z)$ is in L_α^2, and by (6.4), we can write

$$K_{T_\varphi}(\,\cdot\,,z) = \overline{\varphi}K(\,\cdot\,,z) - H(\,\cdot\,,z),$$

where $H(\,\cdot\,,z) \perp F_\alpha^2$. By Corollary 3.12,

$$\operatorname{tr}(T_\varphi S) = \int_{\mathbb{C}} d\lambda_\alpha(w)\int_{\mathbb{C}} \overline{K_{T_\varphi}(z,w)K_S(w,z)}\,d\lambda_\alpha(z)$$

$$= \int_{\mathbb{C}} d\lambda_\alpha(w) \overline{\int_{\mathbb{C}} [\overline{\varphi(z)}K(z,w) - H(z,w)]\overline{K_{S^*}(z,w)}\, d\lambda_\alpha(z)}$$

$$= \int_{\mathbb{C}} d\lambda_\alpha(w) \int_{\mathbb{C}} \overline{\overline{\varphi(z)}K(z,w)}\overline{K_{S^*}(z,w)}\, d\lambda_\alpha(z)$$

$$= \int_{\mathbb{C}} d\lambda_\alpha(w) \int_{\mathbb{C}} \varphi(z)K(w,z)\overline{K_S(w,z)}\, d\lambda_\alpha(z).$$

Hypothesis (4) allows the application of Fubini's theorem, which, along with the reproducing property in F_α^2 and parts (3) and (8) of Proposition 3.9, leads to the desired trace formulas.   □

Taking S to be the identity operator, we obtain the following trace formula for Toeplitz operators on the Fock space.

Corollary 6.3 *Suppose φ satisfies condition (I_2). If T_φ is in the trace class and $\varphi \in L^1(\mathbb{C}, dA)$, then*

$$\mathrm{tr}(T_\varphi) = \int_{\mathbb{C}} \varphi(z)K(z,z)\, d\lambda_\alpha(z) = \frac{\alpha}{\pi}\int_{\mathbb{C}} \varphi(z)\, dA(z). \tag{6.6}$$

Proof. When S is the identity operator, we have $K_S(z,w) = K(z,w)$, so the condition

$$\int_{\mathbb{C}}\int_{\mathbb{C}} |\varphi(z)||K(w,z)||K_S(w,z)|\, d\lambda_\alpha(w)\, d\lambda_\alpha(z) < \infty$$

becomes

$$\int_{\mathbb{C}} |\varphi(z)|K(z,z)\, d\lambda_\alpha(z) = \frac{\alpha}{\pi}\int_{\mathbb{C}} |\varphi(z)|\, dA(z) < \infty.$$

□

Note that there exist symbol functions φ such that T_φ is in the trace class but $\varphi \notin L^1(\mathbb{C}, dA)$. See Exercise 10.

Corollary 6.4 *Suppose φ is bounded and compactly supported in $\mathbb{C}$. Then for any bounded linear operator S on F_α^2, the operator $T_\varphi S$ is trace class and*

$$\mathrm{tr}(T_\varphi S) = \int_{\mathbb{C}} \varphi(z)\overline{K_S(z,z)}\, d\lambda_\alpha(z) = \frac{\alpha}{\pi}\int_{\mathbb{C}} \varphi(z)\tilde{S}(z)\, dA(z). \tag{6.7}$$

Proof. It is easy to see that hypotheses (1)–(3) of Theorem 6.2 are satisfied. To check hypothesis (4) of Theorem 6.2, we write

$$I = \int_{\mathbb{C}} |\varphi(z)|\, d\lambda_\alpha(z) \int_{\mathbb{C}} |K_S(w,z)||K(w,z)|\, d\lambda_\alpha(w).$$

From the definition $K_S(w,z) = S^* K(\,\cdot\,,z)(w)$, we deduce that

$$\int_{\mathbb{C}} |K_S(w,z)|^2 \, d\lambda_\alpha(w) = \|S^* K(\,\cdot\,,z)\|_{2,\alpha}^2.$$

It follows from this and the Cauchy–Schwarz inequality that

$$\int_{\mathbb{C}} |K_S(w,z)| |K(w,z)| \, d\lambda_\alpha(w) \le \|S^* K_z\|_{2,\alpha} \|K_z\|_{2,\alpha} \le \|S\| K(z,z).$$

Thus,

$$I \le \|S\| \int_{\mathbb{C}} |\varphi(z)| K(z,z) \, d\lambda_\alpha(z) = \frac{\alpha \|S\|}{\pi} \int_{\mathbb{C}} |\varphi| \, dA < \infty,$$

as φ is bounded and compactly supported. □

As a consequence of Corollary 6.4, we show that every trace-class operator on F_α^2 can be approximated by trace-class Toeplitz operators in the trace norm, and every compact operator on F_α^2 can be approximated by compact Toeplitz operators in norm.

Theorem 6.5. *Let $\mathcal{C}$ denote the set of all Toeplitz operators T_φ, where φ is continuous and has compact support in $\mathbb{C}$. Then:*

(1) $\mathcal{C}$ is trace-norm dense in the trace class $\mathcal{T}$ of F_α^2.
(2) $\mathcal{C}$ is norm dense in the space $\mathcal{K}$ of all compact operators on F_α^2.

Proof. Let $\mathcal{L}$ denote the space of all bounded linear operators on F_α^2. Then, it is well known that $\mathcal{T}^* = \mathcal{L}$ and $\mathcal{K}^* = \mathcal{T}$, with the duality pairing given by $\langle S, T \rangle = \mathrm{tr}\,(ST)$.

To prove (2), assume that $\mathcal{C}$ is not norm dense in $\mathcal{K}$. By the Hahn–Banach theorem, there must be a nonzero operator S in $\mathcal{T}$ such that

$$\langle T_\varphi, S \rangle = 0, \qquad T_\varphi \in \mathcal{C}.$$

By Corollary 6.4,

$$0 = \langle T_\varphi, S \rangle = \mathrm{tr}\,(T_\varphi S) = \frac{\alpha}{\pi} \int_{\mathbb{C}} \varphi(z) \widetilde{S}(z) \, dA(z)$$

for all continuous functions φ with compact support in $\mathbb{C}$. This implies that $\widetilde{S} = 0$. So $S = 0$, a contradiction which proves (2).

The proof for (1) is similar, and we omit the details here. □

6.2 The Bargmann Transform

The connection between Toeplitz operators on the Fock space and pseudodifferential operators on $L^2(\mathbb{R}, dx)$ is established by the Bargmann transform, and the most elementary way to understand the Bargmann transform is via the classical Hermite polynomials.

Recall that for any nonnegative integer n, the nth Hermite polynomial $H_n(x)$ is defined by

$$H_n(x) = (-1)^n e^{x^2} \frac{d^n}{dx^n} e^{-x^2}.$$

The first five Hermite polynomials are given by

$$H_0(x) = 1,$$
$$H_1(x) = 2x,$$
$$H_2(x) = 4x^2 - 2,$$
$$H_3(x) = 8x^3 - 12x,$$
$$H_4(x) = 16x^4 - 48x^2 + 12.$$

In general, it is easy to check that each H_n has degree n and

$$H_n(x) = 2xH_{n-1}(x) - H'_{n-1}(x), \qquad n \geq 1,$$

which can be used to compute H_n inductively. In particular, the leading term of $H_n(x)$ is $(2x)^n$.

Lemma 6.6. *For nonnegative integers m and n, let*

$$I_{mn} = \int_{\mathbb{R}} H_n(x) H_m(x) e^{-x^2} \, dx.$$

Then $I_{mn} = 0$ for $m \neq n$ and $I_{nn} = 2^n n! \sqrt{\pi}$.

Proof. For any polynomial f, we use integration by parts n times to get

$$\int_{\mathbb{R}} H_n(x) f(x) e^{-x^2} \, dx = (-1)^n \int_{\mathbb{R}} f(x) \frac{d^n}{dx^n} e^{-x^2} \, dx$$
$$= \int_{\mathbb{R}} f^{(n)}(x) e^{-x^2} \, dx.$$

If $m < n$ and $f = H_m$, then $f^{(n)} \equiv 0$ and so $I_{mn} = 0$.

If $f = H_n$, then $f(x) = (2x)^n + \cdots$, and so $f^{(n)} \equiv 2^n n!$. It follows that

$$I_{nn} = 2^n n! \int_{\mathbb{R}} e^{-x^2} \, dx = 2^n n! \sqrt{\pi}.$$

This proves the desired result. $\square$

Theorem 6.7. *For any nonnegative integer n, let*

$$h_n(x) = \left(\frac{2\alpha}{\pi}\right)^{\frac{1}{4}} \frac{1}{\sqrt{2^n n!}} e^{-\alpha x^2} H_n(\sqrt{2\alpha} x).$$

Then $\{h_n\}$ is an orthonormal basis of $L^2(\mathbb{R}, \mathrm{d}x)$.

Proof. It follows from a change of variables and Lemma 6.6 that $\{h_n\}$ is an orthonormal set. In particular, for any positive integer N, the functions

$$h_0(x), h_1(x), \cdots, h_N(x),$$

are linearly independent. It follows that the polynomials

$$H_0(\sqrt{2\alpha} x), H_1(\sqrt{2\alpha} x), \cdots, H_N(\sqrt{2\alpha} x) \tag{6.8}$$

are linearly independent in the vector space of all polynomials of degree less than or equal to N. A dimensionality argument then shows every polynomial of degree less than or equal to N can be written as a linear combination of the polynomials in (6.8). Therefore, the condition

$$\int_{\mathbb{R}} f(x) h_n(x) \, \mathrm{d}x = 0, \qquad n \geq 0,$$

implies that

$$\int_{\mathbb{R}} f(x) P(x) e^{-\alpha x^2} \, \mathrm{d}x = 0$$

for all polynomials P, which, according to Lemma 3.16, implies that $f = 0$ almost everywhere. Thus, the set $\{h_n\}$ is complete in $L^2(\mathbb{R}, \mathrm{d}x)$.   □

We now define the Bargmann transform. Let f be a function on $\mathbb{R}$ satisfying the condition that $f(x) e^{|tx| - \pi x^2}$ is integrable with respect to $\mathrm{d}x$ for any real t. Then for any positive parameter α, we can define an analytic function $\mathcal{B}_\alpha f$ by

$$\mathcal{B}_\alpha f(z) = \left(\frac{2\alpha}{\pi}\right)^{\frac{1}{4}} \int_{\mathbb{R}} f(x) e^{2\alpha x z - \alpha x^2 - \frac{\alpha}{2} z^2} \, \mathrm{d}x, \qquad z \in \mathbb{C}. \tag{6.9}$$

This will be called the (parametrized) Bargmann transform of f.

Theorem 6.8. *For any positive α, the Bargmann transform is an isometry from $L^2(\mathbb{R}, \mathrm{d}x)$ onto F_α^2.*

Proof. It suffices for us to show that for any nonnegative integer n, we have $\mathcal{B}_\alpha h_n = e_n$, where

$$e_n(z) = \sqrt{\frac{\alpha^n}{n!}} z^n.$$

To this end, first observe that if $u = x - z$, where x is fixed, then $d/du = -d/dz$. It follows that

$$\frac{d^n}{dz^n}e^{-(x-z)^2}\bigg|_{z=0} = (-1)^n\frac{d^n}{du^n}e^{-u^2}\bigg|_{u=x} = e^{-u^2}H_n(u)\bigg|_{u=x} = e^{-x^2}H_n(x).$$

Therefore, by Taylor's formula,

$$e^{-(x-z)^2} = \sum_{n=0}^{\infty}e^{-x^2}H_n(x)\frac{z^n}{n!}.$$

Replace x by $\sqrt{2\alpha}\,x$ and replace z by $\sqrt{\alpha/2}\,z$. Then

$$\left(\frac{2\alpha}{\pi}\right)^{\frac{1}{4}}e^{2\alpha xz - \alpha x^2 - \frac{\alpha}{2}z^2} = \sum_{k=0}^{\infty}e_k(z)h_k(x).$$

Multiply both sides by $h_n(x)$ and integrate over the real line. The desired result $\mathcal{B}_\alpha h_n = e_n$ then follows from the fact that $\{h_k\}$ is orthonormal in $L^2(\mathbb{R}, dx)$.   □

Proposition 6.9. *The inverse of the Bargmann transform is given by*

$$[\mathcal{B}_\alpha^{-1}f](x) = \left(\frac{2\alpha}{\pi}\right)^{\frac{1}{4}}\int_{\mathbb{C}}f(z)e^{2\alpha x\bar{z} - \alpha x^2 - \frac{\alpha}{2}\bar{z}^2}\,d\lambda_\alpha(z), \qquad (6.10)$$

where $f \in F_\alpha^2$.

Proof. Fix any polynomial $f \in F_\alpha^2$ and any function $g \in L^2(\mathbb{R}, dx)$ that is compactly supported. Since

$$\mathcal{B}_\alpha : L^2(\mathbb{R}, dx) \to F_\alpha^2 \subset L_\alpha^2$$

is an isometry, we have

$$\langle \mathcal{B}_\alpha^{-1}f, g\rangle_{L^2(\mathbb{R})} = \langle \mathcal{B}_\alpha\mathcal{B}_\alpha^{-1}f, \mathcal{B}_\alpha g\rangle_\alpha = \langle f, \mathcal{B}_\alpha g\rangle_\alpha$$

$$= \left(\frac{2\alpha}{\pi}\right)^{\frac{1}{4}}\int_{\mathbb{C}}f(z)\,d\lambda_\alpha(z)\int_{\mathbb{R}}\overline{g(x)}e^{2\alpha x\bar{z} - \alpha x^2 - \frac{\alpha}{2}\bar{z}^2}\,dx$$

$$= \left(\frac{2\alpha}{\pi}\right)^{\frac{1}{4}}\int_{\mathbb{R}}\overline{g(x)}\,dx\int_{\mathbb{C}}f(z)e^{2\alpha x\bar{z} - \alpha x^2 - \frac{\alpha}{2}\bar{z}^2}\,d\lambda_\alpha(z)$$

$$= \langle F, g\rangle_{L^2(\mathbb{R})},$$

where

$$F(x) = \left(\frac{2\alpha}{\pi}\right)^{\frac{1}{4}}\int_{\mathbb{C}}f(z)e^{2\alpha x\bar{z} - \alpha x^2 - \frac{\alpha}{2}\bar{z}^2}\,d\lambda_\alpha(z).$$

This proves the desired formula for $\mathcal{B}_\alpha^{-1} f$, as the polynomials are dense in F_α^2 and the compactly supported functions in $L^2(\mathbb{R}, dx)$ are dense there.   □

Proposition 6.10. *Let $a = r + is \in \mathbb{C}$ and k_a be the normalized reproducing kernel of F_α^2 at point a. Then*

$$[\mathcal{B}_\alpha^{-1} k_a](x) = \left(\frac{2\alpha}{\pi}\right)^{\frac{1}{4}} e^{-2\alpha i(rD+sX)} e^{-\alpha x^2}, \tag{6.11}$$

where $e^{-2\alpha i(rD+sX)}$ is the pseudodifferential operator defined in (1.21).

Proof. Let $c = (2\alpha/\pi)^{1/4}$. By Proposition 6.9 and the reproducing property in F_α^2,

$$[\mathcal{B}_\alpha^{-1} k_a](x) = c \int_{\mathbb{C}} e^{2\alpha x \bar{z} - \alpha x^2 - \frac{\alpha}{2} \bar{z}^2} k_a(z) \, d\lambda_\alpha(z)$$

$$= c e^{-\frac{\alpha}{2}|a|^2 - \alpha x^2} \overline{\int_{\mathbb{C}} e^{2\alpha x z - \frac{\alpha}{2} z^2} e^{\alpha a \bar{z}} \, d\lambda_\alpha(z)}$$

$$= c e^{-\frac{\alpha}{2}|a|^2 - \alpha x^2 + 2\alpha x \bar{a} - \frac{\alpha}{2} \bar{a}^2}$$

$$= c e^{-\frac{\alpha}{2}(r^2+s^2) - \alpha x^2 + 2\alpha x(r-is) - \frac{\alpha}{2}(r-is)^2}$$

$$= c e^{-2\alpha i x s + \alpha i r s - \alpha x^2 + 2\alpha x r - \alpha r^2}.$$

On the other hand, by (1.21),

$$e^{2\alpha i(-rD-sX)} e^{-\alpha x^2} = e^{-2\alpha i s x + \alpha i r s - \alpha(x-r)^2} = e^{-2\alpha i s x + \alpha i r s - \alpha x^2 + 2\alpha x r - \alpha r^2}.$$

This proves the desired result.   □

Lemma 6.11. *We have*

$$\int_{\mathbb{R}} e^{-2\pi i z x - \pi x^2} \, dx = e^{-\pi z^2} \tag{6.12}$$

for all complex numbers z.

Proof. Recall that

$$h_0(x) = \left(\frac{2\alpha}{\pi}\right)^{\frac{1}{4}} e^{-\alpha x^2}$$

is the first vector in the orthonormal basis $\{h_n\}$ of $L^2(\mathbb{R}, dx)$. By Theorem 6.8 and its proof, $\mathcal{B}_\alpha(h_0) = e_0 = 1$, or equivalently,

$$\sqrt{\frac{2\alpha}{\pi}} \int_{\mathbb{R}} e^{2\alpha x z - 2\alpha x^2} \, dx = e^{\frac{\alpha}{2} z^2}.$$

Replacing z by $-i\sqrt{2\pi/\alpha}\, z$ and changing x to $\sqrt{\pi/(2\alpha)}\, x$, we obtain the desired identity in (6.12).   □

The above lemma simply states that the Fourier transform of $e^{-\pi x^2}$ is $e^{-\pi z^2}$. In what follows, the equivalent form (obtained by suitable changes of variables)

$$\int_{\mathbb{R}} e^{-i\alpha xz - \frac{\alpha}{2}x^2}\, dx = \sqrt{\frac{2\pi}{\alpha}}\, e^{-\frac{\alpha}{2}z^2} \tag{6.13}$$

will be more convenient for us to use.

Theorem 6.12. *Suppose $\sigma(w) = \sigma(u,v)$, with $w = v + iu$, is a symbol function and $\sigma(D,X)$ is the Weyl pseudodifferential operator on $L^2(\mathbb{R}, dx)$ with symbol σ. Let $T = \mathcal{B}_\alpha \sigma(D,X)\mathcal{B}_\alpha^{-1}$ on F_α^2. Then $\widetilde{T}(z) = B_{2\alpha}\sigma(\overline{z})$ for all $z \in \mathbb{C}$.*

Proof. Recall that

$$\sigma(D,X) = \int_{\mathbb{R}}\int_{\mathbb{R}} \widehat{\sigma}(p,q) e^{2\pi i(pD+qX)}\, dp\, dq$$

$$= \left(\frac{\alpha}{\pi}\right)^2 \int_{\mathbb{R}}\int_{\mathbb{R}} \widehat{\sigma}\left(\frac{\alpha}{\pi}p, \frac{\alpha}{\pi}q\right) e^{2\alpha i(pD+qX)}\, dp\, dq.$$

It follows from this and Fubini's theorem that

$$\widetilde{T}(z) = \langle \mathcal{B}_\alpha \sigma(D,X)\mathcal{B}_\alpha^{-1}k_z, k_z\rangle_\alpha$$

$$= \langle \sigma(D,X)\mathcal{B}_\alpha^{-1}k_z, \mathcal{B}_\alpha^{-1}k_z\rangle_{L^2(\mathbb{R})}$$

$$= \left(\frac{\alpha}{\pi}\right)^2 \int_{\mathbb{R}}\int_{\mathbb{R}} \widehat{\sigma}\left(\frac{\alpha}{\pi}p, \frac{\alpha}{\pi}q\right) \langle e^{2\alpha i(pD+qX)}\mathcal{B}_\alpha^{-1}k_z, \mathcal{B}_\alpha^{-1}k_z\rangle_{L^2(\mathbb{R})}\, dp\, dq.$$

To simplify notation, let us write

$$\rho(p,q) = e^{2\alpha i(pD+qX)}$$

for real p and q, and proceed to compute the integral

$$I = \langle e^{2\alpha i(pD+qX)}\mathcal{B}_\alpha^{-1}k_z, \mathcal{B}_\alpha^{-1}k_z\rangle_{L^2(\mathbb{R})}.$$

Let $z = r + is$. By Proposition 6.10, Lemma 1.28, and the fact that each $\rho(-r,-s)$ is a unitary operator on $L^2(\mathbb{R}, dx)$, we have

$$I = c^2\langle \rho(p,q)\rho(-r,-s)e^{-\alpha x^2}, \rho(-r,-s)e^{-\alpha x^2}\rangle_{L^2(\mathbb{R})}$$

$$= c^2 e^{2\alpha i(-ps+qr)}\langle \rho(-r,-s)\rho(p,q)e^{-\alpha x^2}, \rho(-r,-s)e^{-\alpha x^2}\rangle_{L^2(\mathbb{R})}$$

$$= c^2 e^{2\alpha i(-ps+qr)}\langle \rho(p,q)e^{-\alpha x^2}, e^{-\alpha x^2}\rangle_{L^2(\mathbb{R})},$$

where $c^2 = \sqrt{2\alpha/\pi}$.

By (1.21) and the change of variables $x \mapsto x - (p/2)$,

$$I = c^2 e^{2\alpha i(-ps+qr)} \int_{\mathbb{R}} e^{2\alpha iqx + \alpha ipq - \alpha(x+p)^2 - \alpha x^2} \, dx$$

$$= c^2 e^{2\alpha i(-ps+qr)} \int_{\mathbb{R}} e^{2\alpha iqx - \alpha(x+\frac{p}{2})^2 - \alpha(x-\frac{p}{2})^2} \, dx$$

$$= c^2 e^{2\alpha i(-ps+qr) - \frac{\alpha}{2}p^2} \int_{\mathbb{R}} e^{2\alpha iqx - 2\alpha x^2} \, dx.$$

It follows from another change of variables and Lemma 6.11 that

$$\int_{\mathbb{R}} e^{2\alpha iqx - 2\alpha x^2} \, dx = \sqrt{\frac{\pi}{2\alpha}} \int_{\mathbb{R}} e^{-2\pi i \sqrt{\frac{\alpha}{2\pi}} qx - \pi x^2} \, dx$$

$$= \sqrt{\frac{\pi}{2\alpha}} e^{-\pi \cdot \frac{\alpha}{2\pi} q^2} = \sqrt{\frac{\pi}{2\alpha}} e^{-\frac{\alpha}{2} q^2}.$$

Therefore,

$$\widetilde{T}(z) = \left(\frac{\alpha}{\pi}\right)^2 \int_{\mathbb{R}} \int_{\mathbb{R}} \widehat{\sigma}\left(\frac{\alpha}{\pi}p, \frac{\alpha}{\pi}q\right) e^{2\alpha i(-ps+qr) - \frac{\alpha}{2}(p^2+q^2)} \, dp \, dq$$

$$= \int_{\mathbb{R}} \int_{\mathbb{R}} \widehat{\sigma}(p,q) e^{2\pi i(-ps+qr) - \frac{\pi^2}{2\alpha}(p^2+q^2)} \, dp \, dq.$$

We rewrite $\widetilde{T}(z)$ as

$$\int_{\mathbb{R}} \int_{\mathbb{R}} e^{2\pi i(-ps+qr) - \frac{\pi^2}{2\alpha}(p^2+q^2)} \, dp \, dq \int_{\mathbb{R}} \int_{\mathbb{R}} \sigma(u,v) e^{-2\pi i(pu+qv)} \, du \, dv.$$

Interchanging the order of integration above, we see that $\widetilde{T}(z)$ is equal to

$$\int_{\mathbb{R}} \int_{\mathbb{R}} \sigma(u,v) \, du \, dv \int_{\mathbb{R}} \int_{\mathbb{R}} e^{[-2\pi ip(s+u) - \frac{\pi^2}{2\alpha}p^2] + [-2\pi iq(-r+v) - \frac{\pi^2}{2\alpha}q^2]} \, dp \, dq.$$

Evaluate the inner integrals using Lemma 6.11 again. We obtain

$$\widetilde{T}(z) = \frac{2\alpha}{\pi} \int_{\mathbb{R}} \int_{\mathbb{R}} \sigma(u,v) e^{-2\alpha[(s+u)^2 + (-r+v)^2]} \, du \, dv.$$

Since $z = r + is$ and $w = v + iu$, we can rewrite the above formula as

$$\widetilde{T}(z) = \frac{2\alpha}{\pi} \int_{\mathbb{C}} \sigma(w) e^{-2\alpha|w-\bar{z}|^2} \, dA(w) = B_{2\alpha} \sigma(\bar{z}).$$

This completes the proof of the theorem.   □

The rest of this section is devoted to showing that every Toeplitz operator on F_α^2 is unitarily equivalent to an anti-Wick pseudodifferential operator on $L^2(\mathbb{R}, dx)$.

We begin with the unbounded operator A of differentiation on F_α^2 together with its adjoint. Thus,

$$Af(z) = \frac{1}{\alpha}f'(z), \qquad A^*f(z) = zf(z). \tag{6.14}$$

We show that, via the Bargmann transform $\mathcal{B}_\alpha$, these operators are unitarily equivalent to certain familiar operators on $L^2(\mathbb{R}, dx)$.

Lemma 6.13. *For any positive α, we have*

$$\mathcal{B}_\alpha^{-1}A\mathcal{B}_\alpha = X + iD = Z, \quad \mathcal{B}_\alpha^{-1}A^*\mathcal{B}_\alpha = X - iD = Z^*, \tag{6.15}$$

where X, D, and Z are the (unbounded) operators on $L^2(\mathbb{R}, dx)$ defined in Sect. 1.4.

Proof. Let $C_c(\mathbb{R})$ denote the space of continuous functions on $\mathbb{R}$ having compact support. Then $C_c(\mathbb{R})$ is dense in $L^2(\mathbb{R}, dx)$. Given $f \in C_c(\mathbb{R})$, we differentiate

$$\mathcal{B}_\alpha f(z) = \left(\frac{2\alpha}{\pi}\right)^{\frac{1}{4}} \int_\mathbb{R} e^{2\alpha xz - \alpha x^2 - \frac{\alpha}{2}z^2} f(x)\,dx$$

under the integral sign to obtain

$$A\mathcal{B}_\alpha f(z) = \left(\frac{2\alpha}{\pi}\right)^{\frac{1}{4}} \int_\mathbb{R} (2x - z)e^{2\alpha xz - \alpha x^2 - \frac{\alpha}{2}z^2} f(x)\,dx. \tag{6.16}$$

This gives

$$A\mathcal{B}_\alpha f = 2\mathcal{B}_\alpha Xf - A^*\mathcal{B}_\alpha f,$$

and hence

$$\mathcal{B}_\alpha^{-1}A\mathcal{B}_\alpha + \mathcal{B}_\alpha^{-1}A^*\mathcal{B}_\alpha = 2X. \tag{6.17}$$

On the other hand, we can rewrite (6.16) as

$$A\mathcal{B}_\alpha f(z) = -\frac{1}{\alpha}\left(\frac{2\alpha}{\pi}\right)^{\frac{1}{4}} \int_\mathbb{R} f(x)\frac{d}{dx}e^{2\alpha xz - \alpha x^2 - \frac{\alpha}{2}z^2}\,dx + A^*\mathcal{B}_\alpha f(z).$$

Apply integration by parts to the integral above. We obtain

$$A\mathcal{B}_\alpha f = 2i\mathcal{B}_\alpha Df + A^*\mathcal{B}_\alpha f,$$

and hence

$$\mathcal{B}_\alpha^{-1}A\mathcal{B}_\alpha - \mathcal{B}_\alpha^{-1}A^*\mathcal{B}_\alpha = 2iD. \tag{6.18}$$

Solving for $\mathcal{B}_\alpha^{-1}A\mathcal{B}_\alpha$ and $\mathcal{B}_\alpha^{-1}A^*\mathcal{B}_\alpha$ from (6.17) and (6.18), we obtain the desired results. $\qquad\square$

We now establish the relationship between anti-Wick pseudodifferential operators on $L^2(\mathbb{R}, dx)$ and Toeplitz operators on F_α^2.

Theorem 6.14. *Let*

$$\sigma(z) = \sigma(z, \bar{z}) = \sum c_{nm} z^n \bar{z}^m$$

be real analytic and

$$\sigma(Z, Z^*) = \sum c_{nm} Z^n Z^{*m}$$

be the anti-Wick pseudodifferential operator on $L^2(\mathbb{R}, dx)$. We have

$$\mathcal{B}_\alpha \sigma(Z, Z^*) \mathcal{B}_\alpha^{-1} = T_\varphi, \tag{6.19}$$

where T_φ is the Toeplitz operator on F_α^2 with symbol $\varphi(z) = \sigma(\bar{z}, z) = \sigma(\bar{z})$.

Proof. By Lemma 6.13, we have

$$\mathcal{B}_\alpha \sigma(Z, Z^*) \mathcal{B}_\alpha^{-1} = \sum c_{nm} A^n A^{*m}.$$

Thus, for $f \in F_\alpha^2$, we have

$$\mathcal{B}_\alpha \sigma(Z, Z^*) \mathcal{B}_\alpha^{-1} f(z) = \sum c_{nm} \left(\frac{1}{\alpha}\right)^n \frac{\partial^n}{\partial z^n} (z^m f(z)).$$

If f has the property that the function $z^m f(z)$ is also in F_α^2 (all polynomials, which are dense in F_α^2, clearly have this property), then we can write

$$z^m f(z) = \int_{\mathbb{C}} w^m f(w) e^{\alpha z \bar{w}} \, d\lambda_\alpha(w).$$

Differentiating under the integral sign n times, we obtain

$$\frac{\partial^n}{\partial z^n} (z^m f(z)) = \alpha^n \int_{\mathbb{C}} \bar{w}^n w^m f(w) e^{\alpha z \bar{w}} \, d\lambda_\alpha(w).$$

Therefore,

$$\mathcal{B}_\alpha \sigma(Z, Z^*) \mathcal{B}_\alpha^{-1} f(z) = \int_{\mathbb{C}} \left[\sum c_{nm} \bar{w}^n w^m\right] f(w) e^{\alpha z \bar{w}} \, d\lambda_\alpha(w)$$

$$= \int_{\mathbb{C}} \varphi(w) f(w) e^{\alpha z \bar{w}} \, d\lambda_\alpha(w)$$

$$= T_\varphi f(z).$$

This proves the desired relation.   □

6.3 Boundedness

In this section, we obtain necessary and sufficient conditions for the Toeplitz operator T_φ to be bounded on F_α^2. These conditions are based on the Berezin transform or the heat transform at particular time points.

The main results of the section can be summarized as follows:

(a) If T_φ is bounded on F_α^2, then $B_\beta \varphi$ is bounded for all $\beta \in (0, 2\alpha)$.
(b) If $B_\beta \varphi$ is bounded for some $\beta > 2\alpha$, then T_φ is bounded on F_α^2.
(c) If $\varphi \geq 0$, then T_φ is bounded on F_α^2 if and only if $B_\alpha \varphi$ is bounded if and only if $\widehat{\varphi}_r$ is bounded, where r is any fixed radius. Here,

$$\widehat{\varphi}_r(z) = \frac{1}{\pi r^2} \int_{B(z,r)} \varphi(w) \, dA(w)$$

is the averaging function of φ with respect to area measure.
(d) If $\varphi \in \mathrm{BMO}^1$, then T_φ is bounded on F_α^2 if and only if $T_{|\varphi|}$ is bounded on F_α^2 if and only if $B_\alpha \varphi$ is bounded.

The proof of (c) uses the characterizations of Fock–Carleson measures and is almost straightforward. The result in (d) follows from (c) and the translation invariant characterization of BMO^1.

The proof of (a) depends on some general trace estimates and the semigroup property of the weighted Berezin transforms. The proof of (b) requires certain estimates from the theory of pseudodifferential operators.

We now get down to the details.

Recall that the standard orthonormal basis for F_α^2 is given by

$$e_n(z) = \sqrt{\frac{\alpha^n}{n!}} \, z^n, \qquad n = 0, 1, 2, 3, \cdots.$$

For any nonnegative integer n, let P_n denote the rank-one projection from F_α^2 onto the one-dimensional subspace generated by e_n. Thus,

$$P_n f = \langle f, e_n \rangle e_n, \qquad n \geq 0, f \in F_\alpha^2.$$

It follows from (3.3), the definition of $K_S(w, z)$, that

$$K_{P_n}(z, w) = e_n(z) \overline{e_n(w)}, \qquad n \geq 0.$$

For any parameter $t \in (-1, 1)$, we consider the operator

$$T^{(t)} = (1 - t) \sum_{n=0}^{\infty} t^n P_n, \tag{6.20}$$

with the usual convention that $T^{(0)} = P_0$. It is clear that the series above converges in the norm topology of F_α^2. By property (7) of Proposition 3.9, we have

$$K_{T^{(t)}}(z,w) = (1-t) \sum_{n=0}^{\infty} t^n K_{P_n}(z,w)$$

$$= (1-t) \sum_{n=0}^{\infty} t^n e_n(z)\overline{e_n(w)}$$

$$= (1-t)e^{\alpha t z \overline{w}},$$

and the series converges uniformly on compact subsets of $\mathbb{C} \times \mathbb{C}$.

Let $\| \ \|_{S_1}$ denote the norm in the trace class S_1. Since each P_n is a positive trace-class operator with $\|P_n\|_{S_1} = \mathrm{tr}\,(P_n) = 1$, the series in (6.20) also converges in S_1 with

$$\|T^{(t)}\|_{S_1} \le (1-t) \sum_{n=0}^{\infty} |t|^n \|P_n\|_{S_1} = \frac{1-t}{1-|t|}, \qquad (6.21)$$

and

$$\mathrm{tr}\,(T^{(t)}) = (1-t) \sum_{n=0}^{\infty} t^n \mathrm{tr}\,(P_n) = 1. \qquad (6.22)$$

Recall that for each $a \in \mathbb{C}$, we have the Weyl unitary operator W_a on F_α^2, a weighted translation operator, defined by

$$W_a f(z) = f(z-a)k_a(z) = e^{\alpha z \overline{a} - \frac{\alpha}{2}|a|^2} f(z-a).$$

We always have

$$W_a^* = W_{-a}, \qquad W_a T_\varphi W_a^* = T_{\varphi \circ \tau_a}, \qquad W_a^* T_\varphi W_a = T_{\varphi \circ t_a},$$

where $\tau_a(z) = z - a$ and $t_a(z) = z + a$. The translation invariance of the parametrized Berezin transform also gives

$$B_\beta(\varphi \circ \tau_a) = (B_\beta \varphi) \circ \tau_a, \qquad B_\beta(\varphi \circ t_a) = (B_\beta \varphi) \circ t_a.$$

Now for every $t \in (-1,1)$ and every $a \in \mathbb{C}$, we consider the operator

$$T_a^{(t)} = W_a T^{(t)} W_a^*.$$

Thus, $T_0^{(t)} = T^{(t)}$, each $T_a^{(t)}$ is still in the trace class, and it follows from the well-known trace identity $\mathrm{tr}\,(AB) = \mathrm{tr}\,(BA)$ that

$$\mathrm{tr}\,(T_a^{(t)}) = \mathrm{tr}\,[T^{(t)} W_a^* W_a] = \mathrm{tr}\,(T^{(t)}) = 1$$

for all $t \in (-1,1)$ and $a \in \mathbb{C}$.

Theorem 6.15. *Suppose* φ *satisfies condition* (I_2) *and* T_φ *is bounded on* F_α^2. *Then*

$$\operatorname{tr}(T_\varphi T_a^{(t)}) = B_\beta \varphi(a) \tag{6.23}$$

for all $-1 < t < \sqrt{2} - 1$, *where* $\beta = \alpha(1-t)$.

Proof. We first prove the result for $a = 0$. The problem is reduced to checking hypothesis (4) of Theorem 6.2. In fact, it would then follow from (6.5) that

$$
\begin{aligned}
\operatorname{tr}\left(T_\varphi T^{(t)}\right) &= \int_{\mathbb{C}} \varphi(z)\overline{K_{T^{(t)}}(z,z)}\,d\lambda_\alpha(z) \\
&= (1-t)\int_{\mathbb{C}} \varphi(z)e^{\alpha t|z|^2}\,d\lambda_\alpha(z) \\
&= \frac{\alpha(1-t)}{\pi}\int_{\mathbb{C}} \varphi(z)e^{-\alpha(1-t)|z|^2}\,dA(z) \\
&= B_\beta\varphi(0).
\end{aligned}
$$

Thus, we need to estimate the integral

$$
\begin{aligned}
I(t) &= \int_{\mathbb{C}} |\varphi(z)|\,d\lambda_\alpha(z)\int_{\mathbb{C}} |K(z,w)||K_{T^{(t)}}(z,w)|\,d\lambda_\alpha(w) \\
&= (1-t)\int_{\mathbb{C}} |\varphi(z)|\,d\lambda_\alpha(z)\int_{\mathbb{C}} |e^{\alpha(1+t)z\overline{w}}|\,d\lambda_\alpha(w) \\
&= (1-t)\int_{\mathbb{C}} |\varphi(z)|e^{\frac{\alpha(1+t)^2}{4}|z|^2}\,d\lambda_\alpha(z) \\
&= \frac{\alpha(1-t)}{\pi}\int_{\mathbb{C}} |\varphi(z)|e^{\alpha[\frac{(1+t)^2}{4}-1]|z|^2}\,dA(z) \\
&= \frac{\alpha(1-t)}{\pi}\int_{\mathbb{C}} |\varphi(z)|e^{-\frac{\alpha}{2}|z|^2}e^{-\delta(t)|z|^2}\,dA(z),
\end{aligned}
$$

where

$$\delta(t) = \alpha\left[1 - \frac{(1+t)^2}{4}\right] - \frac{\alpha}{2} = \frac{\alpha}{4}(1 - 2t - t^2).$$

By the Cauchy–Schwarz inequality, $I < \infty$ whenever $\delta(t) > 0$. It is elementary that for $t \in (-1,1)$, we have $\delta(t) > 0$ if and only if $-1 < t < \sqrt{2} - 1$. This proves the desired result for $a = 0$.

In general, note that T_φ is bounded if and only if $T_{\varphi \circ t_a}$ is bounded. Thus,

$$
\begin{aligned}
\operatorname{tr}(T_\varphi T_a^{(t)}) &= \operatorname{tr}(T_\varphi W_a T^{(t)} W_a^*) = \operatorname{tr}(W_a^* T_\varphi W_a T^{(t)}) \\
&= \operatorname{tr}(T_{\varphi \circ t_a} T^{(t)}) = B_\beta(\varphi \circ t_a)(0) \\
&= B_\beta \varphi(a),
\end{aligned}
$$

completing the proof of the theorem. $\qquad\square$

As a consequence of the theorem above, we obtain the following necessary condition for a Toeplitz operator to be bounded on F_α^2, one of the main results of this section.

Theorem 6.16. *Suppose φ satisfies condition (I_2) and T_φ is bounded on F_α^2. Then $B_\beta \varphi$ is bounded for all β with $0 < \beta < 2\alpha$.*

Proof. Let $\beta = \alpha(1-t)$ with $-1 < t < 1$. The condition $-1 < t < \sqrt{2} - 1$ is equivalent to $\alpha(2 - \sqrt{2}) < \beta < 2\alpha$. Also, according to the trace-norm estimate in (6.21), we have

$$\|T_a^{(t)}\|_{S_1} = \|W_a T^{(t)} W_a^*\|_{S_1} = \|T^{(t)}\|_{S_1} \leq \frac{1-t}{1-|t|}.$$

Combining this with Theorem 6.15, we obtain

$$|B_\beta \varphi(a)| = |\mathrm{tr}\,(T_\varphi T_a^{(t)})| \leq \|T_\varphi\| \|T_a^{(t)}\|_{S_1} \leq \frac{1-t}{1-|t|} \|T_\varphi\|$$

for all $a \in \mathbb{C}$. This shows that

$$\|B_\beta \varphi\|_\infty \leq \frac{1-t}{1-|t|} \|T_\varphi\| < \infty \tag{6.24}$$

whenever $\alpha(2 - \sqrt{2}) < \beta < 2\alpha$.

If $0 < \beta \leq \alpha(2 - \sqrt{2}) < \alpha$, we can find a positive number γ such that

$$\frac{1}{\beta} = \frac{1}{\gamma} + \frac{1}{\alpha}.$$

By Theorem 3.13, the semigroup property of the heat transform H_t, we have $H_{1/\beta} = H_{1/\gamma} H_{1/\alpha}$. In terms of the parametrized Berezin transforms, we have $B_\beta = B_\gamma B_\alpha$. By what was proved in the previous paragraph, or directly from $B_\alpha \varphi(a) = \langle T_\varphi k_a, k_a \rangle_\alpha$, the boundedness of T_φ on F_α^2 implies $\|B_\alpha \varphi\|_\infty \leq \|T_\varphi\|$. Since B_γ is a contraction on L^∞, we have

$$\|B_\beta \varphi\|_\infty = \|B_\gamma B_\alpha \varphi\|_\infty \leq \|B_\alpha \varphi\|_\infty \leq \|T_\varphi\|.$$

This completes the proof of the theorem. $\square$

Our next goal is to show that if $B_\beta \varphi$ is bounded for some $\beta > 2\alpha$, then T_φ is bounded on F_α^2. This is accomplished with the help of the theory of pseudodifferential operators.

Theorem 6.17. *Suppose g satisfies condition (I_2) and $\sigma(D,X)$ is the pseudodifferential operator on $L^2(\mathbb{R}, dx)$ with symbol*

$$\sigma(\zeta, x) = \sigma(z) = B_{2\alpha} g(\bar{z}), \qquad z = x + i\zeta.$$

Then $T_g = \mathcal{B}_\alpha \sigma(D,X) \mathcal{B}_\alpha^{-1}$ and $B_{2\alpha} \sigma(\bar{z}) = B_\alpha g(z)$.

Proof. Let $T = \mathcal{B}_\alpha \sigma(D,X) \mathcal{B}_\alpha^{-1}$. By Theorem 6.12, we have

$$\widetilde{T}(z) = B_{2\alpha}\sigma(\bar{z}) = B_{2\alpha}B_{2\alpha}g(z).$$

By the semigroup property (Corollary 3.15), we have

$$B_{2\alpha}B_{2\alpha}g = B_\alpha g = \widetilde{T}_g.$$

It follows that the operators T and T_g have the same Berezin symbol. Since the mapping $S \mapsto \widetilde{S}$ is one-to-one, we conclude that $T = T_g$. $\qquad\square$

Theorem 6.18. *Let g be a symbol function on $\mathbb{C}$ that satisfies condition (I_2). If there exists some $\beta \in (2\alpha, \infty)$ such that $B_\beta g \in L^\infty(\mathbb{C})$, then T_g is bounded on F_α^2.*

Proof. Let $\sigma(z) = B_{2\alpha}g(\bar{z})$. In view of Theorem 6.17, the Toeplitz operator T_g on F_α^2 is unitarily equivalent to the pseudodifferential operator $\sigma(D,X)$ on $L^2(\mathbb{R}, dx)$. We proceed to show that the pseudodifferential operator $\sigma(D,X)$ is bounded.

Let γ be the positive number satisfying

$$\frac{1}{2\alpha} = \frac{1}{\beta} + \frac{1}{\gamma}.$$

By the semigroup property of the parametrized Berezin transforms, we have

$$\sigma(z) = B_{2\alpha}g(\bar{z}) = B_\gamma B_\beta g(\bar{z}).$$

Let $\varphi(z) = B_\beta g(\bar{z})$. Then φ is in $L^\infty(\mathbb{C})$, and

$$\sigma(z) = \frac{\gamma}{\pi}\int_{\mathbb{C}} \varphi(w)e^{-\gamma|z-w|^2}\, dA(w).$$

Differentiating under the integral sign, we see that for any nonnegative integers n and m, we have

$$\frac{\partial^{n+m}\sigma}{\partial z^n \partial \bar{z}^m}(z) = \int_{\mathbb{C}} h_{mn}(z-w, \bar{z}-\bar{w})\varphi(w)e^{-\gamma|z-w|^2}\, dA(w),$$

where h_{mn} is a polynomial of degree $m+n$. Thus, for all $z \in \mathbb{C}$, we have

$$\left|\frac{\partial^{n+m}\sigma}{\partial z^n \partial \bar{z}^m}(z)\right| \le \|\varphi\|_\infty \int_{\mathbb{C}} |h_{mn}(z-w, \bar{z}-\bar{w})|e^{-\gamma|z-w|^2}\, dA(w)$$

$$= \|\varphi\|_\infty \int_{\mathbb{C}} |h_{mn}(u, \bar{u})|e^{-\gamma|u|^2}\, dA(u) < \infty.$$

This shows that $\partial^{m+n}\sigma/\partial z^n\partial\bar{z}^m$ is bounded on $\mathbb{C}$ for all nonnegative integer n and m. By Theorem 1.24, the pseudodifferential operator $\sigma(D,X)$ is bounded on $L^2(\mathbb{R},dx)$. $\square$

When the symbol function φ is nonnegative, we have the following characterization for boundedness.

Theorem 6.19. *Suppose $\varphi \geq 0$ satisfies condition (I_1). Then the following conditions are equivalent:*

(a) *T_φ is bounded on F_α^2.*
(b) *$\widetilde{\varphi} = B_\alpha\varphi \in L^\infty(\mathbb{C})$.*
(c) *$B_\beta\varphi \in L^\infty(\mathbb{C})$, where β is any fixed positive weight parameter.*
(d) *$\widehat{\varphi}_r \in L^\infty(\mathbb{C})$, where r is any fixed positive radius.*

Proof. The equivalences of (a), (b), and (d) follow from the characterization of Fock–Carleson measures in Sect. 3.4. In fact, when φ is nonnegative, we have

$$\langle T_\varphi f, f\rangle_\alpha = \int_{\mathbb{C}} |f|^2 \varphi\, d\lambda_\alpha.$$

The densely defined positive operator T_φ is bounded if and only if there exists a constant $C > 0$ such that

$$\langle T_\varphi f, f\rangle_\alpha \leq C\|f\|_{2,\alpha}^2, \quad f \in F_\alpha^2,$$

which is the same as

$$\int_{\mathbb{C}} |f|^2 \varphi\, d\lambda_\alpha \leq C \int_{\mathbb{C}} |f|^2\, d\lambda_\alpha, \quad f \in F_\alpha^2.$$

This condition simply means that the measure $d\mu(z) = \varphi(z)\, dA(z)$ is Fock–Carleson. The equivalence of (b) and (c) follows from Theorem 3.23. $\square$

As a consequence of the above theorem, we obtain the following characterization of bounded Toeplitz operators on F_α^2 induced by symbols from BMO^1.

Theorem 6.20. *Suppose $\varphi \in BMO^1$. Then the following conditions are equivalent:*

(a) *T_φ is bounded on F_α^2.*
(b) *$\widetilde{\varphi} = B_\alpha\varphi \in L^\infty(\mathbb{C})$.*
(c) *$B_\beta\varphi \in L^\infty(\mathbb{C})$, where β is any fixed positive weight parameter.*
(d) *$\widehat{\varphi}_r \in L^\infty(\mathbb{C})$, where r is any fixed positive radius.*

Proof. By (3.22) of Theorem 3.34, there exists a constant $C > 0$ such that

$$\|\varphi \circ \varphi_z - \widetilde{\varphi}(z)\|_{L^1(d\lambda_\alpha)} \leq C$$

for all $z \in \mathbb{C}$, where $\varphi_z(w) = z - w$. By the triangle inequality, we also have

$$\|\varphi \circ \varphi_z\|_{L^1(d\lambda_\alpha)} - |\widetilde{\varphi}(z)| \leq C$$

for all $z \in \mathbb{C}$, which is the same as

$$\widetilde{|\varphi|} - |\widetilde{\varphi}| \in L^\infty(\mathbb{C}).$$

Therefore, $\widetilde{\varphi} \in L^\infty(\mathbb{C})$ if and only if $\widetilde{|\varphi|} \in L^\infty(\mathbb{C})$. It follows from this and the characterization of bounded Toeplitz operators with nonnegative symbols (see Theorem 6.19) that the condition $\widetilde{\varphi} \in L^\infty(\mathbb{C})$ implies that $T_{|\varphi|}$ is bounded on F_α^2.

Let $\varphi = f + ig$, where f and g are the real and imaginary parts of φ, respectively. Since $|f| \leq |\varphi|$ and $|g| \leq |\varphi|$, and nonnegative symbols induce positive operators, we see that the boundedness of $T_{|\varphi|}$ implies that both $T_{|f|}$ and $T_{|g|}$ are bounded on F_α^2.

Since f is real-valued, we can write $f = f^+ - f^-$, where

$$f^+ = \max(f, 0), \qquad f^- = \max(0, -f),$$

are the positive and negative parts of f, respectively. It follows from $0 \leq f^+ \leq |f|$ and $0 \leq f^- \leq |f|$ that T_{f^+} and T_{f^-} are both bounded on F_α^2. Thus, $T_f = T_{f^+} - T_{f^-}$ is bounded. Similarly, T_g is bounded. This shows that the condition $\widetilde{\varphi} \in L^\infty(\mathbb{C})$ implies the boundedness of T_φ on F_α^2. Since the inverse implication is obvious, we have proved the equivalence of (a) and (b).

Recall from the proof of Theorem 3.36 that $B_\beta \varphi - \widehat{\varphi}_r$ is bounded when $\varphi \in \mathrm{BMO}^1$. This shows that conditions (b), (c), and (d) are equivalent whenever $\varphi \in \mathrm{BMO}^1$. This completes the proof of the theorem. $\qquad\square$

6.4 Compactness

In this section, we discuss the compactness of Toeplitz operators on F_α^2. The main results are parallel to those in the previous section. All conditions in this section are in terms of membership in the space $C_0(\mathbb{C})$ which consists of continuous functions f on $\mathbb{C}$ such that $f(z) \to 0$ as $z \to \infty$. In several approximation arguments, we will also need the space $C_c(\mathbb{C})$, consisting of continuous functions f on $\mathbb{C}$ with compact support. It is clear that $C_c(\mathbb{C})$ is dense in $C_0(\mathbb{C})$ in the supremum norm of $C_0(\mathbb{C})$.

Theorem 6.21. *Suppose φ satisfies condition (I_2) and T_φ is compact on F_α^2. Then $B_\beta \varphi \in C_0(\mathbb{C})$ for all $\beta \in (0, 2\alpha)$.*

Proof. Recall from Theorem 6.16 and its proof that, for any $\beta \in (0, 2\alpha)$, there exists a positive constant $C = C(\beta)$ such that $\|B_\beta f\|_\infty \leq C\|T_f\|$ whenever T_f is bounded on F_α^2.

If T_φ is compact on F_α^2, then by Theorem 6.5, there exists a sequence $\{f_n\}$ of functions in $C_c(\mathbb{C})$ such that

$$\|T_\varphi - T_{f_n}\| \leq \frac{1}{n}, \qquad n \geq 1.$$

Therefore,

$$\|B_\beta \varphi - B_\beta f_n\|_\infty \leq C\|T_\varphi - T_{f_n}\| < \frac{1}{n}$$

for all $n \geq 1$. Each f_n has compact support, so $B_\beta f_n \in C_0(\mathbb{C})$. Since $C_0(\mathbb{C})$ is closed in the supremum norm, we conclude that $B_\beta \varphi$ is in $C_0(\mathbb{C})$ as well. $\qquad \square$

Theorem 6.22. *Suppose g is a symbol function that satisfies condition (I_2). If there exists some $\beta \in (2\alpha, \infty)$ such that $B_\beta g \in C_0(\mathbb{C})$, then T_g is compact on F_α^2.*

Proof. As in the proof of Theorem 6.18, the Toeplitz operator T_g on F_α^2 is unitarily equivalent to the pseudodifferential operator $\sigma(D, X)$ on $L^2(\mathbb{R}, dx)$, where $\sigma(z) = B_{2\alpha} g(\bar{z})$. Furthermore, it follows from Theorem 6.18 that T_g and $\sigma(D, X)$ are both bounded operators with

$$\sigma(z) = B_\gamma \varphi(z) = \frac{\gamma}{\pi} \int_{\mathbb{C}} \varphi(w) e^{-\gamma|z-w|^2} \, dA(w),$$

where $\varphi(z) = B_\beta g(\bar{z})$. For any pair of nonnegative integers m and n, there is a polynomial $h_{mn}(z, \bar{z})$ such that

$$\frac{\partial^{m+n} \sigma}{\partial z^m \partial \bar{z}^n}(z) = \int_{\mathbb{C}} h_{mn}(z - w, \bar{z} - \bar{w}) \varphi(w) e^{-\gamma|z-w|^2} \, dA(w). \qquad (6.25)$$

The integral transform T defined by

$$Tf(z) = \int_{\mathbb{C}} h_{mn}(z - w, \bar{z} - \bar{w}) f(w) e^{-\gamma|z-w|^2} \, dA(w)$$

is bounded on $L^\infty(\mathbb{C})$. See the proof of Theorem 6.18. If f is compactly supported, say on $|z| \leq R$, then

$$Tf(z) = \int_{|w| \leq R} h_{mn}(z-w, \bar{z}-\bar{w}) f(w) e^{-\gamma|z-w|^2} \, dA(w)$$

$$= \int_{|w-z| \leq R} h_{mn}(w, \bar{w}) f(z-w) e^{-\gamma|w|^2} \, dA(w),$$

and so

$$|Tf(z)| \leq \|f\|_\infty \int_{|w-z| \leq R} |h_{mn}(w, \bar{w})| e^{-\gamma|w|^2} \, dA(w).$$

The convergence of the integral

$$\int_{\mathbb{C}} |h_{mn}(w, \bar{w})| e^{-\gamma|w|^2} \, dA(w)$$

clearly implies that $Tf(z) \to 0$ as $z \to \infty$. Thus, T maps $C_c(\mathbb{C})$ into $C_0(\mathbb{C})$. Since $C_c(\mathbb{C})$ is dense in $C_0(\mathbb{C})$ in the norm topology of $L^\infty(\mathbb{C})$, we infer from the boundedness of $T: L^\infty(\mathbb{C}) \to L^\infty(\mathbb{C})$ that T maps $C_0(\mathbb{C})$ into $C_0(\mathbb{C})$. This, along with (6.25), shows that $\partial^{m+n}\sigma/\partial z^m \partial \bar{z}^m$ is in $C_0(\mathbb{C})$ for any pair of nonnegative integers m and n. By Theorem 1.25, the pseudodifferential operator $\sigma(D, X)$ is compact on $L^2(\mathbb{R}, dx)$, and hence the Toeplitz operator T_g is compact on F_α^2. $\square$

Theorem 6.23. *Suppose φ is nonnegative and satisfies condition (I_1). Then, the following conditions are equivalent:*

(a) T_φ is compact on F_α^2.
(b) $\widetilde{\varphi} \in C_0(\mathbb{C})$.
(c) $B_\beta \varphi \in C_0(\mathbb{C})$, where β is any fixed positive weight parameter.
(d) $\widehat{\varphi}_r \in C_0(\mathbb{C})$, where r is any fixed positive radius.

Proof. The equivalence of (a), (b), and (d) follow from the characterization of vanishing Fock–Carleson measures in Sect. 3.4. See the proof of Theorem 6.19 for the connection to Fock–Carleson measures. The equivalence of (b) and (c) follows from Theorem 3.23. $\square$

The rest of this section is devoted to the compactness of Toeplitz operators with symbols in BMO^1.

Lemma 6.24. *Suppose $f \in$ BMO^1 and $\widetilde{f} = B_\alpha f$ is bounded. Then*

$$T_f K_z = K_z[P(f \circ \varphi_z)] \circ \varphi_z \qquad (6.26)$$

for all $z \in \mathbb{C}$, where $P: L_\alpha^2 \to F_\alpha^2$ is the orthogonal projection, T_f is the Toeplitz operator on F_α^2, K_z is the reproducing kernel of F_α^2, and $\varphi_z(w) = z - w$.

Proof. Since BMO^1 and the Berezin transform are both translation invariant, we see that for any $z \in \mathbb{C}$, we have

$$f \circ \varphi_z \in \text{BMO}^1, \qquad B_\alpha(f \circ \varphi_z) \in L^\infty(\mathbb{C}).$$

In particular, each side of (6.26) is well defined.

By the definition of Toeplitz operators and a change of variables,

$$T_f K_z(w) = P(fK_z)(w) = \int_\mathbb{C} f(u)K_z(u)\overline{K_w(u)}\,d\lambda_\alpha(u)$$

$$= \int_\mathbb{C} f(\varphi_z(u))K_z(\varphi_z(u))\overline{K_w(\varphi_z(u))}|k_z(u)|^2\,d\lambda_\alpha(u)$$

$$= \int_\mathbb{C} f(\varphi_z(u))e^{\alpha(\bar{z}w-\bar{u}w+z\bar{u})}\,d\lambda_\alpha(u).$$

On the other hand,

$$K_z(w)[P(f \circ \varphi_z)](\varphi_z(w)) = e^{\alpha\bar{z}w}\int_\mathbb{C} f(\varphi_z(u))e^{\alpha(z-w)\bar{u}}\,d\lambda_\alpha(u)$$

$$= \int_\mathbb{C} f(\varphi_z(u))e^{\alpha(\bar{z}w+z\bar{u}-w\bar{u})}\,d\lambda_\alpha(u).$$

This proves the desired identity. $\qquad\qquad\qquad\qquad\qquad\qquad\qquad\qquad\square$

Lemma 6.25. *Suppose $f \in \text{BMO}^1$ and $\widetilde{f}$ is bounded. Then there exists a positive constant C such that*

$$\sup_{z \in \mathbb{C}}|P(f \circ \varphi_z)(w)| \leq Ce^{\alpha|w|^2/4} \qquad\qquad (6.27)$$

for all $w \in \mathbb{C}$.

Proof. Recall from the proof of Theorem 6.20 that if $f \in \text{BMO}^1$ and $\widetilde{f}$ is bounded, then $\widetilde{|f|}$ is bounded as well. By translation invariance of BMO^1 and the Berezin transform, there exists a positive constant C such that

$$B_\alpha(|f \circ \varphi_z|)(w) \leq C, \qquad z, w \in \mathbb{C}.$$

By Theorem 3.29, there exists another positive constant C (independent of z) such that

$$\int_\mathbb{C} |g(u)||f \circ \varphi_z(u)|\,d\lambda_\alpha(u) \leq C \int_\mathbb{C} |g(u)|\,d\lambda_\alpha(u)$$

for all entire functions g. In particular,

$$|P(f \circ \varphi_z)(w)| = \left|\int_\mathbb{C} f \circ \varphi_z(u)e^{\alpha w\bar{u}}\,d\lambda_\alpha(u)\right|$$

$$\leq \int_{\mathbb{C}} |f \circ \varphi_z(u)| |e^{\alpha \overline{w} u}| \, d\lambda_\alpha(u)$$

$$\leq C \int_{\mathbb{C}} |e^{\alpha \overline{w} u}| \, d\lambda_\alpha(u)$$

$$= C e^{\frac{\alpha}{4} |w|^2}.$$

This proves the desired estimate.   □

Lemma 6.26. *Suppose $f \in \mathrm{BMO}^1$ and $\widetilde{f} = B_\alpha f \in C_0(\mathbb{C})$. Then:*

(a) For any $a \in \mathbb{C}$, we have $P(f \circ \varphi_z)(a) = T_{f \circ \varphi_z} 1(a) \to 0$ as $z \to \infty$.
(b) $T_{f \circ \varphi_z} 1 \to 0$ weakly in F_α^2 as $z \to \infty$.

Proof. By Theorem 6.20 and the fact that $T_{f \circ \varphi_z} = U_z T_f U_z$, where $U_z f = f \circ \varphi_z k_z$ is a self-adjoint unitary operator, there exists a constant $C > 0$ such that $\|T_{f \circ \varphi_z}\| \leq C$ for all $z \in \mathbb{C}$. In particular, $\|T_{f \circ \varphi_z} 1\| \leq C$ for all $z \in \mathbb{C}$. Since

$$T_{f \circ \varphi_z} 1(a) = \langle T_{f \circ \varphi_z} 1, K_a \rangle$$

and the set of all finite linear combinations of kernel functions is dense in F_α^2, we see that (a) and (b) are actually equivalent.

To prove part (b), it suffices to show that

$$\lim_{z \to \infty} \langle T_{f \circ \varphi_z} 1, u^n \rangle = 0 \tag{6.28}$$

for every nonnegative integer n because the set of polynomials is dense in F_α^2.

Fix a nonnegative integer n and a point $a \in \mathbb{C}$. Observe that

$$\widetilde{f}(\varphi_z(a)) = \widetilde{f \circ \varphi_z}(a) = e^{-\alpha |a|^2} \langle T_{f \circ \varphi_z} K_a, K_a \rangle,$$

where

$$K_a(u) = e^{\alpha u \overline{a}} = \sum_{k=1}^{\infty} \frac{\alpha^k}{k!} u^k \overline{a}^k.$$

It follows that

$$\widetilde{f}(\varphi_z(a)) = e^{-\alpha |a|^2} \sum_{k,j=0}^{\infty} \frac{\alpha^{k+j}}{k! j!} \langle T_{f \circ \varphi_z} u^k, u^j \rangle \overline{a}^k a^j.$$

Thus, for any positive radius r, the integral

$$I_r(z) = \int_{|u| < r} \widetilde{f}(\varphi_z(u)) \overline{u}^n e^{\alpha |u|^2} \, dA(u)$$

can be written as

$$I_r(z) = \sum_{k,j=0}^{\infty} \frac{\alpha^{k+j}}{k!j!} \left\langle T_{f\circ\varphi_z} u^k, u^j \right\rangle \int_{|v|<r} \bar{v}^{k+n} v^j \, dA(v)$$

$$= \sum_{k=0}^{\infty} \frac{\alpha^{2k+n}}{k!(k+n)!} \langle T_{f\circ\varphi_z} u^k, u^{k+n}\rangle \int_{|v|<r} |v|^{2(k+n)} \, dA(v)$$

$$= \pi \sum_{k=0}^{\infty} \frac{\alpha^{2k+n}}{k!(k+n+1)!} \langle T_{f\circ\varphi_z} u^k, u^{k+n}\rangle r^{2(k+n+1)}$$

$$= \pi r^{2(n+1)} \left[\frac{\alpha^n}{(n+1)!} \langle T_{f\circ\varphi_z} 1, u^n\rangle + \Sigma_{r,n}(z) \right],$$

where

$$\Sigma_{r,n}(z) = \sum_{k=1}^{\infty} \frac{\alpha^{2k+n}}{k!(k+n+1)!} \langle T_{f\circ\varphi_z} u^k, u^{n+k}\rangle r^{2k}.$$

As $z \to \infty$, we have $\widetilde{f}(\varphi_z(u)) \to 0$ for every $u \in \mathbb{C}$. By the dominated convergence theorem,

$$\lim_{z\to\infty} I_r(z) = \lim_{z\to\infty} \int_{|u|<r} \widetilde{f}(\varphi_z(u)) \bar{u}^n e^{\alpha|u|^2} \, dA(u) = 0$$

for any $r > 0$. It follows that

$$\lim_{z\to\infty} \left[\frac{\alpha^n}{(n+1)!} \langle T_{f\circ\varphi_z} 1, u^n\rangle + \Sigma_{r,n}(z) \right] = 0 \tag{6.29}$$

for any fixed $r > 0$. Since $\|T_{f\circ\varphi_z}\| \leq C$ for all $z \in \mathbb{C}$, where C is independent of z, we see that

$$|\Sigma_{r,n}(z)| \leq C \sum_{k=1}^{\infty} \frac{\alpha^{2k+n}\|u^k\|\|u^{k+n}\|}{k!(k+n+1)!} r^{2k}$$

$$= C \sum_{k=1}^{\infty} \frac{\alpha^{2k+n}}{k!(k+n+1)!} \sqrt{\frac{k!(k+n)!}{\alpha^{2k+n}}} r^{2k}$$

$$\leq C\alpha^{\frac{n}{2}} \sum_{k=1}^{\infty} \frac{(\alpha r^2)^k}{k!}$$

$$= C\alpha^{\frac{n}{2}} \left[e^{\alpha r^2} - 1 \right]$$

for all $r > 0$, $n \geq 0$, and $z \in \mathbb{C}$. Given any $\varepsilon > 0$, choose a small enough positive radius r such that

$$C\alpha^{\frac{n}{2}} \left[e^{\alpha r^2} - 1 \right] < \varepsilon.$$

Then by (6.29), we have

$$\limsup_{z\to\infty} |\langle T_{f\circ\varphi_z} 1, u^n\rangle| \le \frac{(n+1)!}{\alpha^n}\varepsilon.$$

This proves (6.28) and completes the proof of the lemma.   □

We can now characterize the compactness of Toeplitz operators with symbols in BMO^1 in terms of the Berezin transform.

Theorem 6.27. *If $f \in \mathrm{BMO}^1$, then T_f is compact on F_α^2 if and only if $\tilde{f} \in C_0(\mathbb{C})$.*

Proof. It suffices to show that the condition $\tilde{f} \in C_0(\mathbb{C})$ implies the compactness of T_f on F_α^2. The other implication is obvious.

So let us assume that $f \in \mathrm{BMO}^1$ and $\tilde{f} \in C_0(\mathbb{C})$. We will actually prove that the operator

$$T_f : F_\alpha^2 \to L_\alpha^2$$

is compact, which clearly implies the desired compactness of $T_f : F_\alpha^2 \to F_\alpha^2$.

For any positive radius R, we consider the operator

$$T_f^R = M_{\chi_R} T_f : F_\alpha^2 \to L_\alpha^2,$$

where χ_R is the characteristic function of the open ball $|z| < R$ and M_{χ_R} is the operator of multiplication on L_α^2 by χ_R. It follows from the boundedness of T_f and a simple normal family argument that each T_f^R is compact. Thus, the compactness of T_f will follow if we can show that

$$\lim_{R\to\infty} \|T_f^R - T_f\|_{F_\alpha^2 \to L_\alpha^2} = 0. \tag{6.30}$$

Given $g \in F_\alpha^2$, we have

$$\begin{aligned}(T_f - T_f^R)g(z) &= (1-\chi_R)T_f g(z)\\ &= (1-\chi_R(z))\langle T_f g, K_z\rangle_\alpha\\ &= (1-\chi_R(z))\langle g, T_{\bar f} K_z\rangle_\alpha\\ &= \int_{\mathbb{C}} g(u)(1-\chi_R(z))\overline{T_{\bar f}K_z}(u)\,d\lambda_\alpha(u).\end{aligned}$$

Thus, $T_f - T_f^R$ is an integral operator with kernel

$$K_f^R(z,u) = (1-\chi_R(z))\overline{T_{\bar f}K_z}(u).$$

By Schur's test (Lemma 2.14), whenever there exists a positive function h on $\mathbb{C}$ such that

$$\int_{\mathbb{C}} |K_f^R(z,u)|h(z)\,d\lambda_\alpha(z) \le C_1 h(u), \qquad u\in\mathbb{C}, \tag{6.31}$$

and

$$\int_{\mathbb{C}} |K_f^R(z,u)| h(u)\, d\lambda_\alpha(u) \le C_2 h(z), \qquad z \in \mathbb{C}, \tag{6.32}$$

we then have

$$\|T_f - T_f^R\|_{F_\alpha^2 \to L_\alpha^2}^2 \le C_1 C_2. \tag{6.33}$$

We will arrive at constants C_1 and C_2 such that the product $C_1 C_2$ tends to 0 as $R \to \infty$, which then implies the compactness of T_f.

Let $h(z) = \sqrt{K(z,z)} = e^{\frac{\alpha}{2}|z|^2}$ and consider the integrals

$$I(z) = \int_{\mathbb{C}} |K_f^R(z,u)| h(u)\, d\lambda_\alpha(u), \qquad z \in \mathbb{C},$$

from (6.32). It is clear that $I(z) = 0$ for $|z| < R$. For $|z| \ge R$, we have

$$I(z) = \int_{\mathbb{C}} |T_{\bar{f}} K_z(u)| \sqrt{K(u,u)}\, d\lambda_\alpha(u),$$

which by Lemma 6.24 can be written as

$$I(z) = \int_{\mathbb{C}} |K_z(u)| |P(\bar{f} \circ \varphi_z)(\varphi_z(u))| \sqrt{K(u,u)}\, d\lambda_\alpha(u).$$

Making the change of variables $u \mapsto \varphi_z(u)$ and simplifying the result, we get

$$I(z) = \frac{\alpha}{\pi} \int_{\mathbb{C}} |P(\bar{f} \circ \varphi_z)(u)| \left| e^{\alpha(z-u)\bar{z}} \right| e^{-\frac{\alpha}{2}|z-u|^2}\, dA(u).$$

Fix $p \in (1, \infty)$ and $\sigma \in (\alpha/4, \alpha/2)$. Let $1/p + 1/q = 1$. By Hölder's inequality,

$$I(z) = \frac{\alpha}{\pi} \int_{\mathbb{C}} \left[|P(\bar{f} \circ \varphi_z(u))| e^{-\sigma|u|^2} \right] \left[e^{\sigma|u|^2} |e^{\alpha(z-u)\bar{z}}| e^{-\frac{\alpha}{2}|z-u|^2} \right] dA(u)$$

$$\le \frac{\alpha}{\pi} \left[\int_{\mathbb{C}} |P(\bar{f} \circ \varphi_z(u))|^p e^{-p\sigma|u|^2}\, dA(u) \right]^{\frac{1}{p}}$$

$$\times \left[\int_{\mathbb{C}} e^{q\sigma|u|^2} |e^{q\alpha(z-u)\bar{z}}| e^{-\frac{q\alpha}{2}|z-u|^2}\, dA(u) \right]^{\frac{1}{q}}.$$

The second integral above can be written as

$$\int_{\mathbb{C}} e^{\frac{q\alpha}{2}|z|^2 + q\sigma|u|^2} \left| e^{-\frac{q\alpha}{2}|z|^2 + \frac{q\alpha}{2}(z-u)\bar{z} + \frac{q\alpha}{2}(\bar{z}-\bar{u})z - \frac{q\alpha}{2}|z-u|^2} \right| dA(u),$$

which is equal to

$$\int_{\mathbb{C}} e^{\frac{q\alpha}{2}|z|^2 + q\sigma|u|^2 - \frac{q\alpha}{2}|u|^2}\, dA(u) = e^{\frac{q\alpha}{2}|z|^2} \int_{\mathbb{C}} e^{-q(\frac{\alpha}{2} - \sigma)|u|^2}\, dA(u).$$

On the other hand, it follows from Lemma 6.25 that for all $z \in \mathbb{C}$, we have

$$|P(\overline{f} \circ \varphi_z)(u)|^p e^{-p\sigma|u|^2} \leq Ce^{-p(\sigma-\frac{\alpha}{4})|u|^2},$$

with the function on the right-hand side above integrable with respect to dA. This, along with Lemma 6.26 and the dominated convergence theorem, shows that the constants

$$C'_{1,R} = \sup_{|z|\geq R} \left[\int_{\mathbb{C}} |P(\overline{f} \circ \varphi_z(u))|^p e^{-p\sigma|u|^2} \, dA(u) \right]^{\frac{1}{p}}$$

tend to 0 as $R \to \infty$. Therefore, we can find constants $C_{1,R}$ such that $C_{1,R} \to 0$ as $R \to \infty$ and $I(z) \leq C_{1,R}h(z)$ for all $z \in \mathbb{C}$. This proves the desired estimate in (6.32).

The integrals

$$J(u) = \int_{\mathbb{C}} |K_f^R(z,u)| h(z) \, d\lambda_\alpha(z)$$

from (6.31) are slightly easier to estimate. In fact, by Lemma 6.24 and a change of variables,

$$
\begin{aligned}
J(u) &\leq \int_{\mathbb{C}} |T_{\overline{f}} K_z(u)| \sqrt{K(z,z)} \, d\lambda_\alpha(z) \\
&= \frac{\alpha}{\pi} \int_{\mathbb{C}} |K_z(u)| |P(\overline{f} \circ \varphi_z)(\varphi_z(u))| e^{-\frac{\alpha}{2}|z|^2} \, dA(z) \\
&= \frac{\alpha}{\pi} \int_{\mathbb{C}} |K_{z+u}(u)| |P(\overline{f} \circ \varphi_{z+u})(z)| e^{-\frac{\alpha}{2}|z+u|^2} \, dA(z) \\
&= \frac{\alpha}{\pi} e^{\frac{\alpha}{2}|u|^2} \int_{\mathbb{C}} |P(\overline{f} \circ \varphi_{z+u})(z)| e^{-\frac{\alpha}{2}|z|^2} \, dA(z).
\end{aligned}
$$

By Lemma 6.25, there is a positive constant C such that

$$J(u) \leq Ce^{\frac{\alpha}{2}|u|^2} \int_{\mathbb{C}} e^{\frac{\alpha}{4}|z|^2 - \frac{\alpha}{2}|z|^2} \, dA(z) = Ce^{\frac{\alpha}{2}|u|^2} \int_{\mathbb{C}} e^{-\frac{\alpha}{4}|z|^2} \, dA(z).$$

This proves the desired estimate in (6.31) and completes the proof of the theorem.

□

Corollary 6.28 *Let* $f \in \mathrm{BMO}^1$, $\alpha > 0$, *and* $\beta > 0$. *Then* $B_\alpha f \in C_0(\mathbb{C})$ *if and only if* $B_\beta f \in C_0(\mathbb{C})$.

Proof. Without loss of generality, assume that $0 < \alpha < \beta$. If $B_\beta f \in C_0(\mathbb{C})$, then by Proposition 3.21, $B_\alpha f \in C_0(\mathbb{C})$. We do not need the assumption that $f \in \mathrm{BMO}^1$ here.

If $B_\alpha f \in C_0(\mathbb{C})$, then by Theorem 6.27, T_f is compact on F_α^2, which, according to Theorem 6.21, implies that $B_\gamma f \in C_0(\mathbb{C})$ for all $0 < \gamma < 2\alpha$. Repeat this process a certain number of times, we will then get $B_\beta f \in C_0(\mathbb{C})$. □

6.5   Toeplitz Operators in Schatten Classes

For $\mu \geq 0$, we are going to determine when the Toeplitz operator T_μ on F_α^2 belongs to the Schatten class S_p. The case when $p \geq 1$ is relatively easy and will be taken up first.

Recall that for any bounded linear operator T on F_α^2 we define the Berezin transform $\widetilde{T}$ by

$$\widetilde{T}(z) = \langle Tk_z, k_z \rangle, \qquad z \in \mathbb{C},$$

where k_z are the normalized reproducing kernels in F_α^2. If T is positive on F_α^2, then

$$\mathrm{tr}\,(T) = \frac{\alpha}{\pi} \int_{\mathbb{C}} \widetilde{T}(z)\,\mathrm{d}A(z).$$

See Proposition 3.3. In particular, T is in the trace-class S_1 if and only if the integral above converges. As a consequence, we obtain the following trace formula for Toeplitz operators on the Fock space.

Proposition 6.29. *Suppose μ is a positive Borel measure on $\mathbb{C}$ and satisfies condition (M). Then T_μ is in the trace-class S_1 if and only if μ is finite on $\mathbb{C}$. Moreover, $\mathrm{tr}\,(T_\mu) = (\alpha/\pi)\mu(\mathbb{C})$.*

Proof. Since all integrands below are nonnegative, we use Fubini's theorem to obtain

$$\mathrm{tr}\,(T_\mu) = \frac{\alpha}{\pi} \int_{\mathbb{C}} \widetilde{\mu}(z)\,\mathrm{d}A(z)$$

$$= \frac{\alpha}{\pi} \int_{\mathbb{C}} e^{\alpha |z|^2}\,\mathrm{d}\lambda_\alpha(z) \int_{\mathbb{C}} |e^{\alpha \bar{z}w}|^2 e^{-\alpha(|z|^2 + |w|^2)}\,\mathrm{d}\mu(w)$$

$$= \frac{\alpha}{\pi} \int_{\mathbb{C}} e^{-\alpha |w|^2}\,\mathrm{d}\mu(w) \int_{\mathbb{C}} |e^{\alpha \bar{z}w}|^2\,\mathrm{d}\lambda_\alpha(z)$$

$$= \frac{\alpha}{\pi} \int_{\mathbb{C}} \mathrm{d}\mu(w) = \frac{\alpha}{\pi}\mu(\mathbb{C}).$$

This also shows that $\mathrm{tr}\,(T_\mu) < \infty$ if and only if $\mu(\mathbb{C}) < \infty$. $\square$

Lemma 6.30. *If $p \geq 1$ and $\varphi \in L^p(\mathbb{C}, \mathrm{d}A)$, then $T_\varphi \in S_p$.*

Proof. If $\varphi \in L^p(\mathbb{C}, \mathrm{d}A)$, then $\varphi \circ t_a \in L^p(\mathbb{C}, \mathrm{d}A)$ by a simple change of variables. It follows that $\varphi \circ t_a \in L^p(\mathbb{C}, \mathrm{d}\lambda_\alpha)$ for every $a \in \mathbb{C}$. Thus, φ satisfies condition (I_p). Since $p \geq 1$, φ also satisfies condition (I_1) so that T_φ is densely defined on F_α^2.

The rest is proved in exactly the same way that Proposition 7.11 in [250] was proved. $\square$

Lemma 6.31. *Suppose $r > 0$, μ is a positive Borel measure on $\mathbb{C}$, and*

$$\widehat{\mu}_r(z) = \frac{\mu(B(z,r))}{\pi r^2}, \qquad z \in \mathbb{C}.$$

If $\widehat{\mu}_r$ is in $L^p(\mathbb{C}, dA)$ for some $0 < p < \infty$, then μ satisfies condition (M), and the Toeplitz operators T_μ and $T_{\widehat{\mu}_r}$ are both bounded on F_α^2. Moreover, there exists a positive constant C (independent of μ) such that $T_\mu \leq CT_{\widehat{\mu}_r}$.

Proof. Let

$$C = \int_{\mathbb{C}} \mu(B(z,r))^p \, dA(z) < \infty.$$

For any $a \in \mathbb{C}$, we have

$$\int_{B(a,r/2)} \mu(B(z,r))^p \, dA(z) \leq C.$$

When $z \in B(a,r/2)$, we have $B(a,r/2) \subset B(z,r)$ by the triangle inequality. It follows that $\mu(B(z,r)) \geq \mu(B(a,r/2))$, and so

$$\frac{\pi r^2}{4} \mu(B(a,r/2))^p \leq C, \qquad a \in \mathbb{C}.$$

This shows that the function $a \mapsto \mu(B(a,r/2))$ is bounded. By Theorem 3.29 (with $p = 2$ there), the measure μ satisfies condition (M), and the Toeplitz operator T_μ is bounded on F_α^2, which in turn implies that the function $z \mapsto \mu(B(z,r))$ is bounded. Thus, $T_{\widehat{\mu}_r}$ is bounded on F_α^2 as well.

Given $f \in F_\alpha^2$, we use Fubini's theorem to obtain

$$\pi r^2 \langle T_{\widehat{\mu}_r} f, f \rangle = \pi r^2 \int_{\mathbb{C}} |f(z)|^2 \widehat{\mu}_r(z) \, d\lambda_\alpha(z)$$

$$= \int_{\mathbb{C}} |f(z)|^2 \mu(B(z,r)) \, d\lambda_\alpha(z)$$

$$= \int_{\mathbb{C}} |f(z)|^2 \, d\lambda_\alpha(z) \int_{\mathbb{C}} \chi_{B(z,r)}(w) \, d\mu(w)$$

$$= \int_{\mathbb{C}} d\mu(w) \int_{\mathbb{C}} |f(z)|^2 \chi_{B(w,r)}(z) \, d\lambda_\alpha(z)$$

$$= \frac{\alpha}{\pi} \int_{\mathbb{C}} d\mu(w) \int_{B(w,r)} |f(z) e^{-\alpha|z|^2/2}|^2 \, dA(z).$$

Combining the above identity with Lemma 2.32, we obtain a positive constant C such that

$$C\langle T_{\widehat{\mu}_r} f, f \rangle \geq \int_{\mathbb{C}} |f(w)|^2 e^{-\alpha|w|^2} \, d\mu(w) = \langle T_\mu f, f \rangle.$$

This proves the desired result.   □

Note that the condition $\widehat{\mu}_r \in L^p(\mathbb{C}, dA)$ for $0 < p < \infty$ implies that

$$\lim_{a \to \infty} \int_{B(a,r/2)} \mu(B(z,r))^p \, dA(z) = 0.$$

Refining the arguments in the above proof then shows that $\mu(B(a,r/2)) \to 0$ as $a \to \infty$, which implies that T_μ is compact on F_α^2 and $\widehat{\mu}_r \in C_0(\mathbb{C})$.

For the remainder of this section, we let $\{a_n\}$ denote any fixed arrangement of the square lattice $r\mathbb{Z}^2$ into a sequence. We are now ready to characterize positive Toeplitz operators in S_p when $p \geq 1$.

Theorem 6.32. *Suppose $\mu \geq 0$, $r > 0$, and $p \geq 1$. If μ satisfies condition (M), then the following conditions are equivalent:*

(a) The operator T_μ is in the Schatten class S_p.
(b) The function $\widetilde{\mu}(z)$ is in $L^p(\mathbb{C}, dA)$.
(c) The function $\mu(B(z,r))$ is in $L^p(\mathbb{C}, dA)$.
(d) The sequence $\{\mu(B(a_n,r))\}$ is in l^p.

Proof. That (a) implies (b) follows from Proposition 3.5. The elementary inequality $\widehat{\mu}_r(z) \leq C\widetilde{\mu}(z)$ (see the proof of Theorem 3.29) shows that condition (b) implies (c).

If the averaging function $\widehat{\mu}_r(z)$, which differs from $\mu(B(z,r))$ by a constant, is in $L^p(\mathbb{C}, dA)$, then it follows from Lemma 6.30 that $T_{\widehat{\mu}_r}$ is in S_p. Combining this with Lemma 6.31, we conclude that T_μ is in S_p. This proves that (c) implies (a). Hence, conditions (a), (b), and (c) are equivalent.

To prove that condition (d) is equivalent to the other conditions, we first assume that condition (b) holds, which implies that the function $\mu(B(z,2r))$ is in $L^p(\mathbb{C}, dA)$. Choose a positive integer m such that each point in the complex plane belongs to at most m of the disks $B(a_n,r)$. Then

$$m \int_{\mathbb{C}} \mu(B(z,2r))^p \, dA(z) \geq \sum_{n=1}^{\infty} \int_{B(a_n,r)} \mu(B(z,2r))^p \, dA(z).$$

For each $z \in B(a_n,r)$, we deduce from the triangle inequality that

$$\mu(B(z,2r)) \geq \mu(B(a_n,r)).$$

Therefore,

$$m \int_{\mathbb{C}} \mu(B(z,2r))^p \, dA(z) \geq \pi r^2 \sum_{n=1}^{\infty} \mu(B(a_n,r))^p.$$

This shows that condition (b) implies (d).

To finish the proof, we assume that condition (d) holds, that is,

$$\sum_{n=1}^{\infty} \mu(B(a_n,r))^p < \infty.$$

It is easy to see that we also have

$$\sum_{n=1}^{\infty} \mu(B(z_n,r))^p < \infty,$$

where $\{z_n\}$ is any arrangement of the lattice $(r/2)\mathbb{Z}^2$. In fact, for each point z_k that is not in the lattice $\{a_n\}$, the disk $B(z_k, r)$ is covered by six adjacent disks $B(a_k, r)$. Therefore,

$$
\int_{\mathbb{C}} \mu(B(z, r/2))^p \, dA(z) \leq \sum_{n=1}^{\infty} \int_{B(z_n, r/2)} \mu(B(z, r/2))^p \, dA(z)
$$

$$
\leq \sum_{n=1}^{\infty} \int_{B(z_n, r/2)} \mu(B(z_n, r))^p \, dA(z)
$$

$$
= \frac{\pi r^2}{4} \sum_{n=1}^{\infty} \mu(B(z_n, r))^p < \infty.
$$

This shows that condition (d) implies (c), as the equivalence of (c) to (b) implies that if condition (c) holds for one positive radius, then it will hold for any other positive radius. This completes the proof of the theorem.   □

Specializing to the case when

$$
d\mu(z) = \frac{\alpha}{\pi} \varphi(z) \, dA(z),
$$

we obtain the following corollary concerning Toeplitz operators induced by non-negative functions.

Corollary 6.33 *Suppose* $\varphi \geq 0$, $p \geq 1$, *and* $r > 0$. *If* φ *satisfies condition* (I_1), *then the following conditions are equivalent:*

(a) The Toeplitz operator T_φ *belongs to* S_p.
(b) The Berezin transform $\widetilde{\varphi}$ *belongs to* $L^p(\mathbb{C}, dA)$.
(c) The averaging function

$$
\widehat{\varphi}_r(z) = \frac{1}{\pi r^2} \int_{B(z, r)} \varphi(w) \, dA(w)
$$

belongs to $L^p(\mathbb{C}, dA)$.
(d) The averaging sequence $\{\widehat{\varphi}_r(a_n)\}$ *belongs to* l^p.

We now turn our attention to the case $0 < p \leq 1$, which requires new ideas and techniques.

Lemma 6.34. *Suppose* $\mu \geq 0$, $r > 0$, *and* $0 < p \leq 1$. *If* μ *satisfies condition* (M), *then the following conditions are equivalent:*

(a) The function $\widetilde{\mu}(z)$ *is in* $L^p(\mathbb{C}, dA)$.
(b) The function $\mu(B(z, r))$ *is in* $L^p(\mathbb{C}, dA)$.
(c) The sequence $\{\mu(B(a_n, r))\}$ *is in* l^p.

Proof. We begin with the inequality

$$\widetilde{\mu}(z) = \frac{\alpha}{\pi} \int_{\mathbb{C}} e^{-\alpha|z-w|^2} \, d\mu(w) \leq \frac{\alpha}{\pi} \sum_{n=1}^{\infty} \int_{B(a_n,r)} e^{-\alpha|z-w|^2} \, d\mu(w).$$

For $w \in B(a_n, r)$, we have

$$|z - w|^2 \geq (|z - a_n| - |a_n - w|)^2 \geq |z - a_n|^2 - 2r|z - a_n|.$$

It follows that

$$\widetilde{\mu}(z) \leq \frac{\alpha}{\pi} \sum_{n=1}^{\infty} e^{-\alpha|z-a_n|^2 + 2\alpha r|z-a_n|} \mu(B(a_n, r)).$$

Since $0 < p \leq 1$, Hölder's inequality gives

$$\widetilde{\mu}(z)^p \leq \left(\frac{\alpha}{\pi}\right)^p \sum_{n=1}^{\infty} e^{-p\alpha|z-a_n|^2 + 2pr\alpha|z-a_n|} \mu(B(a_n, r))^p.$$

It follows from this and Fubini's theorem that

$$\int_{\mathbb{C}} \widetilde{\mu}(z)^p \, dA(z) \leq \left(\frac{\alpha}{\pi}\right)^p \sum_{n=1}^{\infty} \mu(B(a_n, r))^p \int_{\mathbb{C}} e^{-p\alpha|z-a_n|^2 + 2pr\alpha|z-a_n|} \, dA(z).$$

By an obvious change of variables, the integral above equals

$$\int_{\mathbb{C}} e^{-p\alpha|z|^2 + 2pr\alpha|z|} \, dA(z),$$

which is easily seen to be convergent. Thus, the condition $\{\mu(B(a_n, r))\} \in l^p$ implies $\widetilde{\mu} \in L^p(\mathbb{C}, dA)$.

On the other hand, there exists a positive integer m such that every point in the complex plane belongs to at most m of the disks $B(a_n, r)$. Thus,

$$m \int_{\mathbb{C}} \widetilde{\mu}(z)^p \, dA(z) \geq \sum_{n=1}^{\infty} \int_{B(a_n,r)} \widetilde{\mu}(z)^p \, dA(z).$$

For any $z \in B(a_n, r)$, we have

$$\widetilde{\mu}(z) = \frac{\alpha}{\pi} \int_{\mathbb{C}} e^{-\alpha|z-w|^2} \, d\mu(w) \geq \frac{\alpha}{\pi} \int_{B(a_n,r)} e^{-\alpha|z-w|^2} \, d\mu(w)$$

$$\geq \frac{\alpha}{\pi} e^{-4\alpha r^2} \mu(B(a_n, r)).$$

It follows that

$$m \int_{\mathbb{C}} \widetilde{\mu}(z)^p \, \mathrm{d}A(z) \geq \alpha r^2 e^{-4p\alpha r^2} \sum_{n=1}^{\infty} \mu(B(a_n, r))^p.$$

Thus, $\widetilde{\mu} \in L^p(\mathbb{C}, \mathrm{d}A)$ implies that $\{\mu(B(a_n, r))\} \in l^p$, which proves the equivalence of conditions (a) and (c).

That condition (a) implies (b) follows from the inequality $\mu(B(z, r)) \leq C\widetilde{\mu}(z)$ observed in the proof of Theorem 3.29.

To prove that condition (b) implies (c), we assume that the function $\mu(B(z, r))$ is in $L^p(\mathbb{C}, \mathrm{d}A)$. Consider the lattice $(r/2)\mathbb{Z}^2$ and arrange it into a sequence $\{z_n\}$. There exists a positive integer m such that every point in the complex plane belongs to at most m of the disks $B(z_n, r/2)$. Therefore,

$$m \int_{\mathbb{C}} \mu(B(z, r))^p \, \mathrm{d}A(z) \geq \sum_{n=1}^{\infty} \int_{B(z_n, r/2)} \mu(B(z, r))^p \, \mathrm{d}A(z).$$

For each $z \in B(z_n, r/2)$, the triangle inequality gives us that

$$\mu(B(z, r)) \geq \mu(B(z_n, r/2)).$$

Thus,

$$m \int_{\mathbb{C}} \mu(B(z, r))^p \, \mathrm{d}A(z) \geq \frac{\pi r^2}{4} \sum_{n=1}^{\infty} \mu(B(z_n, r/2))^p.$$

By the equivalence of conditions (a) and (c), the function $\widetilde{\mu}$ belongs to $L^p(\mathbb{C}, \mathrm{d}A)$. Applying the equivalence of (a) and (c) once more, we conclude that $\{\mu(B(a_n, r))\} \in l^p$. This completes the proof of the lemma. $\qquad\square$

Lemma 6.35. *Suppose $\mu \geq 0$, $0 < p \leq 1$, and μ satisfies condition (M). If the function $\widetilde{\mu}$ belongs to $L^p(\mathbb{C}, \mathrm{d}A)$, then the operator T_μ belongs to S_p.*

Proof. Since $\widetilde{\mu}$ belongs to $L^p(\mathbb{C}, \mathrm{d}A)$ and $\widetilde{\mu}$ dominates $\widehat{\mu}_r$, Lemma 6.31 shows that T_μ is bounded. Thus, $\widetilde{T_\mu} = \widetilde{\mu}$ and the desired result follows from Proposition 3.6. $\qquad\square$

We will need the following lemma, which can be found as Proposition 1.29 in [250].

Lemma 6.36. *If $0 < p \leq 2$, then for any orthonormal basis $\{e_n\}$ of a separable Hilbert space H and any compact operator T on H, we have*

$$\|T\|_{S_p}^p \leq \sum_{n=1}^{\infty} \sum_{k=1}^{\infty} |\langle Te_n, e_k \rangle|^p.$$

We are now ready to characterize Toeplitz operators T_μ in S_p when $0 < p \leq 1$.

Theorem 6.37. *Suppose $\mu \geq 0$, $r > 0$, $0 < p \leq 1$, and μ satisfies condition (M). Then the following conditions are equivalent:*

(a) T_μ belongs to the Schatten class S_p.
(b) $\widetilde{\mu}$ belongs to $L^p(\mathbb{C}, dA)$.
(c) $\widehat{\mu}_r$ belongs to $L^p(\mathbb{C}, dA)$.
(d) $\{\widehat{\mu}_r(a_n)\}$ belongs to l^p.

Proof. The equivalence of (b), (c), and (d) was proved in Lemma 6.34. That condition (b) implies condition (a) was proved in Lemma 6.35. Therefore, to finish the proof, we will show that condition (a) implies (d).

To this end, fix some large R with $R > 2r$ and use Lemma 1.14 to partition $\{a_n\}$ into N sublattices such that the Euclidean distance between any two points in each sublattice is at least R. Let $\{\zeta_n\}$ be such a sublattice and let

$$v = \sum_{n=1}^{\infty} \mu \chi_n,$$

where χ_n is the characteristic function of $B(\zeta_n, r)$. Since $T_\mu \in S_p$ and $\mu \geq v$, we have $T_v \leq T_\mu$, and so $T_v \in S_p$ with $\|T_v\|_{S_p} \leq \|T_\mu\|_{S_p}$.

Let $\{e_n\}$ be an orthonormal basis for F_α^2 and define a linear operator A on F_α^2 by $Ae_n = k_{\zeta_n}$, $n \geq 1$, where k_ζ is the normalized reproducing kernel of F_α^2 at ζ. By the proof of Theorem 2.34, the operator A is bounded. Let $T = A^* T_v A$. Then $\|T\|_{S_p} \leq \|A\|^2 \|T_\mu\|_{S_p}$.

We split the operator T as $T = D + E$, where D is the diagonal operator defined on F_α^2 by

$$Df = \sum_{n=1}^{\infty} \langle Te_n, e_n \rangle \langle f, e_n \rangle e_n,$$

and $E = T - D$. Since $0 < p \leq 1$, it follows from the triangle inequality that

$$\|T\|_{S_p}^p \geq \|D\|_{S_p}^p - \|E\|_{S_p}^p. \tag{6.34}$$

Also, D is a positive diagonal operator, so

$$\|D\|_{S_p}^p = \sum_{n=1}^{\infty} \langle Te_n, e_n \rangle^p = \sum_{n=1}^{\infty} \langle T_v k_{\zeta_n}, k_{\zeta_n} \rangle^p \tag{6.35}$$

$$= \left(\frac{\alpha}{\pi}\right)^p \sum_{n=1}^{\infty} \left(\int_{\mathbb{C}} e^{-\alpha|z-\zeta_n|^2} dv(z)\right)^p$$

$$\geq \left(\frac{\alpha}{\pi}\right)^p \sum_{n=1}^{\infty} \left(\int_{B(\zeta_n,r)} e^{-\alpha|z-\zeta_n|^2} dv(z)\right)^p$$

$$\geq C_1 \sum_{n=1}^{\infty} v(B(\zeta_n, r))^p.$$

On the other hand, by Lemma 6.36, we have

$$\|E\|_{S_p}^p \le \sum_{n=1}^{\infty}\sum_{k=1}^{\infty} |\langle Ee_n, e_k\rangle|^p = \sum_{n\ne k} |\langle T_v k_{\zeta_n}, k_{\zeta_k}\rangle|^p$$

$$= \left(\frac{\alpha}{\pi}\right)^p \sum_{n\ne k} \left| \int_{\mathbb{C}} k_{\zeta_n}(z)\overline{k_{\zeta_k}(z)}e^{-\alpha|z|^2}\,d\nu(z)\right|^p. \qquad (6.36)$$

A straightforward calculation shows that

$$\left|k_{\zeta_n}(z)\overline{k_{\zeta_k}(z)}e^{-\alpha|z|^2}\right| = e^{-\frac{\alpha|z-\zeta_n|^2}{2}}e^{-\frac{\alpha|z-\zeta_k|^2}{2}},$$

so (6.36) gives us

$$\|E\|_{S_p}^p \le \left(\frac{\alpha}{\pi}\right)^p \sum_{n\ne k}\left(\int_{\mathbb{C}} e^{-\frac{\alpha|z-\zeta_n|^2}{2}}e^{-\frac{\alpha|z-\zeta_k|^2}{2}}\,d\nu(z)\right)^p. \qquad (6.37)$$

If $n\ne k$, then $|\zeta_n - \zeta_k| \ge R$. Thus, for $|z - \zeta_n| \le \frac{R}{2}$, the triangle inequality gives us $|z - \zeta_k| \ge \frac{R}{2}$. Therefore, for each $z \in \mathbb{C}$, at least one of $|z - \zeta_n| \ge \frac{R}{2}$ and $|z - \zeta_k| \ge \frac{R}{2}$ must hold. From this, we deduce that

$$e^{-\frac{\alpha|z-\zeta_n|^2}{2}}e^{-\frac{\alpha|z-\zeta_k|^2}{2}} \le e^{-\frac{\alpha R^2}{16}}e^{-\frac{\alpha|z-\zeta_n|^2}{4}}e^{-\frac{\alpha|z-\zeta_k|^2}{4}}.$$

Plugging this into (6.37), we obtain

$$\|E\|_{S_p}^p \le \left(\frac{\alpha}{\pi}\right)^p e^{-\frac{p\alpha R^2}{16}} \sum_{n\ne k}\left(\int_{\mathbb{C}} e^{-\frac{\alpha|z-\zeta_n|^2}{4}}e^{-\frac{\alpha|z-\zeta_k|^2}{4}}\,d\nu(z)\right)^p. \qquad (6.38)$$

Since the measure ν is supported on $\cup_j B(\zeta_j, r)$, we have

$$\int_{\mathbb{C}} e^{-\frac{\alpha}{4}|z-\zeta_n|^2-\frac{\alpha}{4}|z-\zeta_k|^2}\,d\nu(z) = \sum_{j=1}^{\infty}\int_{B(\zeta_j,r)} e^{-\frac{\alpha}{4}|z-\zeta_n|^2-\frac{\alpha}{4}|z-\zeta_k|^2}\,d\mu(z)$$

$$= \sum_{j=1}^{\infty} e^{-\frac{\alpha}{4}|z_*-\zeta_n|^2-\frac{\alpha}{4}|z_*-\zeta_k|^2}\mu(B(\zeta_j,r)).$$

The last step above follows from the mean value theorem with

$$z_* = z_*(n,k,j) \in B(\zeta_j, r).$$

Since $0 < p \le 1$, it follows from Hölder's inequality that

$$\left[\int_{\mathbb{C}} e^{-\frac{\alpha}{4}|z-\zeta_n|^2-\frac{\alpha}{4}|z-\zeta_k|^2}\,d\nu(z)\right]^p \le \sum_{j=1}^{\infty} \mu(B(\zeta_j,r))^p e^{-\frac{p\alpha}{4}|z_*-\zeta_n|^2-\frac{p\alpha}{4}|z_*-\zeta_k|^2},$$

and so

$$\|E\|_{S_p}^p \le \left(\frac{\alpha}{\pi}\right)^p e^{-\frac{p\alpha}{16}R^2} \sum_{n,k=1}^{\infty} \sum_{j=1}^{\infty} \mu(B(\zeta_j,r))^p e^{-\frac{p\alpha}{4}|z_*-\zeta_n|^2 - \frac{p\alpha}{4}|z_*-\zeta_k|^2}$$

$$= \left(\frac{\alpha}{\pi}\right)^p e^{-\frac{p\alpha}{16}R^2} \sum_{j=1}^{\infty} \mu(B(\zeta_j,r))^p \sum_{n,k=1}^{\infty} e^{-\frac{p\alpha}{4}|z_*-\zeta_n|^2 - \frac{p\alpha}{4}|z_*-\zeta_k|^2}.$$

If $n \ne j$, then $|\zeta_j - \zeta_n| \ge R > 2r$, so by the triangle inequality,

$$|z_* - \zeta_n| \ge |\zeta_j - \zeta_n| - r = |\zeta_j - \zeta_n|\left[1 - \frac{r}{|\zeta_j - \zeta_n|}\right] > \frac{1}{2}|\zeta_j - \zeta_n|.$$

This holds trivially for $n = j$ as well. Thus,

$$\|E\|_{S_p}^p \le \left(\frac{\alpha}{\pi}\right)^p e^{-\frac{p\alpha}{16}R^2} \sum_{j=1}^{\infty} \mu(B(\zeta_j,r))^p \sum_{n,k=1}^{\infty} e^{-\frac{p\alpha}{16}|\zeta_j-\zeta_n|^2 - \frac{p\alpha}{16}|\zeta_j-\zeta_k|^2}$$

$$= \left(\frac{\alpha}{\pi}\right)^p e^{-\frac{p\alpha}{16}R^2} \sum_{j=1}^{\infty} \mu(B(\zeta_j,r))^p \left[\sum_{n=1}^{\infty} e^{-\frac{p\alpha}{16}|\zeta_j-\zeta_n|^2}\right]^2$$

$$\le \left(\frac{\alpha}{\pi}\right)^p e^{-\frac{p\alpha}{16}R^2} \sum_{j=1}^{\infty} \mu(B(\zeta_j,r))^p \left[\sum_{n=1}^{\infty} e^{-\frac{p\alpha}{16}|\zeta_j-a_n|^2}\right]^2$$

$$= \left(\frac{\alpha}{\pi}\right)^p e^{-\frac{p\alpha}{16}R^2} \sum_{j=1}^{\infty} \mu(B(\zeta_j,r))^p \left[\sum_{n=1}^{\infty} e^{-\frac{p\alpha}{16}|a_n|^2}\right]^2.$$

The last series above is clearly convergent. So we can find a positive constant C_2, independent of R, such that

$$\|E\|_{S_p}^p \le C_2 e^{-\frac{p\alpha}{16}R^2} \sum_{j=1}^{\infty} \mu(B(\zeta_j,r))^p.$$

Going back to (6.34) and (6.35), we deduce that

$$\|T\|_{S_p}^p \ge \|D\|_{S_p}^p - \|E\|_{S_p}^p \ge \left(C_1 - C_2 e^{-\frac{p\alpha}{16}R^2}\right) \sum_{j=1}^{\infty} \mu(B(\zeta_j,r))^p.$$

Since C_1 and C_2 do not depend on R, setting $R > 0$ large enough gives us

$$\sum_{j=1}^{\infty} \mu(B(\zeta_j,r))^p \le C_3 \|T_\mu\|_{S_p}^p,$$

where C_3 is another positive constant. Since this holds for each of the N subsequences of $\{a_n\}$, we obtain

$$\sum_{n=1}^{\infty} \mu(B(a_n,r))^p \leq C_3 N \|T_\mu\|_{S_p}^p \tag{6.39}$$

for all positive Borel measures μ such that

$$\sum_{n=1}^{\infty} \mu(B(a_n,r))^p < \infty.$$

Finally, an easy approximation argument shows that (6.39) holds for all positive Borel measures μ with $T_\mu \in S_p$. This proves that condition (a) implies (d), and thus completes the proof of Theorem 6.37.   □

Again, specializing to the case when

$$d\mu(z) = \frac{\alpha}{\pi}\varphi(z)\,dA(z),$$

we obtain the following corollary concerning Toeplitz operators induced by nonnegative functions:

Corollary 6.38 *Suppose $\varphi \geq 0$, $0 < p \leq 1$, $r > 0$, and φ satisfies condition (I_1). Then the following conditions are equivalent:*

(a)   The Toeplitz operator T_φ belongs to S_p.
(b)   The Berezin transform $\widetilde{\varphi}$ belongs to $L^p(\mathbb{C}, dA)$.
(c)   The averaging function

$$\widehat{\varphi}_r(z) = \frac{1}{\pi r^2} \int_{B(z,r)} \varphi(w)\,dA(w)$$

belongs to $L^p(\mathbb{C}, dA)$.
(d)   The sequence $\{\widehat{\varphi}_r(a_n)\}$ belongs to l^p.

6.6 Finite Rank Toeplitz Operators

In this section, we consider the following problem: when does a Toeplitz operator T_μ have finite rank on the Fock space F_α^2? It turns out the problem is pretty tricky. If μ has compact support in $\mathbb{C}$, we will be able to determine exactly when T_μ has finite rank. But on the other hand, we will also construct a radial function φ, not identically zero, such that $T_\varphi = 0$ in a natural way on the Fock space. This is something unique for the Fock space setting. In particular, in the Fock space setting, the Berezin transform $\varphi \mapsto \widetilde{\varphi}$ is not one-to-one if no additional assumptions are made about φ.

Let n be a positive integer and denote by $P(\mathbb{C}^n)$ the algebra of all holomorphic polynomials on $\mathbb{C}^n$. For any tuple $k = (k_1, \cdots, k_n)$ of nonnegative integers, we write

$$z^k = z_1^{k_1} \cdots z_n^{k_n}, \qquad |k| = k_1 + \cdots + k_n.$$

These are the monomials in $P(\mathbb{C}^n)$.

Given a permutation σ on $\{1, \cdots, n\}$, we write

$$\sigma(z) = (z_{\sigma(1)}, \cdots, z_{\sigma(n)}), \qquad z = (z_1, \cdots, z_n) \in \mathbb{C}^n.$$

A function $f : \mathbb{C}^n \to \mathbb{C}$ is called symmetric if $f(\sigma(z)) = f(z)$ for all $z \in \mathbb{C}^n$ and all permutations σ on $\{1, \cdots, n\}$. We say that $f : \mathbb{C}^n \to \mathbb{C}$ is antisymmetric if $f(\sigma(z)) = \mathrm{sgn}(\sigma)f(z)$ for all $z \in \mathbb{C}^n$ and all permutations σ on $\{1, \cdots, n\}$.

A set $U \subset \mathbb{C}^n$ is called permutation-invariant if $\sigma(z) \in U$ for all $z \in U$ and all permutations σ on $\{1, \cdots, n\}$. Obviously, the notions of symmetric and antisymmetric functions can also be defined on any permutation-invariant subset of $\mathbb{C}^n$. In particular, if R is any positive radius, we let $C_S(R)$ denote the space of all symmetric, complex-valued, and continuous functions f on the closed ball $\overline{B}(0, R)$ in $\mathbb{C}^n$.

For any complex-valued function f on a permutation-invariant subset U of $\mathbb{C}^n$, we can define two functions, called the symmetrization and antisymmetrization of f, respectively, as follows:

$$f_s(z) = \frac{1}{n!} \sum_\sigma f(\sigma(z)), \qquad z \in U,$$

and

$$f_a(z) = \frac{1}{n!} \sum_\sigma \mathrm{sgn}(\sigma)f(\sigma(z)), \qquad z \in U,$$

where the sums are taken over all permutations on $\{1, \cdots, n\}$.

Let $P_s(\mathbb{C}^n)$ denote the subspace of $P(\mathbb{C}^n)$ consisting of all symmetric polynomials. Similarly, let $P_a(\mathbb{C}^n)$ denote the subspace of $P(\mathbb{C}^n)$ consisting of all antisymmetric polynomials.

Let $P^*(\mathbb{C}^n)$ denote the vector space of all conjugate linear functionals on $P(\mathbb{C}^n)$. If μ is a finite complex Borel measure with compact support in $\mathbb{C}$, then the Toeplitz operator T_μ is well defined on the dense set $P(\mathbb{C})$ in F_α^2. Furthermore, for any $f \in P(\mathbb{C})$, we have $T_\mu(f) \in P^*(\mathbb{C})$ in the sense that

$$T_\mu(f)(g) = \frac{\alpha}{\pi} \int_{\mathbb{C}} f(z)\overline{g(z)} e^{-\alpha|z|^2} \, d\mu(z), \qquad g \in P(\mathbb{C}).$$

Therefore, when restricted to polynomials, we can think of the Toeplitz operator T_μ as a mapping from $P(\mathbb{C})$ to $P^*(\mathbb{C})$. If $T_\mu : F_\alpha^2 \to F_\alpha^2$ has finite rank, then so does $T_\mu : P(\mathbb{C}) \to P^*(\mathbb{C})$.

Lemma 6.39. *Suppose μ is a finite complex Borel measure on $\mathbb{C}$ with compact support. If T_μ has rank less than n, then*

$$\det \begin{pmatrix} \mu(f_1\overline{g}_1) & \cdots & \mu(f_n\overline{g}_1) \\ \vdots & \vdots & \vdots \\ \mu(f_1\overline{g}_n) & \cdots & \mu(f_n\overline{g}_n) \end{pmatrix} = 0 \qquad (6.40)$$

for all complex polynomials f_k and g_k in $P(\mathbb{C})$. Here,

$$\mu(f\overline{g}) = T_\mu(f)(g) = \frac{\alpha}{\pi} \int_{\mathbb{C}} f(z)\overline{g(z)} e^{-\alpha|z|^2} \, d\mu(z).$$

Proof. Given one-variable polynomials $f_1, \cdots, f_n$, the functionals $T_\mu(f_1), \cdots, T_\mu(f_n)$ are linearly dependent because T_μ has rank less than n. So there are coefficients $c_1, \cdots, c_n$, not all 0, such that

$$c_1 T_\mu(f_1) + \cdots + c_n T_\mu(f_n) = 0. \qquad (6.41)$$

If $\{g_1, \cdots, g_n\}$ is another collection of polynomials of one complex variable, we take the inner product of g_k with both sides of (6.41) to obtain

$$\begin{pmatrix} \mu(f_1\overline{g}_1) & \cdots & \mu(f_n\overline{g}_1) \\ \vdots & \vdots & \vdots \\ \mu(f_1\overline{g}_n) & \cdots & \mu(f_n\overline{g}_n) \end{pmatrix} \begin{pmatrix} c_1 \\ \vdots \\ c_n \end{pmatrix} = \begin{pmatrix} 0 \\ \vdots \\ 0 \end{pmatrix}.$$

Since the c_k's are not all 0, we see that the determinant of the matrix above must be 0. $\quad\square$

Lemma 6.40. *Suppose μ is a finite complex Borel measure on $\mathbb{C}$ with compact support. If T_μ has rank less than n and*

$$d\mu_n(z_1, \cdots, z_n) = e^{-\alpha(|z_1|^2 + \cdots + |z_n|^2)} \, d\mu(z_1) \cdots d\mu(z_n)$$

is the product measure on $\mathbb{C}^n$, then

$$\int_{\mathbb{C}^n} f\bar{g}\,d\mu_n = 0 \tag{6.42}$$

for all polynomials $f \in P(\mathbb{C}^n)$ and all antisymmetric polynomials $g \in P(\mathbb{C}^n)$.

Proof. Since the determinant is linear in each column, we can rephrase (6.40) as follows:

$$\int_{\mathbb{C}^n} f_1(z_1)\cdots f_n(z_n)\overline{\Delta(g_1,\cdots,g_n)(z)}\,d\mu_n(z) = 0, \tag{6.43}$$

where $z = (z_1,\cdots,z_n)$ and

$$\Delta(g_1,\cdots,g_n)(z) = \det \begin{pmatrix} g_1(z_1) & \cdots & g_1(z_n) \\ \vdots & \vdots & \vdots \\ g_n(z_1) & \cdots & g_n(z_n) \end{pmatrix}.$$

Inserting monomials f_k into (6.43) and then taking finite linear combinations, we see that (6.43) remains valid if the product $f_1(z_1)\cdots f_n(z_n)$ is replaced by any polynomial $f \in P(\mathbb{C}^n)$. In other words,

$$\int_{\mathbb{C}^n} f(z)\overline{\Delta(g_1,\cdots,g_n)(z)}\,d\mu_n(z) = 0 \tag{6.44}$$

for all $f \in P(\mathbb{C}^n)$ and $g_k \in P(\mathbb{C})$, $1 \le k \le n$.

If each g_k is a monomial in $P(\mathbb{C})$, then the function $\Delta(g_1,\cdots,g_n)(z)$ is an antisymmetric polynomial in $P(\mathbb{C}^n)$. On the other hand, it follows from the elementary identities

$$[g_1(z_1)\cdots g_n(z_n)]_a = \frac{1}{n!}\sum_\sigma(\mathrm{sgn}\sigma)g_1(z_{\sigma(1)})\cdots g_n(z_{\sigma(n)})$$

$$= \frac{1}{n!}\Delta(g_1,\cdots,g_n)(z)$$

that any antisymmetric polynomial in $P(\mathbb{C}^n)$ is a finite linear combination of functions of the form $\Delta(g_1,\cdots,g_n)(z)$. This proves the desired result. $\quad\square$

Lemma 6.41. *Let K be a permutation invariant compact set in $\mathbb{C}^n$, let Φ_s denote the algebra consisting of all finite linear combinations of functions of the form $\psi\bar{\varphi}$, where ψ and φ are symmetric polynomials in $P(\mathbb{C}^n)$, and let $C_s(K)$ denote the space of symmetric continuous functions on K. Then Φ_s is dense in $C_s(K)$ in the sense of uniform convergence.*

Proof. It is clear that Φ_s is an algebra that contains the constant functions and is closed under complex conjugation. If it also separated points in K, the desired result would then follow from the Stone–Weierstrass approximation theorem. But it is easy to see that Φ_s does not separate points in K. In fact, if $z \in K$ and $w = \sigma(z) = (z_{\sigma(1)}, \cdots, z_{\sigma(n)})$, where σ is a permutation not equal to the identity, then $z \neq w$ but $f(z) = f(w)$ for all $f \in \Phi_s$.

To overcome this obstacle, we define an equivalence relation $\sim$ on K as follows: $z \sim w$ if and only if $w = \sigma(z)$ for some permutation σ. Let $K' = K/\sim$ be the quotient space equipped with the standard quotient topology. It is clear that every function in $C_s(K)$ induces a function in $\mathbb{C}(K')$, the space of complex-valued continuous functions on the compact Hausdorff space K', and conversely, every function in $\mathbb{C}(K')$ can be lifted to a function in $C_s(K)$. Also, it is easy to see that Φ_s separates points in K'. In fact, if the cosets of $z = (z_1, \cdots, z_n)$ and $w = (w_1, \cdots, w_n)$ are two different points in K' (in other words, if w is not a permutation of z), then the two one-variable polynomials

$$p(u) = \prod_{k=1}^n (u - z_k), \qquad q(u) = \prod_{k=1}^n (u - w_k),$$

either have different zeros or they have the same zeros with different multiplicities. It follows that at least one Taylor coefficient of p differs from the corresponding coefficient of q. Thus, there exists an elementary symmetric polynomial whose values at z and w are different.

We can now apply the Stone–Weierstrass approximation theorem to conclude that every function in $C_s(K)$ can be uniformly approximated by a sequence of functions in Φ_s. □

The main result of this section is the following:

Theorem 6.42. *Suppose μ is a compactly supported finite complex Borel measure on $\mathbb{C}$ such that the rank of T_μ is less than n, where n is a positive integer. Then μ is supported on less than n points in $\mathbb{C}$.*

Proof. Recall that for $z = (z_1, \cdots, z_n)$,

$$V(z) = \det \begin{pmatrix} 1 & 1 & \cdots & 1 \\ z_1 & z_2 & \cdots & z_n \\ \vdots & \vdots & \vdots & \vdots \\ z_1^{n-1} & z_2^{n-1} & \cdots & z_n^{n-1} \end{pmatrix} = \prod_{i>j}(z_i - z_j)$$

is called the Vandermonde determinant, which is an antisymmetric polynomial in $P(\mathbb{C}^n)$.

Fix a compact set $E \subset \mathbb{C}$ that contains the support of μ. Suppose the support of μ contains n distinct points $a_1, \cdots, a_n$. We will obtain a contradiction. To this end, we choose a one-variable polynomial $p \in P(\mathbb{C})$ such that $p(a_i) \neq p(a_j)$ for all $i \neq j$ and consider the multiple-variable polynomial

$$V_p(z_1, \cdots, z_n) = V(p(z_1), \cdots, p(z_n)).$$

The choice of p ensures that $V_p(a_1, \cdots, a_n) \neq 0$.

It is easy to see that V_p is an antisymmetric polynomial in $P(\mathbb{C}^n)$. Since the product of a symmetric function and an antisymmetric function is antisymmetric, an application of Lemma 6.40 to the functions $\psi = \psi_1 V_p$ and $\varphi = \varphi_1 V_p$, where both ψ_1 and φ_1 are symmetric polynomials in $P(\mathbb{C}^n)$, shows that

$$\int_{\mathbb{C}^n} F|V_p|^2 \, d\mu_n = 0 \tag{6.45}$$

for all $F \in \Phi_s$. Since μ_n is supported on the permutation invariant compact set $E^n = E \times \cdots \times E$, it follows from Lemma 6.41 that (6.45) holds for all $F \in C_s(E^n)$.

The measure $|V_p|^2 \, d\mu_n$ is permutation invariant, which implies that

$$\int_{\mathbb{C}^n} F|V_p|^2 \, d\mu_n = \int_{\mathbb{C}^n} F_s |V_p|^2 \, d\mu_n$$

for all $F \in \mathbb{C}(E^n)$, where F_s is the symmetrization of F. Thus, (6.45) holds for all $F \in \mathbb{C}(E^n)$. Consequently, $|V_p|^2 \, d\mu_n$ is the zero measure so that the support of μ_n is contained in the zero variety of V_p. Since $a = (a_1, \cdots, a_n)$ is contained in the support of μ_n, we must have $V_p(a_1, \cdots, a_n) = 0$, which is a contradiction. This shows that μ is supported on less than n distinct points in $\mathbb{C}$. $\qquad\square$

Corollary 6.43 *Let φ be a compactly supported and locally integrable function on $\mathbb{C}$. Then the Toeplitz operator T_φ on F_α^2 has finite rank if and only if $\varphi = 0$.*

In the rest of this section, we present an example to show that it is necessary to assume that the measure μ in Theorem 6.42 and φ in Corollary 6.43 are compactly supported. These results will be false without this assumption. To better understand the intricacy of the problem, we note that if φ is bounded, then it follows easily from the integral representation of the projection P_α and Fubini's theorem that

$$\langle T_\varphi f, g \rangle = \int_{\mathbb{C}} \varphi(z) f(z) \overline{g(z)} \, d\lambda_\alpha(z)$$

for all polynomials f and g. A limit argument then shows that the above also holds for all functions f and g in F_α^2.

Proposition 6.44. *There exists a radial function φ, not identically zero, such that $T_\varphi = 0$ on F_α^2 in the sense that*

$$\int_{\mathbb{C}} \varphi(z) f(z) \overline{g(z)} \, d\lambda_\alpha(z) = 0$$

for all polynomials f and g.

Proof. We start with two constants ρ and c satisfying

$$c = \exp\left(\frac{\pi i}{2}(2 - \rho)\right), \qquad 0 < \rho < 1.$$

Let $z^{\pm \rho}$ denote the branches given by

$$z^{\pm \rho} = |z|^{\pm \rho} e^{\pm i \rho \theta}, \qquad \theta \in \left[-\frac{\pi}{2}, \frac{3\pi}{2} \right).$$

Define a function f on the closed upper half-plane by $f(0) = 0$ and

$$f(z) = \exp \left(\bar{c} z^{-\rho} + c z^{\rho} \right), \qquad \mathrm{Im}\,(z) \geq 0, z \neq 0.$$

Obviously, f is analytic in the upper half-plane.

For $\theta \in [0, \pi]$, we have

$$-\frac{\pi}{2} < -\frac{\pi\rho}{2} \leq -\frac{\pi\rho}{2} + \rho\theta \leq \frac{\pi\rho}{2} < \frac{\pi}{2}.$$

Thus, for $z = |z| e^{i\theta}$ with $\theta \in [0, \pi]$ and $|z| > 0$, we have

$$0 < \cos \frac{\pi\rho}{2} < \cos \left(-\frac{\pi\rho}{2} + \rho\theta \right) \leq 1, \tag{6.46}$$

and

$$
\begin{aligned}
f(z) &= \exp \left[|z|^{-\rho} e^{-\frac{\pi i}{2}(2-\rho) - \rho\theta i} + |z|^{\rho} e^{\frac{\pi i}{2}(2-\rho) + \rho\theta i} \right] \\
&= \exp \left[-(|z|^{-\rho} + |z|^{\rho}) \cos \left(-\frac{\pi\rho}{2} + \rho\theta \right) \right. \\
&\qquad \left. + \mathrm{i}(|z|^{-\rho} - |z|^{\rho}) \sin \left(-\frac{\pi\rho}{2} + \rho\theta \right) \right].
\end{aligned}
$$

In particular,

$$|f(z)| = \exp \left[-(|z|^{-\rho} + |z|^{\rho}) \cos \left(-\frac{\pi\rho}{2} + \rho\theta \right) \right]$$

for $z = |z| e^{i\theta}$ with $\theta \in [0, \pi]$ and $|z| > 0$. This together with (6.46) shows that

$$\lim_{z \to 0} f(z) = 0 = f(0), \qquad \lim_{z \to \infty} f(z) = 0,$$

where z is restricted to the closed upper half-plane, so f is continuous on the closed upper half-plane. Similarly, we can show that

$$\lim_{z \to 0} f^{(k)}(z) = 0, \qquad \lim_{z \to \infty} f^{(k)}(z) = 0, \tag{6.47}$$

where k is any nonnegative integer and z is restricted to the closed upper half-plane.

By the formula for $|f(z)|$, the restriction of f to the real line belongs to $L^1(\mathbb{R}, dx)$. In particular, the Fourier transform of f is well defined. Let g be the function $e^{\alpha x^2}$ times the Fourier transform of f, namely,

$$g(x)e^{-\alpha x^2} = \int_{-\infty}^{\infty} f(t)e^{-2\pi itx}\,dt, \quad -\infty < x < \infty.$$

Since f is analytic in the upper half-plane and continuous on the closed upper half-plane, it follows from (6.47) and contour integration around the semicircle $|z| = R$, $\mathrm{Im}(z) \geq 0$, that $g(x) = 0$ whenever $x \in (-\infty, 0]$. So the function g is supported on $(0, \infty)$.

By the Fourier inversion formula, we can write

$$f(x) = \int_{-\infty}^{\infty} g(t)e^{-\alpha t^2 + 2\pi itx}\,dt = \int_{0}^{\infty} g(t)e^{-\alpha t^2 + 2\pi itx}\,dt$$

for $-\infty < x < \infty$. Differentiating under the integral sign, we obtain

$$f^{(k)}(0) = (2\pi i)^k \int_{0}^{\infty} g(t)t^k e^{-\alpha t^2}\,dt, \quad k = 0, 1, 2, 3, \cdots.$$

Since all derivatives of f vanish at the origin, we have

$$\int_{0}^{\infty} g(t)t^k e^{-\alpha t^2}\,dt = 0, \quad k = 0, 1, 2, 3, \cdots.$$

Set $\varphi(z) = g(|z|)$. Then φ is a radial function, so

$$\int_{\mathbb{C}} \varphi(z)z^k \overline{z^m}\,d\lambda_\alpha(z) = 0$$

whenever $k \neq m$. On the other hand,

$$\int_{\mathbb{C}} \varphi(z)z^k \overline{z^k}\,d\lambda_\alpha(z) = 2\alpha \int_{0}^{\infty} g(r)r^{2k+1}e^{-\alpha r^2}\,dr = 0$$

for all $k \geq 0$. This shows that

$$\int_{\mathbb{C}} \varphi(z)f(z)\overline{g(z)}\,d\lambda_\alpha(z) = 0$$

for all polynomials f and g.   □

6.7 Notes

The systematic study of Toeplitz operators on the Fock space started in [28, 29], where several important techniques were introduced that remain useful up to today. For example, the use of the Berezin transform in function theoretic operator theory began in [28].

The material in Sect. 6.1 is mostly from [30]. The Bargmann transform between the Fock space F_α^2 and $L^2(\mathbb{R}, dx)$ has been a well-known and very useful tool in analysis. Our presentation in Sect. 6.2 follows Folland's book [92] closely. Theorems 6.12 and 6.14 are well known in the theory of pseudodifferential operators.

The idea of using the operators $T^{(t)}$ to study trace-class properties of Toeplitz operators first appeared in [30]. Theorems 6.15–6.18, as well as their compactness counterparts in Sect. 6.4, are all from [30]. The characterization of bounded and compact Toeplitz operators with nonnegative symbols is very similar to the Bergman space setting, and details are worked out in [132].

For Toeplitz operators with bounded symbols, the characterization of compactness in terms of the Berezin transform is also analogous to the Bergman space setting, which was first obtained by Axler and Zheng in [6] and later generalized to BMO symbols by Zorborska in [259]. Our presentation here follows [15, 61] closely.

When $1 \le p < \infty$, the characterization of Toeplitz operators in the Schatten class S_p of the Fock space F_α^2 is relatively easy and follows the Bergman space theory very closely. However, if $0 < p < 1$, there is a critical difference between the Fock and Bergman space theories. More specifically, in the Bergman space theory, there is a cutoff point when Schatten class Toeplitz operators are characterized using the Berezin transform, while the cutoff disappears in the Fock space setting. The proof of Theorem 6.37 here is simpler than the one first constructed in [132].

Theorem 6.42, the characterization of finite-rank Toeplitz operators induced by compactly supported measures, is due to Luecking [153]. The proof in [153] is purely algebraic and works in several different contexts, including Toeplitz operators on the Bergman space of various domains. The example in Proposition 6.44 was constructed in [105]. Note that Proposition 6.44 does not contradict with Proposition 3.17 because the function in Proposition 6.44 is far worse than the functions permitted in Proposition 3.17.

6.8 Exercises

1. Let μ be a positive Borel measure on $\mathbb{C}$ satisfying condition (M). Then the following conditions are equivalent:

   (a) μ is a vanishing Fock–Carleson measure.
   (b) $\|\mu - \mu_R\| \to 0$ as $R \to \infty$, where μ_R is the truncation of μ on the disk $B(0,R)$.
   (c) There exists a sequence of finite Borel measures μ_n, each with compact support, such that $\|\mu - \mu_n\| \to 0$ as $n \to \infty$.

2. Suppose $p > 1$. Show that there exists $\varphi \geq 0$ such that $T_\varphi \in S_p$ but $\varphi \notin L^p(\mathbb{C}, dA)$.

3. Suppose $0 < p < 1$. Show that there exists $\varphi \geq 0$ such that $\varphi \in L^p(\mathbb{C}, dA)$ but $T_\varphi \notin S_p$.

4. Suppose

$$\varphi(z) = e^{\left(\frac{1}{3} + \frac{2}{3}i\right)|z|^2}.$$

   Show that the Toeplitz operator T_φ is unitary on the Fock space F_1^2 ($\alpha = 1$) and the Berezin transform $\widetilde{\varphi}$ vanishes at ∞ and belongs to $L^p(\mathbb{C}, dA)$ for all $0 < p < \infty$.

5. Recall that for any $z \in \mathbb{C}$, we have the self-adjoint unitary operator U_z defined by $U_z f(w) = f(z - w) k_z(w)$. Show that if T_φ is bounded, then

$$\int_{\mathbb{C}} U_z T_\varphi U_z \, d\lambda_\alpha(z) = T_\psi,$$

   where $\psi(w) = \widetilde{\varphi}(-w)$ and the integral converges in the strong operator topology.

6. If T_φ is bounded, show that

$$\int_{\mathbb{C}} W_z T_\varphi W_z^* \, d\lambda_\alpha(z) = T_{\widetilde{\varphi}}.$$

7. Show that there exist functions φ such that $\widetilde{\varphi} \in L^\infty(\mathbb{C})$ but T_φ is not bounded on F_α^2.

8. Show that there exist functions φ such that $\widetilde{\varphi}(z) \to 0$ as $z \to \infty$ but T_φ is not compact on F_α^2.

9. Suppose φ is radial, that is, $\varphi(z) = \varphi(|z|)$ for all $z \in \mathbb{C}$. If φ satisfies condition (I_1), show that the densely defined Toeplitz operator T_φ is diagonal with respect to the standard basis of F_α^2. Characterize boundedness, compactness, and membership in the Schatten classes for such Toeplitz operators in terms of the moments of φ.

10. Suppose $\varphi(z) = e^{i|z|^2}$. Show that T_φ is in the trace class, but $\int_{\mathbb{C}} |\varphi| \, dA = \infty$.

11. If φ is bounded and compactly supported, then T_φ belongs to S_p for all $0 < p < \infty$.
12. Show that the set of bounded Toeplitz operators on F_α^2 is not norm-dense in the space of all bounded linear operators on F_α^2. See [30].
13. Show that there exists no positive constant C such that $\|B_{2\alpha}\varphi\|_\infty \leq C\|T_\varphi\|$ for all φ. See [30].
14. Let T_φ^α denote the Toeplitz operator defined on F_α^2 using the orthogonal projection $P_\alpha : L_\alpha^2 \to F_\alpha^2$. Show that

$$T_{\varphi_r}^\alpha f(z) = T_\varphi^{\alpha/r^2} f_{1/r}(rz)$$

for all polynomials f.
15. Show that the operator

$$T_{\varphi_r}^\alpha : F_\alpha^2 \to F_\alpha^2$$

is unitarily equivalent to the operator

$$T_\varphi^{\alpha/r^2} : F_{\alpha/r^2}^2 \to F_{\alpha/r^2}^2.$$

16. Suppose $1 \leq p < \infty$ and $B_\beta \varphi \in L^p(\mathbb{C}, dA)$ for some $\beta > 2\alpha$. Then the Toeplitz operator $T_\varphi : F_\alpha^2 \to F_\alpha^2$ belongs to the Schatten class S_p. See [30] and [61].
17. Let c be a complex constant and $\varphi(z) = e^{c|z|^2}$. Show that T_φ is bounded on F_α^2 if and only if $B_{2\alpha}\varphi \in L^\infty(C)$, T_φ is compact on F_α^2 if and only if $B_{2\alpha}\varphi \in C_0(\mathbb{C})$, and T_φ belongs to the Schatten class S_p if and only if $B_{2\alpha}\varphi \in L^p(\mathbb{C}, dA)$. See [30].
18. Define $T : F_\alpha^2 \to F_\alpha^2$ by $Tf(z) = f(-z)$. Show that $\|T_\varphi - T\| \geq 1$ for any bounded Toeplitz operator T_φ on F_α^2. See [30].
19. Suppose T is a finite sum of finite products of Toeplitz operators on F_α^2 induced by bounded symbols. Show that T is compact on F_α^2 if and only if $\tilde{T} \in C_0(\mathbb{C})$.
20. Suppose $\varphi(z) = |f(z)|e^{-\sigma|z|^2}$, where f is entire and $\sigma > 0$. Show that T_φ is bounded on F_α^2 if and only if $\varphi \in L^\infty(\mathbb{C})$, T_φ is compact on F_α^2 if and only if $\varphi \in C_0(\mathbb{C})$, and T_φ belongs to the Schatten class S_p if and only if $\varphi \in L^p(\mathbb{C}, dA)$.
21. Show that $H_n(x) = 2xH_{n-1}(x) - H'_{n-1}(x)$ for all $n \geq 1$.

Chapter 7
Small Hankel Operators

In this chapter, we study small Hankel operators on the Fock space F_α^2. Problems considered in the chapter include boundedness, compactness, and membership in the Schatten class S_p. We will also determine when a small Hankel operator has finite rank.

K. Zhu, *Analysis on Fock Spaces*, Graduate Texts in Mathematics 263,   267
DOI 10.1007/978-1-4419-8801-0_7,
© Springer Science+Business Media New York 2012

7.1 Small Hankel Operators

Recall that

$$P : L_\alpha^2 \to F_\alpha^2$$

is the orthogonal projection. Let

$$\overline{F}_\alpha^2 = \left\{ \overline{f} : f \in F_\alpha^2 \right\}$$

and use $\overline{P}$ to denote the orthogonal projection from L_α^2 onto $\overline{F}_\alpha^2$.

Suppose φ is a function on $\mathbb{C}$ that satisfies condition (I_1). Using the integral representation for P (and hence $\overline{P}$) we can define an operator h_φ on a dense subset of F_α^2 by

$$h_\varphi f(z) = \overline{P}(\varphi f)(z) = \int_{\mathbb{C}} K(w,z)\varphi(w)f(w) \, d\lambda_\alpha(w).$$

In fact, as in the definition of Toeplitz operators, the assumption that φ satisfy condition (I_1) ensures that $h_\varphi f$ is well defined whenever

$$f(z) = \sum_{k=1}^n c_k K(z, a_k)$$

is a finite linear combination of reproducing kernels. The set of all such f is a dense subspace of F_α^2.

The operator h_φ is traditionally called the small (or little) Hankel operator with symbol φ. We say that h_φ is bounded on F_α^2 if there exists a constant $C > 0$ such that $\|h_\varphi(f)\|_\alpha \le C\|f\|_\alpha$ whenever f is a finite linear combination of reproducing kernels. In this case, the domain of h_φ can be extended to the whole space F_α^2.

7.2 Boundedness and Compactness

In this section, we determine when the small Hankel operator h_φ is bounded or compact on the Fock space F_α^2. We will focus on the case when φ belongs to L_α^2. In this case, we can further assume that φ is conjugate analytic. In fact, if $\varphi \in L_\alpha^2$, then φ satisfies condition (I_1), and it is easy to check that $h_\varphi = h_{\overline{P(\overline{\varphi})}}$, with $P(\overline{\varphi}) \in F_\alpha^2$.

Theorem 7.1. *Suppose* $\varphi \in F_\alpha^2$. *Then,* $h_{\overline{\varphi}}$ *is bounded on* F_α^2 *if and only if* $\varphi \in F_{\alpha/2}^\infty$, *that is, there exists a constant* $C > 0$ *such that*

$$|\varphi(z)| \le C e^{\alpha|z|^2/4}, \qquad z \in \mathbb{C}.$$

Moreover, we always have

$$\frac{1}{2}\|\varphi\|_{F_{\alpha/2}^\infty} \le \|h_{\overline{\varphi}}\| \le \|\varphi\|_{F_{\alpha/2}^\infty}.$$

Proof. First, suppose that $h_{\overline{\varphi}}$ is bounded on F_α^2. Then there exists a positive constant C such that

$$|\langle h_{\overline{\varphi}} f, \overline{g}\rangle| \le \|h_{\overline{\varphi}}\|\|f\|\|g\|, \qquad f, g \in F_\alpha^2,$$

or

$$\left| \int_{\mathbb{C}} f(w)g(w)\overline{\varphi(w)} \, d\lambda_\alpha(w) \right| \le \|h_{\overline{\varphi}}\|\|f\|\|g\|, \qquad f, g \in F_\alpha^2.$$

Let $f = g = k_z$ be the normalized reproducing kernels in F_α^2. Then

$$\left| \int_{\mathbb{C}} k_z^2(w)\overline{\varphi(w)} \, d\lambda_\alpha(w) \right| \le \|h_{\overline{\varphi}}\|, \qquad z \in \mathbb{C}. \tag{7.1}$$

Rewrite this as

$$e^{-\alpha|z|^2} \left| \int_{\mathbb{C}} e^{(2\alpha z)\overline{w}} \varphi(w) \, d\lambda_\alpha(w) \right| \le \|h_{\overline{\varphi}}\|, \qquad z \in \mathbb{C}.$$

By the reproducing property in F_α^2, the integral above equals $\varphi(2z)$, so

$$e^{-\alpha|z|^2}|\varphi(2z)| \le \|h_{\overline{\varphi}}\|, \qquad z \in \mathbb{C}.$$

Replacing z by $z/2$ shows that $\varphi \in F_{\alpha/2}^\infty$ and $\|\varphi\|_{F_{\alpha/2}^\infty} \le \|h_{\overline{\varphi}}\|$.

Next, we suppose that $\varphi \in F_{\alpha/2}^\infty$ so that the function

$$\psi(w) = 2\varphi(2w)e^{-\alpha|w|^2}$$

is bounded on $\mathbb{C}$ with $\|\psi\|_\infty = 2\|\varphi\|_{F^\infty_{\alpha/2}}$. According to the reproducing formula in $F^2_{2\alpha}$,

$$\varphi(z) = \varphi\left(2 \cdot \frac{z}{2}\right) = \int_{\mathbb{C}} e^{2\alpha(z/2)\overline{w}} \varphi(2w)\, d\lambda_{2\alpha}(w)$$

$$= \int_{\mathbb{C}} e^{\alpha z\overline{w}} \varphi(2w)\, d\lambda_{2\alpha}(w) = P_\alpha(\psi)(z).$$

Therefore, if f and g are polynomials (which are dense in F^2_α), then

$$\langle h_{\overline{\varphi}} f, \overline{g}\rangle = \int_{\mathbb{C}} fg\,\overline{\varphi}\, d\lambda_\alpha = \langle fg, P_\alpha(\psi)\rangle = \langle fg, \psi\rangle = \int_{\mathbb{C}} fg\,\overline{\psi}\, d\lambda_\alpha.$$

Thus, by Hölder's inequality,

$$|\langle h_{\overline{\varphi}} f, \overline{g}\rangle| \le \|\psi\|_\infty \int_{\mathbb{C}} |fg|\, d\lambda_\alpha \le 2\|\varphi\|_{F^\infty_{\alpha/2}} \|f\|\, \|g\|.$$

This shows that the small Hankel operator $h_{\overline{\varphi}}$ is bounded, and we have the norm estimate $\|h_{\overline{\varphi}}\| \le 2\|\varphi\|_{F^\infty_{\alpha/2}}$. $\square$

Theorem 7.2. *Suppose $\varphi \in F^2_\alpha$. Then $h_{\overline{\varphi}}$ is compact on F^2_α if and only if $f \in f^\infty_{\alpha/2}$, that is,*

$$\lim_{z \to \infty} e^{-\alpha|z|^2/4} \varphi(z) = 0. \tag{7.2}$$

Proof. First, assume that φ is an entire function that satisfies condition (7.2). Then there exists a sequence of polynomials $\{p_k\}$ such that

$$\lim_{k \to \infty} \|p_k - \varphi\|_{F^\infty_{\alpha/2}} = 0.$$

By Theorem 7.1, we have $\|h_{\overline{\varphi}} - h_{\overline{p_k}}\| \to 0$ as $k \to \infty$. It is easy to see that each $h_{\overline{p_k}}$ has finite rank and hence is compact. So $h_{\overline{\varphi}}$ is compact.

On the other hand, if $h_{\overline{\varphi}}$ is compact, then it follows from the proof of Theorem 7.1 that

$$\lim_{z \to \infty} e^{-\alpha|z|^2} \varphi(2z) = 0,$$

because $k_z \to 0$ weakly in F^2_α as $z \to \infty$. Replacing z by $z/2$ shows that condition (7.2) must hold. $\square$

Corollary 7.3. *Suppose f is an entire function. Then $f = P_\alpha(g)$ for some $g \in L^\infty(\mathbb{C})$ if and only if $f \in F^\infty_{\alpha/2}$. Similarly, $f = P_\alpha(g)$ for some $g \in C_0(\mathbb{C})$ if and only if $f \in f^\infty_{\alpha/2}$.*

Proof. If $f = P_\alpha(g)$ for some $g \in L^\infty(\mathbb{C})$, then $h_{\bar{f}} = h_{\bar{g}}$ is bounded, so by Theorem 7.1, $f \in F^\infty_{\alpha/2}$.

If $f = P_\alpha(g)$ for some $g \in C_0(\mathbb{C})$, then we can approximate g in $L^\infty(\mathbb{C})$ by a sequence $\{g_k\}$ of functions with compact support in $\mathbb{C}$. Each $h_{\bar{g}_k}$ is obviously compact and

$$\|h_{\bar{f}} - h_{\bar{g}_k}\| = \|h_{\bar{g} - \bar{g}_k}\| \le \|g - g_k\|_\infty \to 0$$

as $k \to \infty$. It follows that $h_{\bar{f}}$ is compact. By Theorem 7.2, $f \in f^\infty_{\alpha/2}$.

On the other hand, if we define $g(z) = 2f(2z)e^{-\alpha|z|^2}$, it follows from the proof of Theorem 7.1 that $f = P_\alpha(g)$. If $f \in F^\infty_{\alpha/2}$, then $g \in L^\infty(\mathbb{C})$. Similarly, if $f \in f^\infty_{\alpha/2}$, then g is in $C_0(\mathbb{C})$. This completes the proof of the corollary. $\square$

7.3 Membership in Schatten Classes

Our next goal is to characterize small Hankel operators induced by entire functions that belong to the Schatten classes S_p. As usual, the cases $1 \le p < \infty$ and $0 < p < 1$ require different treatments. More specifically, we use complex interpolation for the case $1 \le p < \infty$, and we use atomic decomposition for the case $0 < p < 1$.

Theorem 7.4. *Suppose* $1 \le p \le \infty$, $\beta = \alpha/2$, *and* φ *is an entire function satisfying condition* (I_1). *Then* $h_{\overline{\varphi}}$ *is in the Schatten class* S_p *if and only if* $\varphi \in F_\beta^p$.

Proof. By Theorem 7.1, the mapping $F : F_\beta^1 + F_\beta^\infty \to S_\infty$ defined by $F(\varphi) = h_{\overline{\varphi}}$ is bounded (and conjugate linear) because F_β^1 is continuously contained in F_β^∞.

If $\varphi \in F_\beta^1$, then it follows from the reproducing formula in F_β^2 that

$$\varphi(z) = \int_{\mathbb{C}} e^{\beta z \overline{w}} \varphi(w) \, d\lambda_\beta(w).$$

If we write $K_w^\beta(z) = e^{\beta z \overline{w}}$ for the reproducing kernel in F_β^2, then it follows from Fubini's theorem that for polynomials f and g we have

$$
\begin{aligned}
\langle h_{\overline{\varphi}} f, \overline{g} \rangle &= \int_{\mathbb{C}} f(z) g(z) \overline{\varphi(z)} \, d\lambda_\alpha(z) \\
&= \int_{\mathbb{C}} \overline{\varphi(w)} \, d\lambda_\beta(w) \int_{\mathbb{C}} f(z) g(z) \overline{K_w^\beta(z)} \, d\lambda_\alpha(z) \\
&= \int_{\mathbb{C}} \overline{\varphi(w)} \langle h_{\overline{K_w^\beta}} f, \overline{g} \rangle \, d\lambda_\beta(w).
\end{aligned}
$$

In the sense of Banach space valued integrals, we can rewrite the above as

$$h_{\overline{\varphi}} = \int_{\mathbb{C}} \overline{\varphi(w)} h_{\overline{K_w^\beta}} \, d\lambda_\beta(w). \tag{7.3}$$

It is easy to see that each $h_{\overline{K_w^\beta}}$ is an operator of rank one, so by Theorem 7.1,

$$\| h_{\overline{K_w^\beta}} \|_{S_1} = \| h_{\overline{K_w^\beta}} \| \le 2 \| K_w^\beta \|_{F_\beta^\infty} = 2 e^{\beta |w|^2 / 2}.$$

Therefore, it follows from (7.3) that

$$\| h_{\overline{\varphi}} \|_{S_1} \le 2 \int_{\mathbb{C}} |\varphi(w)| e^{\beta |w|^2 / 2} \, d\lambda_\beta(w) = \frac{2\beta}{\pi} \int_{\mathbb{C}} \left| \varphi(w) e^{-\frac{\beta}{2} |w|^2} \right| \, dA(w).$$

This shows that $h_{\overline{\varphi}}$ belongs to the trace-class S_1 whenever φ is in F_β^1. On the other hand, we have already shown in the previous section that $h_{\overline{\varphi}}$ is in S_∞ whenever $\varphi \in F_\beta^\infty$. An application of complex interpolation then shows that, for $1 \le p \le \infty$, the small Hankel operator $h_{\overline{\varphi}}$ is in the Schatten class S_p whenever $\varphi \in F_\beta^p$.

On the other hand, if the small Hankel operator $h_{\overline{\varphi}}$ belongs to the Schatten class S_p, where $1 \le p < \infty$, then according to Proposition 3.5 and its proof, the function

$$\Phi(z) = \langle h_{\overline{\varphi}} k_z, \overline{k_z} \rangle$$

is in $L^p(\mathbb{C}, dA)$, where k_z are the normalized reproducing kernels of F_α^2. We compute that

$$\Phi(z) = \int_{\mathbb{C}} \overline{\varphi(w)} k_z^2(w) \, d\lambda_\alpha(w)$$

$$= e^{-\alpha|z|^2} \int_{\mathbb{C}} \overline{\varphi(w)} e^{2\alpha \overline{z} w} \, d\lambda_\alpha(w)$$

$$= e^{-\alpha|z|^2} \overline{\varphi(2z)}.$$

Obviously, the condition that

$$e^{-\alpha|z|^2} \varphi(2z) \in L^p(\mathbb{C}, dA)$$

is equivalent to the condition that

$$e^{-\alpha|z|^2/4} \varphi(z) \in L^p(\mathbb{C}, dA),$$

which in turn is equivalent to $\varphi \in F_\beta^p$. This completes the proof of the theorem.   □

Note that if $\varphi \in F_\beta^1$, we can also use atomic decomposition to prove that the operator $h_{\overline{\varphi}}$ is in S_1. See the first part of the proof of the next theorem.

Theorem 7.5. *Suppose $0 < p < 1$, $\beta = \alpha/2$, and φ is an entire function satisfying condition (I_1). Then $h_{\overline{\varphi}}$ is in the Schatten class S_p if and only if $\varphi \in F_\beta^p$.*

Proof. First, assume that $\varphi \in F_\beta^p$. By Theorem 2.34, we can write

$$\varphi(z) = \sum_{k=1}^{\infty} c_k \varphi_k(z),$$

where $\{c_k\} \in l^p$ and

$$\varphi_k(z) = e^{-\frac{\beta}{2}|z_k|^2 + \beta \overline{z}_k z}, \qquad k \ge 1.$$

We may also assume that the sequence $\{z_k\}$ is dense enough to be a sampling sequence for F_β^p. Moreover, there is a constant $C > 0$, independent of φ, such that

$$\sum_{k=1}^{\infty} |c_k|^p \le C \|\varphi\|_{p,\beta}^p.$$

It follows that

$$\|h_{\overline{\varphi}}\|_{S_p}^p = \left\| \sum_{k=1}^{\infty} c_k h_{\overline{\varphi}_k} \right\|_{S_p}^p \leq \sum_{k=1}^{\infty} |c_k|^p \|h_{\overline{\varphi}_k}\|_{S_p}^p.$$

Each operator $h_{\overline{\varphi}_k}$ is a rank-one operator. In fact, if we use K^α and K^β to denote the reproducing kernels of F_α^2 and F_β^2, respectively, then for any $f \in F_\alpha^2$, we have

$$\begin{aligned}
h_{\overline{\varphi}_k} f(z) = \overline{P}(\overline{\varphi}_k f)(z) &= \langle \overline{\varphi}_k f, \overline{K_z^\alpha} \rangle_\alpha \\
&= e^{-\beta |z_k|^2/2} \left\langle f K_z^\alpha, K_{z_k}^\beta \right\rangle_\alpha = e^{-\beta |z_k|^2/2} \left\langle f K_z^\alpha, K_{\beta z_k/\alpha}^\alpha \right\rangle_\alpha \\
&= e^{-\beta |z_k|^2/2} f(\beta z_k/\alpha) K_z^\alpha(\beta z_k/\alpha) = f(z_k/2) \overline{\varphi}_k(z) \\
&= \left\langle f, K_{z_k/2}^\alpha \right\rangle_\alpha \overline{\varphi}_k(z).
\end{aligned}$$

Therefore,

$$\|h_{\overline{\varphi}_k}\|_{S_p} = \|h_{\overline{\varphi}_k}\| \leq \|K_{z_k/2}^\alpha\|_{2,\alpha} \|\varphi_k\|_{2,\alpha} = 1,$$

and so

$$\|h_{\overline{\varphi}}\|_{S_p}^p \leq \sum_{k=1}^{\infty} |c_k|^p \leq C \|\varphi\|_{p,\beta}^p.$$

On the other hand, if $h_{\overline{\varphi}}$ is in S_p, we are going to show that $\varphi \in F_\beta^p$. To this end, we fix a square lattice $Z = \{z_k\}$ in $\mathbb{C}$ such that atomic decomposition holds on Z for both F_β^p and F_α^2. We also assume that $2Z$ is a sampling sequence for F_β^p. Fix a sufficiently large R and use Lemma 1.14 to decompose $Z = Z_1 \cup \cdots \cup Z_N$ into N square lattices such that for each $1 \leq k \leq N$ and each pair $\{w_1, w_2\}$ of distinct points in Z_k, we have $|w_1 - w_2| > R$.

Fix an orthonormal basis $\{e_k\}$ for F_α^2 and define an operator A on F_α^2 as follows:

$$A \left(\sum_{k=1}^{\infty} c_k e_k \right)(z) = \sum_{k=1}^{\infty} c_k e^{\alpha z \overline{z}_k - \frac{\alpha}{2} |z_k|^2}.$$

By the atomic decomposition for F_α^2, the operator A is bounded and onto. Clearly, we have $A = A_1 + \cdots + A_N$, where

$$A_j \left(\sum_{k=1}^{\infty} c_k e_k \right)(z) = \sum_{z_k \in Z_j} c_k e^{\alpha z \overline{z}_k - \frac{\alpha}{2} |z_k|^2}$$

for $1 \leq j \leq N$. Each operator A_j is also bounded on F_α^2.

We also consider the companion operators

$$B_j : \overline{F_\alpha^2} \to \overline{F_\alpha^2}$$

defined by

$$B_j(\overline{f}) = \overline{A_j f}, \qquad f \in F_\alpha^2, 1 \le j \le N.$$

Since $h_{\overline{\varphi}}$ is in S_p, so is the operator $T = T_1 + \cdots + T_N$, where

$$T_j = B_j^* h_{\overline{\varphi}} A_j, \qquad 1 \le j \le N.$$

Write $T = D + E$, where D is diagonal with respect to the basis $\{e_k\}$ and satisfies $\langle De_k, \overline{e}_k \rangle_\alpha = \langle Te_k, \overline{e}_k \rangle_\alpha$ for all $k \ge 1$. If we write

$$f_k(z) = e^{\alpha z \overline{z}_k - \frac{\alpha}{2}|z_k|^2},$$

then

$$\|D\|_{S_p}^p = \sum_{k=1}^{\infty} |\langle De_k, \overline{e}_k \rangle|^p = \sum_{k=1}^{\infty} |\langle Te_k, \overline{e}_k \rangle|^p$$

$$= \sum_{k=1}^{\infty} |\langle h_{\overline{\varphi}} f_k, \overline{f}_k \rangle|^p = \sum_{k=1}^{\infty} \left| \varphi(2z_k) e^{-\alpha|z_k|^2} \right|^p$$

$$\ge C \|\varphi\|_{p,\beta}^p,$$

where C is a positive constant independent of φ. Note that the last inequality above follows from the assumption that $\{2z_k\}$ is a sampling sequence for F_β^p.

On the other hand, since $0 < p < 1$, it follows from Lemma 6.36 that

$$\|E\|_{S_p}^p \le \sum_{k,l} |\langle Ee_k, \overline{e}_l \rangle|^p = \sum_{k \ne l} |\langle Te_k, \overline{e}_l \rangle|^p$$

$$= \sum_{j=1}^{N} \sum_{k \ne l} |\langle h_{\overline{\varphi}} A_j e_k, \overline{A_j e_l} \rangle|^p.$$

Since

$$\langle h_{\overline{\varphi}} A_j e_k, \overline{A_j e_l} \rangle = 0$$

unless both z_k and z_l are in Z_j, we see that

$$\|E\|_{S_p}^p \le \sum_{j=1}^{N} \sum \left\{ |\langle h_{\overline{\varphi}} f_k, \overline{f}_l \rangle|^p : k \ne l, z_k \in Z_j, z_l \in Z_j \right\}.$$

If φ is already in F_β^p, we can write

$$\varphi(z) = \sum_{i=1}^{\infty} c_i \varphi_i(z),$$

where

$$\varphi_i(z) = e^{\beta z \bar{z}_i - \frac{\beta}{2}|z_i|^2}$$

and

$$\sum_{i=1}^{\infty} |c_i|^p \leq C \|\varphi\|_{\beta,p}^p.$$

Here, C is a positive constant independent of φ. By Hölder's inequality,

$$\|E\|_{S_p}^p \leq \sum_{i=1}^{\infty} |c_i|^p \sum_{j=1}^{N} \sum \left\{ |\langle h_{\overline{\varphi}_i} f_k, \overline{f}_l \rangle|^p : k \neq l, z_k \in Z_j, z_l \in Z_j \right\}.$$

It is easy to see that

$$|\langle h_{\overline{\varphi}_i} f_k, \overline{f}_l \rangle \alpha| = e^{-\beta|z_k - (z_i/2)|^2 - \beta|z_l - (z_i/2)|^2}.$$

Therefore,

$$\|E\|_{S_p}^p \leq \sum_{i=1}^{\infty} |c_i|^p \sum_{|z_k - z_l| \geq R} e^{-p\beta|z_k - (z_i/2)|^2 - p\beta|z_l - (z_i/2)|^2}.$$

If 2δ is the separation constant for the sequence Z, then by Lemma 2.32, there exists a positive constant $C = C(\delta, \alpha, p)$ such that

$$e^{-p\beta[|z_k - \frac{z_i}{2}|^2 + |z_l - \frac{z_i}{2}|^2]} \leq C \int_{B\left(z_k - \frac{z_i}{2}, \delta\right) \times B\left(z_l - \frac{z_i}{2}, \delta\right)} e^{-p\beta[|z|^2 + |w|^2]} \, dA(z) dA(w).$$

If $(k,l) \neq (k',l')$, then

$$B\left(z_k - \frac{z_i}{2}, \delta\right) \times B\left(z_l - \frac{z_i}{2}, \delta\right) \cap B\left(z_{k'} - \frac{z_i}{2}, \delta\right) \times B\left(z_{l'} - \frac{z_i}{2}, \delta\right) = \emptyset.$$

Also,

$$B\left(z_k - \frac{z_i}{2}, \delta\right) \times B\left(z_l - \frac{z_i}{2}, \delta\right) \subset \left\{ (z,w) \in \mathbb{C}^2 : |z - w| \geq R - 2\delta \right\}.$$

It follows that there exists a positive constant C, independent of large R, such that

$$\sum_{|z_k - z_l| \geq R} e^{-p\beta|z_k - (z_i/2)|^2 - p\beta|z_l - (z_i/2)|^2}$$

is less than or equal to

$$C \int_{|z - w| \geq R - 2\delta} e^{-p\beta(|z|^2 + |w|^2)} \, dA(z) \, dA(w).$$

The above double integral tends to 0 as $R \to \infty$. This, along with the fact that

$$\|D\|_{S_p}^p \le 2^p \left(\|T\|_{S_p}^p + \|E\|_{S_p}^p \right),$$

shows that we can find a positive constant σ such that

$$\sigma \|\varphi\|_{p,\beta} \le \|h_{\overline{\varphi}}\|_{S_p}, \tag{7.4}$$

where $\varphi \in F_\beta^p$ and σ is independent of φ.

The inequality in (7.4) was proved under the assumption that φ is already in F_β^p. The general case then follows from an easy approximation argument. In fact, if φ is any entire function such that $h_{\overline{\varphi}}$ is in S_p, then by Theorem 7.1, φ must be in F_β^p. We consider the functions φ_r, $0 < r < 1$, defined by $\varphi_r(z) = \varphi(rz)$. Each $\varphi_r \in F_\beta^p$, so by (7.4),

$$\sigma \|\varphi_r\|_{p,\beta} \le \|h_{\overline{\varphi}_r}\|_{S_p} \le \|h_{\overline{\varphi}}\|_{S_p}, \quad 0 < r < 1.$$

Let $r \to 1$. We obtain

$$\sigma \|\varphi\|_{p,\beta} \le \|h_{\overline{\varphi}}\|_{S_p} < \infty.$$

This completes the proof of the theorem.   □

7.4   Finite Rank Small Hankel Operators

In this section, we characterize small Hankel operators on F_α^2 whose range is finite dimensional. Such operators are called finite rank operators.

We begin with an example. Suppose $\varphi(z) = K(a,z)$ for some point $a \in \mathbb{C}$. Then for any function $f \in F_\alpha^2$, we have

$$h_\varphi(f)(z) = \overline{P}(\varphi f)(z) = \int_{\mathbb{C}} K(w,z)K(a,w)f(w)\,d\lambda_\alpha(w)$$
$$= f(a)K(a,z).$$

So, in this case, the range of h_φ is the one-dimensional subspace spanned by the function $z \mapsto K(a,z)$. More generally, if

$$\varphi(z) = \sum_{k=0}^{N} c_k \frac{\partial^k}{\partial a^k} K(a,z)$$

for some point $a \in \mathbb{C}$ and some nonnegative integer N, then

$$h_\varphi(f)(z) = \sum_{k=0}^{N} c_k \frac{\partial^k}{\partial a^k} \int_{\mathbb{C}} K(w,z)K(a,w)f(w)\,d\lambda_\alpha(w)$$
$$= \sum_{k=0}^{N} c_k \frac{\partial^k}{\partial a^k} [f(a)K(a,z)],$$

which shows that h_φ is a finite rank operator whose range is spanned by the following functions of z:

$$\frac{\partial^k}{\partial a^k} K(a,z), \qquad 0 \le k \le N.$$

We are going to show that these are essentially all the finite rank small Hankel operators on F_α^2. But we first need the following elementary result from algebra.

Lemma 7.6. *Let $P(\mathbb{C})$ denote the ring of all complex polynomials of the variable z. If J is an ideal in $P(\mathbb{C})$ containing at least one nonzero polynomial, then there are a finite number of complex numbers a_k, $1 \le k \le N$, and for each k, there exists a nonnegative integer N_k, such that J consists of all polynomials φ with the property that*

$$\varphi^{(i)}(a_k) = 0, \qquad 1 \le k \le N, 0 \le i \le N_k.$$

Proof. By a well-known fact in abstract algebra (see [146] for example), every ideal $J \ne (0)$ of $P(\mathbb{C})$ is generated by a polynomial, that is, there exists a polynomial q such that $J = \{pq : p \in P(\mathbb{C})\}$. If $a_1, \cdots, a_N$ are the zeros of q, and each zero a_k has multiplicity $1 + N_k$, then J has the desired form. $\square$

Theorem 7.7. *A bounded small Hankel operator has finite rank if and only if it can be written as* h_φ, *where*

$$\varphi(z) = \sum_{k=1}^{N} \sum_{i=0}^{N_k} c_{ki} \varphi_{ki}(z). \tag{7.5}$$

Here, $\varphi_{ki}(z)$ *denotes the function*

$$\frac{\partial^i}{\partial a^i} K(a, z)$$

evaluated at the point $a = a_k$.

Proof. We have already proved that h_φ has finite rank if φ is given by (7.5).

To prove the other direction, we write the small Hankel operator as h_φ, where φ is conjugate analytic. If h_φ has finite rank, then the restriction of h_φ on $P(\mathbb{C})$ also has finite rank. Consider the kernel of h_φ on $P(\mathbb{C})$:

$$J = \left\{ f \in P(\mathbb{C}) : h_\varphi(f) = 0 \right\}.$$

It is easy to check that J is an ideal in $P(\mathbb{C})$. In fact, if $h_\varphi(f) = 0$, then $\langle \varphi f, \overline{g} \rangle = 0$ for all polynomials g (which are dense in F_α^2). If p is any polynomial, then $\langle \varphi f, \overline{pg} \rangle = 0$ for all polynomials g. This can be rewritten as $\langle \varphi p f, \overline{g} \rangle = 0$ for all polynomials g, which shows that $h_\varphi(pf) = 0$ as well.

By Lemma 7.6, there exist points $a_k \in \mathbb{C}$, $1 \le k \le N$, and for each k, there exists a nonnegative integer N_k, such that

$$J = \left\{ f \in P(\mathbb{C}) : f^{(i)}(a_k) = 0, 1 \le k \le N, 0 \le i \le N_k \right\}.$$

In other words, J is the intersection of the kernels of finitely many linear functionals on $P(\mathbb{C})$.

Let $g = \overline{\varphi} \in F_\alpha^2$. Then the linear functional on $P(\mathbb{C})$ defined by

$$f \mapsto \langle f, g \rangle = \langle h_\varphi(f), 1 \rangle$$

vanishes on J. Combining this with the conclusion from the previous paragraph, we can find constants c_{ki} such that

$$\langle f, g \rangle = \sum_{k=1}^{N} \sum_{i=0}^{N_k} c_{ki} f^{(i)}(a_k) = \left\langle f, \sum_{k=1}^{N} \sum_{i=0}^{N_k} \overline{c}_{ki} \frac{\partial^i}{\partial \overline{a}^i} K(\cdot, a_k) \right\rangle$$

for all polynomials f. This shows that

$$\varphi(z) = \overline{g}(z) = \sum_{k=1}^{N} \sum_{i=0}^{N_k} c_{ki} \frac{\partial^i}{\partial a^i} K(a_k, z),$$

completing the proof of the theorem. $\square$

7.5 Notes

Small Hankel operators on the Fock space were first studied in [138], where the boundedness, compactness, and membership in Schatten classes S_p for $1 \leq p < \infty$ were characterized. The case when $0 < p < 1$ was taken up and settled in [231]. Our presentation here follows [138] and [231] very closely.

7.6 Exercises

1. For a symbol function φ, define a conjugate linear operator

$$\widetilde{h}_\varphi : F_\alpha^2 \to F_\alpha^2$$

by $\widetilde{h}_\varphi(f) = P(\varphi \overline{f})$. Show that h_φ is bounded if and only if $\widetilde{h}_{\overline{\varphi}}$ is bounded, h_φ is compact if and only if $\widetilde{h}_{\overline{\varphi}}$ is compact, and h_φ is in the Schatten class S_p if and only if $\widetilde{h}_{\overline{\varphi}}$ is in the Schatten class S_p.

2. Suppose φ is an entire function. Define a bilinear form

$$\Phi : F_\alpha^2 \times F_\alpha^2 \to \mathbb{C}$$

by

$$\Phi(f,g) = \langle \varphi f, g \rangle_\alpha = \int_{\mathbb{C}} \overline{\varphi} f g \, d\lambda_\alpha.$$

Show that $h_{\overline{\varphi}}$ is bounded on F_α^2 if and only if there exists a constant $C > 0$ such that $|\Phi(f,g)| \leq C \|f\|_{2,\alpha} \|g\|_{2,\alpha}$ for all f and g in F_α^2.

3. Formulate conditions for compactness and membership in Schatten classes for $h_{\overline{\varphi}}$ on F_α^2 in terms of the bilinear form Φ in the previous problem, where φ is any entire function.

4. Suppose $\varphi \in L_\alpha^2$. Show that $h_\varphi = 0$ if and only if $\varphi \perp \overline{F_\alpha^2}$.

5. Consider the integral transform

$$V_\varphi(z) = \langle h_\varphi k_z, \overline{k}_z \rangle_\alpha = \int_{\mathbb{C}} \varphi(w) k_z(w)^2 \, d\lambda_\alpha(w).$$

Show that h_φ is bounded if and only if V_φ is bounded, h_φ is compact if and only if $V_\varphi \in C_0(\mathbb{C})$, and h_φ belongs to the Schatten class S_p if and only if $V_\varphi \in L^p(\mathbb{C}, dA)$.

6. If φ is entire, show that

$$V_{\overline{\varphi}}(z) = e^{-\alpha|z|^2} \overline{\varphi}(2z)$$

for all $z \in \mathbb{C}$.

7. If $\varphi \in L^p(\mathbb{C}, d\lambda_\alpha)$ for some $1 < p < \infty$, then φ satisfies condition (I_1). In particular, every function in F_α^2 satisfies condition (I_1).

8. Show that if φ satisfies condition (I_1) with respect to the weight parameter $\beta = 3\alpha/4$, then $P_\alpha(\varphi)$ satisfies condition (I_1) with respect to the weight parameter α.

9. Show that Theorems 7.1 and 7.2 remain valid with the weaker assumption that φ is entire and satisfies condition (I_1).

10. Verify directly that $h_{\overline{\varphi}}$ has finite rank when φ is a polynomial.

11. Show that $\|h_{\varphi_r}\|_{S_p} \leq \|h_\varphi\|_{S_p}$ for all $0 < r < 1$.

Chapter 8
Hankel Operators

In this chapter, we study (big) Hankel operators H_φ on the Fock space F_α^2. Problems considered include, again, boundedness, compactness, and membership in the Schatten classes. There are basically two theories here: one concerns the simultaneous size estimates for both H_φ and $H_{\overline{\varphi}}$, and one concerns the size estimates for the single operator H_φ. The former is similar to the situations in the more classical Hardy and Bergman space settings, while the latter is unique to the Fock space setting.

K. Zhu, *Analysis on Fock Spaces*, Graduate Texts in Mathematics 263,
DOI 10.1007/978-1-4419-8801-0_8,
© Springer Science+Business Media New York 2012

8.1 Boundedness and Compactness

Suppose $\varphi \in L^\infty(\mathbb{C})$. We can then define an operator H_φ on F_α^2 by

$$H_\varphi(f) = (I - P)(\varphi f),$$

where I is the identity operator on L_α^2 and

$$P : L_\alpha^2 \to F_\alpha^2$$

is the orthogonal projection. It is obvious that H_φ is a bounded linear operator from F_α^2 into $L_\alpha^2 \ominus F_\alpha^2$ and $\|H_\varphi\| \leq \|\varphi\|_\infty$. We call H_φ the (big) Hankel operator with symbol φ. By the integral representation of the projection P, we have

$$\begin{aligned} H_\varphi(f)(z) &= \varphi(z)f(z) - P(\varphi f)(z) \\ &= \int_{\mathbb{C}} (\varphi(z) - \varphi(w))K(z,w)f(w)\,\mathrm{d}\lambda_\alpha(w) \end{aligned}$$

for all $f \in F_\alpha^2$ and $z \in \mathbb{C}$.

Using this integral representation for Hankel operators with bounded symbols, we can extend the definition of H_φ to the case in which φ is not necessarily bounded. In particular, if φ satisfies condition (I_1), then $H_\varphi(f)$ will always be defined whenever f is a finite linear combination of reproducing kernels in F_α^2. A natural question arises: for which symbol functions φ is the Hankel operator H_φ bounded?

In this section, we answer the above question when φ is real-valued. Equivalently, we characterize those symbol functions φ such that both H_φ and $H_{\overline{\varphi}}$ are bounded. A similar characterization will be given for the simultaneous compactness of H_φ and $H_{\overline{\varphi}}$.

We begin with Hankel operators induced by symbol functions that are Lipschitz in the Euclidean metric.

Lemma 8.1. *If there exists a positive constant C such that*

$$|\varphi(z) - \varphi(w)| \leq C|z - w|$$

for all complex numbers z and w. Then φ satisfies condition (I_1) and $\|H_\varphi\| \leq \sqrt{2\pi/\alpha}\,C$.

Proof. It is easy to check that any Lipschitz function satisfies condition (I_1) and hence induces a well-defined Hankel operator. To estimate the norm of H_φ, consider the integrals

$$I(z) = \int_{\mathbb{C}} |z - w||K(z,w)||K(w,w)|^{1/2}\,\mathrm{d}\lambda_\alpha(w), \qquad z \in \mathbb{C}.$$

By a change of variables,

$$
\begin{aligned}
I(z) &= \frac{\alpha}{\pi} \int_{\mathbb{C}} |z - w| \left| e^{\alpha z \bar{w} - \frac{\alpha}{2}|w|^2} \right| dA(w) \\
&= \frac{\alpha}{\pi} e^{\frac{\alpha}{2}|z|^2} \int_{\mathbb{C}} |z - w| e^{-\frac{\alpha}{2}|z - w|^2} dA(w) \\
&= \frac{\alpha}{\pi} e^{\frac{\alpha}{2}|z|^2} \int_{\mathbb{C}} |w| e^{-\frac{\alpha}{2}|w|^2} dA(w) \\
&= \sqrt{\frac{2\pi}{\alpha}} e^{\frac{\alpha}{2}|z|^2}.
\end{aligned}
$$

Thus,

$$
\int_{\mathbb{C}} |\varphi(z) - \varphi(w)| |K(z,w)| K(w,w)^{1/2} d\lambda_\alpha(w) \le \sqrt{\frac{2\pi}{\alpha}} C K(z,z)^{1/2}
$$

for all $z \in \mathbb{C}$. The desired norm estimate then follows from Lemma 2.14 and the integral representation of the Hankel operator H_φ. $\square$

Recall that for any $a \in \mathbb{C}$, we define a unitary operator U_a on L_α^2 by $U_a f = f \circ \varphi_a k_a$, where $\varphi_a(z) = a - z$ and k_a is the normalized reproducing kernel of F_α^2 at a. It is easy to check that $U_a^2 = I$, so $U_a^* = U_a^{-1} = U_a$. Since U_a leaves the Fock space F_α^2 invariant, we have $U_a P = P U_a$.

Lemma 8.2. *Suppose f satisfies condition (I_2). Then the operators T_f and H_f are both densely defined on F_α^2. Moreover, we have*

$$
T_f k_z = U_z P(f \circ \varphi_z) = P(f \circ \varphi_z) \circ \varphi_z k_z \tag{8.1}
$$

and

$$
H_f k_z = U_z (I - P)(f \circ \varphi_z) = [f - P(f \circ \varphi_z) \circ \varphi_z] k_z \tag{8.2}
$$

for all $z \in \mathbb{C}$.

Proof. Since each U_z commutes with the projection P, we have

$$
T_f k_z = P(f k_z) = P U_z(f \circ \varphi_z) = U_z P(f \circ \varphi_z).
$$

This proves the desired results. $\square$

Proposition 8.3. *Suppose f satisfies condition (I_2). Then*

$$
\max \left\{ \|H_f k_z\|, \|H_{\bar{f}} k_z\| \right\} \le MO(f)(z) \le \|H_f k_z\| + \|H_{\bar{f}} k_z\|
$$

for all $z \in \mathbb{C}$, *where* $MO(f) = \sqrt{\widetilde{|f|^2} - |\tilde{f}|^2}$.

Proof. Since each k_z is a unit vector, it follows from the Cauchy–Schwarz inequality that

$$
\begin{aligned}
MO(f)^2(z) &= \|fk_z\|^2 - |\langle fk_z, k_z\rangle|^2 \\
&= \|fk_z\|^2 - |\langle P(fk_z), k_z\rangle|^2 \\
&\geq \|fk_z\|^2 - \|P(fk_z)\|^2 \\
&= \|(I-P)(fk_z)\|^2 = \|H_f k_z\|^2.
\end{aligned}
$$

Replacing f by $\bar{f}$, we also have $MO(f)(z) \geq \|H_{\bar{f}} k_z\|$. Thus,

$$
MO(f)(z) \geq \max\left\{ \|H_f k_z\|, \|H_{\bar{f}} k_z\| \right\}.
$$

On the other hand, it follows from Lemma 8.2 that

$$
\begin{aligned}
\|H_f k_z\| &= \|U_z(I-P)(f \circ \varphi_z)\| = \|(I-P)(f \circ \varphi_z)\| \\
&= \|f \circ \varphi_z - P(f \circ \varphi_z)\|.
\end{aligned}
$$

Similarly, we have

$$
\|H_{\bar{f}} k_z\| = \|\bar{f} \circ \varphi_z - P(\bar{f} \circ \varphi_z)\| = \|f \circ \varphi_z - \overline{P(\bar{f} \circ \varphi_z)}\|.
$$

Since $\tilde{f}(z) = P(f \circ \varphi_z)(0)$ and $P\bar{g}(z) = \bar{g}(0)$ whenever $g \in F_\alpha^2$, we have

$$
\begin{aligned}
MO(f)(z) &= \|f \circ \varphi_z - P(f \circ \varphi_z)(0)\| \\
&\leq \|f \circ \varphi_z - P(f \circ \varphi_z)\| + \|P(f \circ \varphi_z) - P(f \circ \varphi_z)(0))\| \\
&= \|H_f k_z\| + \|P(f \circ \varphi_z) - \overline{P(\bar{f} \circ \varphi_z)(0)}\| \\
&= \|H_f k_z\| + \|P[f \circ \varphi_z - \overline{P(\bar{f} \circ \varphi_z)}]\| \\
&\leq \|H_f k_z\| + \|f \circ \varphi_z - \overline{P(\bar{f} \circ \varphi_z)}\| \\
&= \|H_f k_z\| + \|\bar{f} \circ \varphi_z - P(\bar{f} \circ \varphi_z)\| \\
&= \|H_f k_z\| + \|H_{\bar{f}} k_z\|.
\end{aligned}
$$

This completes the proof of the proposition. $\square$

We can now prove the main result of this section.

Theorem 8.4. *Suppose φ satisfies condition (I_2). Then the following two conditions are equivalent:*

1. *Both H_φ and $H_{\overline{\varphi}}$ are bounded on F_α^2.*
2. *The function φ belongs to* BMO^2.

Proof. First, assume that $\varphi \in$ BMO^2. Then, by Corollary 3.37, we can write $\varphi = \varphi_1 + \varphi_2$, where the function φ_1 satisfies the Lipschitz estimate

$$|\varphi_1(z) - \varphi_1(w)| \le C|z - w|$$

and the Toeplitz operator $T_{|\varphi_2|^2}$ is bounded. By Lemma 8.1, the Hankel operator H_{φ_1} is bounded on F_α^2. On the other hand, it follows from the identity

$$H_{\varphi_2}^* H_{\varphi_2} = T_{|\varphi_2|^2} - T_{\overline{\varphi_2}} T_{\varphi_2}$$

and the boundedness of $T_{|\varphi_2|^2}$ that H_{φ_2} is also bounded on F_α^2. Therefore, H_φ is bounded. Since BMO^2 is closed under complex conjugation, the assumption $\varphi \in$ BMO^2 implies that $\overline{\varphi}$ is also in BMO^2 so that $H_{\overline{\varphi}}$ is bounded on F_α^2 as well.

Next, assume that both H_φ and $H_{\overline{\varphi}}$ are bounded on F_α^2. Then, it follows from the inequality (see Proposition 8.3)

$$MO(\varphi)(z) = \left[\widetilde{|\varphi|^2}(z) - |\widetilde{\varphi}(z)|^2\right]^{1/2} \le \|H_\varphi k_z\| + \|H_{\overline{\varphi}} k_z\|$$

that the function φ is in BMO^2. $\square$

A companion result for the simultaneous compactness of H_φ and $H_{\overline{\varphi}}$ is the following:

Theorem 8.5. *Suppose φ satisfies condition (I_2). Then the following two conditions are equivalent:*

1. *Both H_φ and $H_{\overline{\varphi}}$ are compact on F_α^2.*
2. *The function φ belongs to* VMO^2.

Proof. If $\varphi \in$ VMO^2, then $\|\varphi - \varphi_r\|_{\text{BMO}^2} \to 0$ as $r \to \infty$, where φ_r is φ times the characteristic function of the Euclidean ball $B(0, r)$. It is easy to see that both H_{φ_r} and $H_{\overline{\varphi}_r}$ are compact on F_α^2. Since

$$\|H_\varphi - H_{\varphi_r}\| + \|H_{\overline{\varphi}} - H_{\overline{\varphi}_r}\| \sim \|\varphi - \varphi_r\|_{\text{BMO}^2},$$

we conclude that both H_φ and $H_{\overline{\varphi}}$ can be approximated by compact operators in the norm topology and so must be compact themselves.

Conversely, if H_φ and $H_{\overline{\varphi}}$ are both compact on F_α^2, then it follows from the second inequality in Proposition 8.3 that φ is in VMO^2, as the normalized reproducing kernels k_z tend to 0 weakly in F_α^2. $\square$

Corollary 8.6. *If φ is entire, then $H_{\overline{\varphi}}$ is bounded if and only if φ is a linear polynomial and $H_{\overline{\varphi}}$ is compact if and only if φ is constant.*

8.2 Compact Hankel Operators with Bounded Symbols

The purpose of this section is to show that for bounded symbol functions φ, the Hankel operator H_φ is compact on F_α^2 if and only if $H_{\overline{\varphi}}$ is compact on F_α^2. This striking result probably reflects the lack of bounded analytic functions (except the constants) in the complex plane, as the direct analogs for Hankel operators on the more classical Hardy and Bergman spaces are false.

Lemma 8.7. *If $f \in L^\infty(\mathbb{C})$, then $|Pf(z)| \leq \|f\|_\infty e^{\alpha|z|^2/4}$ for all $z \in \mathbb{C}$.*

Proof. This follows directly from Corollary 2.5.   □

Lemma 8.8. *Suppose $F(w,z)$ is a nonnegative measurable function on $\mathbb{C} \times \mathbb{C}$ with the property that there is a constant $B > 0$ such that*

$$F(w,z) \leq B e^{\frac{\alpha}{4}|z|^2}, \qquad z, w \in \mathbb{C}. \tag{8.3}$$

Then there exists another positive constant C such that

$$\int_{\mathbb{C}} F(w, \varphi_w(z)) |e^{\alpha z \overline{w} + \frac{1}{2}\alpha|z|^2}| \, d\lambda_\alpha(z) \leq C e^{\frac{1}{2}\alpha|w|^2} \left[\int_{\mathbb{C}} F(w,z)^2 \, d\lambda_\alpha(z) \right]^{\frac{1}{4}}$$

for all $w \in \mathbb{C}$.

Proof. We make an obvious change of variables to rewrite the integral on the left-hand side as

$$\frac{\alpha}{\pi} e^{\frac{\alpha}{2}|w|^2} \int_{\mathbb{C}} F(w,u) e^{-\frac{\alpha}{2}|u|^2} \, dA(u).$$

Denote the integral above by I, apply Hölder's inequality with exponents 4 and $4/3$, and use the assumption in (8.3). We obtain

$$I = \int_{\mathbb{C}} F(w,u) e^{-\frac{3\alpha}{8}|u|^2} e^{-\frac{\alpha}{8}|u|^2} \, dA(u)$$

$$\leq \left[\int_{\mathbb{C}} F(w,u)^4 e^{-\frac{3\alpha}{2}|u|^2} \, dA(u) \right]^{\frac{1}{4}} \left[\int_{\mathbb{C}} e^{-\frac{\alpha}{6}|u|^2} \, dA(u) \right]^{\frac{3}{4}}$$

$$\leq C \left[\frac{\alpha}{\pi} \int_{\mathbb{C}} F(w,u)^2 e^{-\alpha|u|^2} \, dA(u) \right]^{\frac{1}{4}}$$

$$= C \left[\int_{\mathbb{C}} F(w,z)^2 \, d\lambda_\alpha(z) \right]^{\frac{1}{4}}.$$

This proves the desired result.   □

Lemma 8.9. *Suppose* $f \in L^\infty(\mathbb{C})$. *For any* $z \in \mathbb{C}$, *we have*

$$\int_{\mathbb{C}} |P(f \circ \varphi_w)(\varphi_w(z))| |K_w(z)| K_w(w)^{\frac{1}{2}} \, d\lambda_\alpha(w) \le 4 \|f\|_\infty K_z(z)^{\frac{1}{2}},$$

and

$$\int_{\mathbb{C}} |f(z) - P(f \circ \varphi_w)(\varphi_w(z))| |K_w(z)| K_w(w)^{\frac{1}{2}} \, d\lambda_\alpha(w) \le 6 \|f\|_\infty K_z(z)^{\frac{1}{2}}.$$

Proof. It follows from (8.1) that

$$|P(f \circ \varphi_w)(\varphi_w(z))| |K_w(z)| = |P(fK_w)(z)|$$

$$= \left| \int_{\mathbb{C}} f(u) K_w(u) K(z, u) \, d\lambda_\alpha(u) \right|$$

$$\le \|f\|_\infty \int_{\mathbb{C}} |K(u, w)| |K(z, u)| \, d\lambda_\alpha(u)$$

$$= \|f\|_\infty e^{\frac{\alpha}{4}|z + w|^2}.$$

Thus the integral

$$I = \int_{\mathbb{C}} |P(f \circ \varphi_w)(\varphi_w(z))| |e^{\alpha z \overline{w} + \frac{1}{2}\alpha|w|^2}| \, d\lambda_\alpha(w)$$

satisfies the following estimates:

$$I \le \|f\|_\infty \int_{\mathbb{C}} e^{\frac{\alpha}{4}|z + w|^2 + \frac{\alpha}{2}|w|^2} \, d\lambda_\alpha(w)$$

$$= \frac{\alpha}{\pi} \|f\|_\infty e^{\frac{\alpha}{4}|z|^2} \int_{\mathbb{C}} \left| e^{\frac{\alpha}{4} z \overline{w}} \right|^2 e^{-\frac{\alpha}{4}|w|^2} \, dA(w)$$

$$= 4 \|f\|_\infty e^{\frac{\alpha}{4}|z|^2} \int_{\mathbb{C}} \left| e^{\frac{\alpha}{4} z \overline{w}} \right|^2 \, d\lambda_{\frac{\alpha}{4}}(w)$$

$$= 4 \|f\|_\infty e^{\frac{\alpha}{4}|z|^2} e^{\frac{\alpha}{4}|z|^2} = 4 \|f\|_\infty e^{\frac{\alpha}{2}|z|^2}.$$

This proves the first estimate. The second estimate follows from the triangle inequality, the first estimate, and the fact that

$$\int_{\mathbb{C}} |K_w(z)| K_w(w)^{\frac{1}{2}} \, d\lambda_\alpha(w) = 2K_z(z)^{\frac{1}{2}}. \qquad \square$$

Theorem 8.10. *Suppose* $f \in L^\infty(\mathbb{C})$. *Then*

(a) T_f *is compact if and only if* $\|P(f \circ \varphi_a)\| \to 0$ *as* $a \to \infty$.
(b) H_f *is compact if and only if* $\|f \circ \varphi_a - P(f \circ \varphi_a)\| \to 0$ *as* $a \to \infty$.

Proof. The proof of part (a) is exactly the same as the proof of part (b). The only difference is in the projection that is used in the definitions of Toeplitz operators and Hankel operators: $T_f = PM_f$ and $H_f = (I - P)M_f$. Therefore, we use Q to denote either P or $I - P$ in the rest of the proof.

Since $k_a \to 0$ weakly in F_α^2 as $a \to \infty$, the compactness of

$$QM_f : F_\alpha^2 \to Q(L_\alpha^2)$$

implies that $QM_f(k_a) \to 0$ in norm as $a \to \infty$. By Lemma 8.2,

$$\|QM_f(k_a)\|^2 = \|Q(f \circ \varphi_a)\|^2, \qquad a \in \mathbb{C}.$$

Thus the compactness of QM_f implies $\|Q(f \circ \varphi_a)\| \to 0$ as $a \to \infty$.

Next, we assume that $\|Q(f \circ \varphi_a)\| \to 0$ as $a \to \infty$ and proceed to show that the operator QM_f is compact. Obviously, it is equivalent for us to show that the operator

$$(QM_f)^* : Q(L_\alpha^2) \to F_\alpha^2 \subset L_\alpha^2$$

is compact.

Given $h \in Q(L_\alpha^2)$ and $w \in \mathbb{C}$, we use Lemma 8.2 to write

$$
\begin{aligned}
(QM_f)^*h(w) &= \langle (QM_f)^*h, K_w \rangle = \langle h, QM_f K_w \rangle \\
&= \langle h, Q(f \circ \varphi_w) \circ \varphi_w K_w \rangle \\
&= \int_{\mathbb{C}} h(z)\overline{Q(f \circ \varphi_w)(\varphi_w(z))K_w(z)}\, d\lambda_\alpha(z).
\end{aligned}
$$

For each positive number R, define an operator

$$S_R : Q(L_\alpha^2) \to L_\alpha^2$$

by

$$S_R h(w) = \chi_R(w)(QM_f)^*h(w), \qquad w \in \mathbb{C},$$

where χ_R is the characteristic function of the ball $\{u \in \mathbb{C} : |u| \le R\}$.

By Fubini's theorem and a change of variables,

$$
\begin{aligned}
\int_{\mathbb{C}} \int_{\mathbb{C}} &\chi_R(w)|Q(f \circ \varphi_w)(\varphi_w(z))|^2 |K_w(z)|^2 \, d\lambda_\alpha(z)d\lambda_\alpha(w) \\
&= \int_{|w|\le R} K_w(w)\|Q(f \circ \varphi_w)\|^2 \, d\lambda_\alpha(w) \\
&= \frac{\alpha}{\pi} \int_{|w|\le R} \|QM_f k_w\|^2 \, dA(w) \\
&\le \alpha R^2 \|QM_f\|^2 < \infty.
\end{aligned}
$$

It follows that the operator S_R is Hilbert–Schmidt. In particular, S_R is compact.

We write

$$[(QM_f)^* - S_R]\,g(w) = \int_{\mathbb{C}} H(w,z)g(z)\,d\lambda_\alpha(z), \qquad g \in Q(L_\alpha^2),$$

where

$$H(w,z) = (1 - \chi_R(w))\overline{Q(f \circ \varphi_w)(\varphi_w(z))K_w(z)}.$$

We are going to apply Schur's test to obtain an estimate on the norm of $(QM_f)^* - S_R$. To this end, we let $h(z) = \sqrt{K(z,z)}$. It follows from Lemma 8.9 that

$$\int_{\mathbb{C}} |H(w,z)|h(w)\,d\lambda_\alpha(w) \le 6\|f\|_\infty h(z)$$

for all $z \in \mathbb{C}$. On the other hand, if we write

$$F(w,z) = (1 - \chi_R(w))|Q(f \circ \varphi_w)(z)|,$$

then by Lemma 8.7,

$$F(w,z) \le 2\|f\|_\infty e^{\frac{\alpha}{4}|z|^2},$$

so we can apply Lemma 8.8. In fact, since

$$|H(w,z)| = F(w,\varphi_w(z))|K_w(z)|,$$

an application of Lemma 8.8 tells us that there exists a positive constant C, depending on f only, such that

$$\int_{\mathbb{C}} |H(w,z)|h(z)\,d\lambda_\alpha(z) \le Ch(w)\left[\int_{\mathbb{C}} F(w,z)^2\,d\lambda_\alpha(z)\right]^{\frac{1}{4}}$$

$$= Ch(w)(1 - \chi_R(w))\|Q(f \circ \varphi_w)\|^{\frac{1}{2}}.$$

By Schur's test, there exists a positive constant C such that

$$\|(QM_f)^* - S_R\| \le C\sup\left\{\|Q(f \circ \varphi_w)\|^{1/4} : |w| > R\right\}.$$

This shows that the condition

$$\lim_{a \to \infty} \|Q(f \circ \varphi_a)\| = 0$$

implies that

$$\lim_{R \to \infty} \|(QM_f)^* - S_R\| = 0.$$

In other words, $(QM_f)^*$ can be approximated in norm by compact operators, and so it must be compact as well. This completes the proof of the theorem. $\qquad\square$

Lemma 8.11. *For any $f \in L^\infty(\mathbb{C})$, there exists a positive constant C such that*

$$\|\widetilde{f} \circ \varphi_a - P(f \circ \varphi_a)\| \leq C \|f \circ \varphi_a - P(f \circ \varphi_a)\|^{\frac{1}{4}}$$

for all $a \in \mathbb{C}$.

Proof. It follows from Corollary 2.5 that

$$|\widetilde{f}(w) - Pf(w)| \leq 2\|f\|_\infty e^{\frac{\alpha}{4}|w|^2} \tag{8.4}$$

for all $w \in \mathbb{C}$. Since the Berezin transform fixes entire functions, we have

$$\widetilde{f}(w) - Pf(w) = \int_{\mathbb{C}} (f(z) - Pf(z))|k_w(z)|^2 \, d\lambda_\alpha(z)$$

so that

$$|\widetilde{f}(w) - Pf(w)| \leq e^{-\alpha|w|^2} \int_{\mathbb{C}} |f(z) - Pf(z)||K_w(z)|^2 \, d\lambda_\alpha(z) \tag{8.5}$$

for all $w \in \mathbb{C}$. By (8.4),

$$\|\widetilde{f} - Pf\|^2 = \frac{\alpha}{\pi} \int_{\mathbb{C}} |\widetilde{f}(w) - Pf(w)|^2 e^{-\alpha|w|^2} \, dA(w)$$

$$\leq \frac{2\alpha}{\pi} \|f\|_\infty \int_{\mathbb{C}} |\widetilde{f}(w) - Pf(w)| e^{-\frac{3}{4}\alpha|w|^2} \, dA(w).$$

Using (8.5), Fubini's theorem, and Corollary 2.5, we arrive at

$$\|\widetilde{f} - Pf\|^2 = \frac{8}{7} \|f\|_\infty \int_{\mathbb{C}} |f(z) - Pf(z)| e^{\frac{4}{7}\alpha|z|^2} \, d\lambda_\alpha(z)$$

$$= \frac{8\alpha}{7\pi} \|f\|_\infty \int_{\mathbb{C}} |f(z) - Pf(z)| e^{-\frac{3}{8}\alpha|z|^2} e^{-\frac{3}{56}\alpha|z|^2} \, dA(z).$$

Applying Hölder's inequality (with exponents 4 and 4/3) and Lemma 8.7, we obtain

$$\|\widetilde{f} - Pf\|^2 \leq C_1 \|f\|_\infty \left[\int_{\mathbb{C}} |f(z) - Pf(z)|^4 e^{-\frac{3}{2}\alpha|z|^2} \, dA(z) \right]^{\frac{1}{4}}$$

$$\leq C_2 \|f\|_\infty^{\frac{3}{2}} \left[\int_{\mathbb{C}} |f(z) - Pf(z)|^2 \, d\lambda_\alpha(z) \right]^{\frac{1}{4}}$$

$$= C_2 \|f\|_\infty^{\frac{3}{2}} \|f - Pf\|^{1/2}.$$

This shows that

$$\|\tilde{f} - Pf\| \le C\|f\|_{\infty}^{\frac{3}{4}}\|f - Pf\|^{1/4},$$

where the constant C is independent of f. Replacing f by $f \circ \varphi_a$ and using the translation invariance of the Berezin transform, we obtain the desired estimate. □

Lemma 8.12. *If* $f \in L^{\infty}(\mathbb{C})$ *and* H_f *is compact, then both* $H_{\tilde{f}}$ *and* $T_{f-\tilde{f}}$ *are compact.*

Proof. By Theorem 8.10,

$$\lim_{a \to \infty} \|f \circ \varphi_a - P(f \circ \varphi_a)\| = 0,$$

which, according to Lemma 8.11, implies that

$$\lim_{a \to \infty} \|\tilde{f} \circ \varphi_a - P(f \circ \varphi_a)\| = 0.$$

Since the projection P is bounded on L_{α}^2, we also have

$$\lim_{a \to \infty} \|P(\tilde{f} \circ \varphi_a) - P(f \circ \varphi_a)\| = 0.$$

By part (a) of Theorem 8.10, the Toeplitz operator $T_{f-\tilde{f}}$ is compact. Since

$$\|\tilde{f} \circ \varphi_a - P(\tilde{f} \circ \varphi_a)\| \le \|\tilde{f} \circ \varphi_a - P(f \circ \varphi_a)\| + \|P(f \circ \varphi_a) - P(\tilde{f} \circ \varphi_a)\|,$$

we see that

$$\lim_{a \to \infty} \|\tilde{f} \circ \varphi_a - P(\tilde{f} \circ \varphi_a)\| = 0,$$

which, in view of part (b) of Theorem 8.10, shows that $H_{\tilde{f}}$ is compact. □

Theorem 8.13. *Suppose* $f \in L^{\infty}(\mathbb{C})$. *Then* H_f *is compact if and only if* $H_{\tilde{f}}$ *is compact.*

Proof. Let $g = \tilde{f}$ and assume that H_g is compact. By Theorem 8.10,

$$\lim_{a \to \infty} \|g \circ \varphi_a - P(g \circ \varphi_a)\| = 0.$$

Combining this with Lemma 8.11, we see that

$$\lim_{a \to \infty} \|\tilde{g} \circ \varphi_a - P(g \circ \varphi_a)\| = 0,$$

and so by the triangle inequality,

$$\lim_{a \to \infty} \|g \circ \varphi_a - \tilde{g} \circ \varphi_a\| = 0.$$

Since complex conjugation commutes with the Berezin transform, we also have

$$\lim_{a \to \infty} \| f \circ \varphi_a - \widetilde{f} \circ \varphi_a \| = 0,$$

which implies that $H_{f-\widetilde{f}}$ is compact. Using Lemma 8.12 and iteration, we conclude that $H_{f-\widetilde{f}^{(m)}}$ is compact for every positive integer m.

On the other hand, Theorem 3.25 shows that $f \in L^\infty(\mathbb{C})$ implies

$$|\widetilde{f}^{(m)}(z) - \widetilde{f}^{(m)}(w)| \le \frac{C}{\sqrt{m}} |z - w|,$$

which, along with Lemma 8.1, shows that $\| H_{\widetilde{f}^{(m)}} \| \to 0$ as $m \to \infty$. This, combined with the fact that each $H_{f-\widetilde{f}^{(m)}}$ is compact, shows that H_f is compact. □

8.3 Membership in Schatten Classes

In this section, we characterize when the Hankel operators H_f and $H_{\bar{f}}$ belong to the Schatten class S_p simultaneously. Throughout the section, we fix a positive radius r and write

$$MO_r(f)(z) = \left[\widetilde{|f|^2}_r(z) - |\widehat{f}_r(z)|^2\right]^{\frac{1}{2}},$$

and

$$MO(f)(z) = \left[\widetilde{|f|^2}(z) - |\widetilde{f}(z)|^2\right]^{\frac{1}{2}}.$$

Lemma 8.14. *Let* $2 \leq p < \infty$. *If* H_f *and* $H_{\bar{f}}$ *are both in the Schatten class* S_p, *then* $MO(f) \in L^p(\mathbb{C}, dA)$.

Proof. If H_f is in S_p, then $(H_f^* H_f)^{p/2}$ is in the trace class S_1, so by Proposition 3.3,

$$\int_{\mathbb{C}} \langle (H_f^* H_f)^{p/2} k_z, k_z \rangle \, dA(z) < \infty,$$

where k_z are the normalized reproducing kernels of F_α^2. By Lemma 3.4,

$$\int_{\mathbb{C}} \langle H_f^* H_f k_z, k_z \rangle^{p/2} \, dA(z) < \infty,$$

or

$$\int_{\mathbb{C}} \|H_f k_z\|^p \, dA(z) < \infty.$$

Similarly, if $H_{\bar{f}}$ is in S_p, then

$$\int_{\mathbb{C}} \|H_{\bar{f}} k_z\|^p \, dA(z) < \infty.$$

The desired result then follows from Proposition 8.3. $\qquad\qquad\qquad\square$

Lemma 8.15. *Let* $0 < p \leq 2$. *If* $MO(f) \in L^p(\mathbb{C}, dA)$, *then both* H_f *and* $H_{\bar{f}}$ *are in the Schatten class* S_p.

Proof. By Proposition 8.3, the condition $MO(f) \in L^p(\mathbb{C}, dA)$ implies that the function $z \mapsto \|H_f k_z\|$ is in $L^p(\mathbb{C}, dA)$. This, along with Proposition 3.3 and Lemma 3.4, shows that

$$\mathrm{tr}\left[(H_f^* H_f)^{p/2}\right] = \frac{\alpha}{\pi} \int_{\mathbb{C}} \langle (H_f^* H_f)^{p/2} k_z, k_z \rangle \, dA(z)$$

$$\leq \frac{\alpha}{\pi} \int_{\mathbb{C}} \langle H_f^* H_f k_z, k_z \rangle^{p/2} \, dA(z)$$

$$= \frac{\alpha}{\pi} \int_{\mathbb{C}} \|H_f k_z\|^p \, dA(z) < \infty.$$

Therefore, $H_f \in S_p$. Since the condition $MO(f) \in L^p(\mathbb{C}, dA)$ is closed under complex conjugation, we also have $H_{\bar{f}} \in S_p$. $\square$

Lemma 8.16. *Suppose $2 \leq p < \infty$ and T is the integral operator defined by*

$$Tf(z) = \int_{\mathbb{C}} G(z,w) K(z,w) f(w) \, d\lambda_\alpha(w),$$

where G is a measurable function on $\mathbb{C} \times \mathbb{C}$ and $K(z,w)$ is the reproducing kernel of F_α^2. If

$$\int_{\mathbb{C}} \int_{\mathbb{C}} |G(z,w)|^p |K(z,w)|^2 \, d\lambda_\alpha(z) \, d\lambda_\alpha(w) < \infty,$$

then T is in the Schatten class S_p of L_α^2.

Proof. The case $p = 2$ follows from the classical characterization of Hilbert–Schmidt integral operators on L^2 spaces; see [113]. If $G \in L^\infty(\mathbb{C} \times \mathbb{C})$, then T is dominated by the bounded operator Q_α considered in Sect. 2.2, so the operator T is bounded on L_α^2 as well. The case $2 < p < \infty$ then follows from complex interpolation. $\square$

Lemma 8.17. *Let $1 \leq p < \infty$. There exists a positive constant $C = C_p$ such that*

$$\int_{\mathbb{C}} |\tilde{f}(z) - \tilde{f}(0)|^p \, d\lambda_\alpha(z) \leq C \int_{\mathbb{C}} \frac{1 + |z|^{p-1}}{|z|} [MO(f)(z)]^p \, d\lambda_\alpha(z) \qquad (8.6)$$

for all f.

Proof. Recall from the proof of Theorem 3.35 that there exists a positive constant $C = C(\alpha)$ such that

$$\left| \frac{d}{dt} \tilde{f}(tz/|z|) \right| \leq C MO(f)(tz/|z|)$$

for all $t \geq 0$ and $z \in \mathbb{C} - \{0\}$. Thus,

$$|\tilde{f}(z) - \tilde{f}(0)| = \left| \int_0^{|z|} \frac{d}{dt} \tilde{f}(tz/|z|) \, dt \right|$$

$$\leq C \int_0^{|z|} MO(f)(tz/|z|) \, dt$$

$$= C|z| \int_0^1 MO(f)(tz) \, dt.$$

Since $p \geq 1$, an application of Hölder's inequality gives

$$|\widetilde{f}(z) - \widetilde{f}(0)|^p \leq C^p |z|^p \int_0^1 MO(f)^p(tz) \, dt.$$

This, along with Fubini's theorem, shows that the integral

$$I = \int_{\mathbb{C}} |\widetilde{f}(z) - \widetilde{f}(0)|^p \, d\lambda_\alpha(z)$$

satisfies

$$
\begin{aligned}
I &\leq C^p \int_{\mathbb{C}} |z|^p \, d\lambda_\alpha(z) \int_0^1 MO(f)^p(tz) \, dt \\
&= C^p \int_0^1 dt \int_{\mathbb{C}} |z|^p MO(f)^p(tz) \, d\lambda_\alpha(z) \\
&= C' \int_0^1 dt \int_{\mathbb{C}} |z|^p e^{-\alpha|z|^2} MO(f)^p(tz) \, dA(z) \\
&= C' \int_0^1 \frac{dt}{t^{2+p}} \int_{\mathbb{C}} |z|^p e^{-\alpha|z|^2/t^2} MO(f)^p(z) \, dA(z) \\
&= C' \int_{\mathbb{C}} |z|^p MO(f)^p(z) \, dA(z) \int_0^1 t^{-(2+p)} e^{-\alpha|z|^2/t^2} \, dt \\
&= C' \int_{\mathbb{C}} |z|^p MO(f)^p(z) \, dA(z) \int_1^\infty t^p e^{-\alpha t^2 |z|^2} \, dt \\
&= C' \int_{\mathbb{C}} \frac{MO(f)^p(z)}{|z|} \, dA(z) \int_{|z|}^\infty t^p e^{-\alpha t^2} \, dt,
\end{aligned}
$$

where $C' = C\alpha/\pi$. By L'Hôpital's rule,

$$\lim_{|z| \to \infty} \frac{\displaystyle\int_{|z|}^\infty t^p e^{-\alpha t^2} \, dt}{|z|^{p-1} e^{-\alpha|z|^2}} = \frac{1}{2\alpha}.$$

It follows that there exists another constant $C > 0$, independent of z, such that

$$\int_{|z|}^\infty t^p e^{-\alpha t^2} \, dt \leq C \left(1 + |z|^{p-1}\right) e^{-\alpha|z|^2}$$

for all $z \in \mathbb{C}$. This proves the desired estimate. $\qquad\square$

Lemma 8.18. *Suppose $2 \leq p < \infty$ and $MO(f) \in L^p(\mathbb{C}, dA)$. Then both H_f and $H_{\overline{f}}$ are in the Schatten class S_p.*

Proof. First, consider the integral

$$I = \int_{\mathbb{C}} \int_{\mathbb{C}} |\tilde{f}(z) - \tilde{f}(w)|^p |K(z,w)|^2 \, d\lambda_\alpha(z) \, d\lambda_\alpha(w).$$

By Fubini's theorem and the change of variables formula, we have

$$
\begin{aligned}
I &= \frac{\alpha}{\pi} \int_{\mathbb{C}} dA(z) \int_{\mathbb{C}} |\tilde{f}(z) - \tilde{f}(w)|^p |k_z(w)|^2 \, d\lambda_\alpha(w) \\
&= \frac{\alpha}{\pi} \int_{\mathbb{C}} dA(z) \int_{\mathbb{C}} |\tilde{f}(z) - \tilde{f}(z-w)|^p \, d\lambda_\alpha(w) \\
&= \frac{\alpha}{\pi} \int_{\mathbb{C}} dA(z) \int_{\mathbb{C}} |\widetilde{f \circ \varphi_z}(0) - \widetilde{f \circ \varphi_z}(w)|^p \, d\lambda_\alpha(w),
\end{aligned}
$$

where $\varphi_z(w) = z - w$. By Lemma 8.17 and the invariance of the Berezin transform under the action of φ_z, there exists a positive constant C, independent of f, such that

$$
\begin{aligned}
I &\leq C \int_{\mathbb{C}} dA(z) \int_{\mathbb{C}} \varphi(w) MO(f \circ \varphi_z)^p(w) \, d\lambda_\alpha(w) \\
&= C \int_{\mathbb{C}} dA(z) \int_{\mathbb{C}} \varphi(w) MO(f)^p(\varphi_z(w)) \, d\lambda_\alpha(w),
\end{aligned}
$$

where $\varphi(w) = (1 + |w|^{p-1})/|w|$. Changing variables again and applying Fubini's theorem, we obtain

$$
\begin{aligned}
I &\leq C \int_{\mathbb{C}} dA(z) \int_{\mathbb{C}} \varphi(\varphi_z(w)) MO(f)^p(w) |k_z(w)|^2 \, d\lambda_\alpha(w) \\
&= C \int_{\mathbb{C}} MO(f)^p(w) \, dA(w) \int_{\mathbb{C}} \varphi(\varphi_z(w)) |k_w(z)|^2 \, d\lambda_\alpha(z) \\
&= C \int_{\mathbb{C}} MO(f)^p(w) \, dA(w) \int_{\mathbb{C}} \varphi(u) \, d\lambda_\alpha(u).
\end{aligned}
$$

It is clear that the integral

$$\int_{\mathbb{C}} \varphi(u) \, d\lambda_\alpha(u) = \frac{\alpha}{\pi} \int_{\mathbb{C}} \frac{1 + |z|^{p-1}}{|z|} e^{-\alpha|z|^2} \, dA(z)$$

converges. It follows that $I < \infty$, and by Lemma 8.16, the Hankel operator $H_{\tilde{f}}$ belongs to S_p.

Next, we consider the function $g = f - \tilde{f}$. By the triangle inequality,

$$\left[\widetilde{|g|^2}(z) \right]^{\frac{1}{2}} = \left[\int_{\mathbb{C}} |f(w) - \tilde{f}(w)|^2 |k_z(w)|^2 \, d\lambda_\alpha(w) \right]^{\frac{1}{2}}$$

$$\leq \left[\int_{\mathbb{C}} |f(w) - \widetilde{f}(z)|^2 |k_z(w)|^2 \, d\lambda_\alpha(w) \right]^{\frac{1}{2}}$$

$$+ \left[\int_{\mathbb{C}} |\widetilde{f}(z) - \widetilde{f}(w)|^2 |k_z(w)|^2 \, d\lambda_\alpha(w) \right]^{\frac{1}{2}}$$

$$= MO(f)(z) + \left[\int_{\mathbb{C}} |\widetilde{f \circ \varphi_z}(0) - \widetilde{f \circ \varphi_z}(w)|^2 \, d\lambda_\alpha(w) \right]^{\frac{1}{2}}.$$

By assumption, the first term above is in $L^p(\mathbb{C}, dA)$. The second term is also in $L^p(\mathbb{C}, dA)$. In fact, since $p \geq 2$, an application of Hölder's inequality gives

$$\int_{\mathbb{C}} \left[\int_{\mathbb{C}} |\widetilde{f \circ \varphi_z}(w) - \widetilde{f \circ \varphi_z}(0)|^2 \, d\lambda_\alpha(w) \right]^{\frac{p}{2}} \, dA(z)$$

$$\leq \int_{\mathbb{C}} dA(z) \int_{\mathbb{C}} |\widetilde{f \circ \varphi_z}(w) - \widetilde{f \circ \varphi_z}(0)|^p \, d\lambda_\alpha(w)$$

$$\leq C \int_{\mathbb{C}} MO(f)^p(w) \, dA(w).$$

The last inequality above was proved in the previous paragraph. We conclude that the function $\sqrt{|g|^2}$ belongs to $L^p(\mathbb{C}, dA)$. In other words, the function $\widetilde{|g|^2}$ belongs to $L^{p/2}(\mathbb{C}, dA)$. By Corollary 6.33, the Toeplitz operator $T_{|g|^2}$ belongs to the Schatten class $S_{p/2}$. Since

$$H_g^* H_g = T_{|g|^2} - T_{\bar{g}} T_g \leq T_{|g|^2},$$

the operator $H_g^* H_g$ belongs to the Schatten class $S_{p/2}$. This shows that H_g belongs to S_p, and consequently, $H_f = H_{\widetilde{f}} + H_{f-\widetilde{f}}$ belongs to S_p. The condition $MO(f) \in L^p(\mathbb{C}, dA)$ is closed under complex conjugation, so we must have $H_{\bar{f}} \in S_p$ as well. $\qquad \square$

Recall that $\mathbb{Z}$ denotes the additive integer group and

$$\mathbb{Z}^2 = \{n + im : n, m \in \mathbb{Z}\}$$

is the lattice of integers in the complex plane. Throughout this section, we fix a positive integer N and consider the finer lattice

$$\frac{1}{N} \mathbb{Z}^2 = \left\{ \frac{n + im}{N} : n, m \in \mathbb{Z} \right\}.$$

We also consider the following two special squares in the complex plane:

$$S_N = \left\{ x + iy : 0 \leq x < \frac{1}{N}, 0 \leq y < \frac{1}{N} \right\},$$

and

$$Q_N = \left\{ x + iy : -\frac{1}{N} \le x < \frac{2}{N}, -\frac{1}{N} \le y < \frac{2}{N} \right\}.$$

If f is a Lebesgue measurable function on the complex plane, we write

$$J_N(f) = \int_{Q_N} \int_{Q_N} |f(u) - f(v)|^2 \, dA(u) \, dA(v).$$

If E is a measurable set in $\mathbb{C}$ with $0 < A(E) < \infty$ and f is integrable on E, we use

$$f_E = \frac{1}{A(E)} \int_E f \, dA$$

to denote the average (mean) of f over the set E.

Lemma 8.19. *Suppose f is locally square integrable and $v \in \mathbb{Z}^2/N$. Then*

$$\int_{S_N} |f \circ t_v - f_{S_N}|^2 \, dA \le \left(N^2 + \frac{4N^4 |\gamma(v)|}{9} \right) \sum_{a \in \gamma(v)} J_N(f \circ t_a),$$

where $t_a(z) = z + a$ is the translation by a and $\gamma(v)$ is the canonical path in $\mathbb{Z}^2/N$ from v to 0 (see Sect. 1.2).

Proof. The case $v = 0$ is trivial. If $v \neq 0$, we write

$$\gamma(v) = \{a_0, a_1, \ldots, a_l\}$$

in the order in which $\gamma(v)$ is defined, where $l + 1 = |\gamma(v)|$ is the length of the path $\gamma(v)$. It is clear that

$$(S_N + a_{j-1}) \cup (S_N + a_j) \subset Q_N + a_{j-1}, \qquad 1 \le j \le l.$$

We will estimate the integral

$$I = \int_{S_N} |f \circ t_v - f_{S_N}|^2 \, dA$$

using the elementary inequality

$$|z_1 + \cdots + z_k|^2 \le k(|z_1|^2 + \cdots + |z_k|^2)$$

along with several natural "telescoping" decompositions.

We begin with the estimate

$$I = \int_{S_N} |f \circ t_{a_l} - (f \circ t_{a_0})_{S_N}|^2 \, dA$$

$$\leq 2 \int_{S_N} \left[|f \circ t_{a_l} - (f \circ t_{a_l})_{S_N}|^2 + |(f \circ t_{a_l})_{S_N} - (f \circ t_{a_0})_{S_N}|^2 \right] dA.$$

It is easy to see that

$$2 \int_{S_N} |f \circ t_{a_l} - (f \circ t_{a_l})_{S_N}|^2 \, dA$$

$$= \frac{1}{A(S_N)} \int_{S_N} \int_{S_N} |f \circ t_{a_l}(u) - f \circ t_{a_l}(v)|^2 \, dA(u) \, dA(v)$$

$$\leq N^2 J_N(f \circ t_{a_l}).$$

On the other hand,

$$2 \int_{S_N} |(f \circ t_{a_l})_{S_N} - (f \circ t_{a_0})_{S_N}|^2 \, dA$$

$$\leq 2l \sum_{j=1}^{l} \int_{S_N} |(f \circ t_{a_j})_{S_N} - (f \circ t_{a_{j-1}})_{S_N}|^2 \, dA$$

$$\leq 4l \sum_{j=1}^{l} \int_{S_N} \left[|(f \circ t_{a_j})_{S_N} - (f \circ t_{a_{j-1}})_{Q_N}|^2 \right.$$

$$\left. + |(f \circ t_{a_{j-1}})_{Q_N} - (f \circ t_{a_{j-1}})_{S_N}|^2 \right] dA.$$

Thus the quantity

$$D = |(f \circ t_{a_j})_{S_N} - (f \circ t_{a_{j-1}})_{Q_N}|^2$$

can be estimated as follows:

$$D = \left| \frac{1}{A(S_N)} \int_{S_N} [f \circ t_{a_j} - (f \circ t_{a_{j-1}})_{Q_N}] \, dA \right|^2$$

$$\leq N^2 \int_{S_N + a_j} |f - (f \circ t_{a_{j-1}})_{Q_N}|^2 \, dA$$

$$\leq N^2 \int_{Q_N + a_{j-1}} |f - (f \circ t_{a_{j-1}})_{Q_N}|^2 \, dA$$

$$= N^2 \int_{Q_N} |f \circ t_{a_{j-1}} - (f \circ t_{a_{j-1}})_{Q_N}|^2 \, dA$$

$$= \frac{N^4}{18} J_N(f \circ t_{a_{j-1}}).$$

Similarly,

$$|(f \circ t_{a_{j-1}})_{Q_N} - (f \circ t_{a_{j-1}})_{S_N}|^2 \leq \frac{N^4}{18} J_N(f \circ t_{a_{j-1}}).$$

Therefore,

$$2\int_{S_N} |(f \circ t_{a_l})_{S_N} - (f \circ t_{a_0})_{S_N}|^2 \, dA \leq \frac{4lN^4}{9} \sum_{j=1}^{l} J_N(f \circ t_{a_{j-1}}).$$

This proves the desired result. $\square$

Lemma 8.20. *Suppose f satisfies condition (I_2). There exists a positive constant $C = C_N$ (depending on N) such that*

$$\sup_{z \in S_N} MO(f)^2(z) \leq C \sum_{v \in \mathbb{Z}^2/N} \sum_{a \in \gamma(v)} e^{-\alpha|v|^2/3} J_N(f \circ t_a).$$

Proof. For any constant c, we have

$$\int_{\mathbb{C}} |f \circ t_z - c|^2 \, d\lambda_\alpha = \widetilde{|f|^2}(z) - \overline{c}\widetilde{f}(z) - c\overline{\widetilde{f}(z)} + |c|^2$$

$$= \widetilde{|f|^2}(z) - |\widetilde{f}(z)|^2 + |\widetilde{f}(z) - c|^2$$

$$\geq \widetilde{|f|^2}(z) - |\widetilde{f}(z)|^2.$$

Thus, for any $z \in \mathbb{C}$, we have

$$MO(f)^2(z) \leq \int_{\mathbb{C}} |f \circ t_z - f_{S_N}|^2 \, d\lambda_\alpha$$

$$= \sum_{v \in \mathbb{Z}^2/N} \int_{S_N+v+z} |f(w+z) - f_{S_N}|^2 \, d\lambda_\alpha(w)$$

$$= \frac{\alpha}{\pi} \sum_{v \in \mathbb{Z}^2/N} \int_{S_N+v} |f(w) - f_{S_N}|^2 e^{-\alpha|w-z|^2} \, dA(w)$$

$$= \frac{\alpha}{\pi} \sum_{v \in \mathbb{Z}^2/N} \int_{S_N} |f \circ t_v(w) - f_{S_N}|^2 e^{-\alpha|w-z+v|^2} \, dA(w).$$

For w and z in S_N, we have

$$|w - z + v|^2 \geq |v|^2 + |w - z|^2 - 2|w - z||v|$$

$$\geq |v|^2/2 - |w - z|^2$$

$$\geq |v|^2/2 - N^{-2}.$$

It follows from this and Lemma 8.19 that

$$MO(f)^2(z) \le \frac{\alpha}{\pi} e^{\frac{\alpha}{N^2}} \sum_{v \in \mathbb{Z}^2/N} e^{-\frac{\alpha}{2}|v|^2} \int_{S_N} |f \circ t_v - f_{S_N}|^2 \, dA$$

$$\le \frac{\alpha}{\pi} e^{\frac{\alpha}{N^2}} \sum_{v \in \mathbb{Z}^2/N} e^{-\frac{\alpha}{2}|v|^2} \left[N^2 + \frac{4N^4|\gamma(v)|}{9} \right] \sum_{a \in \gamma(v)} J_N(f \circ t_a).$$

Since the length of $\gamma(v)$ is comparable to $|v|$, it is clear that we can find a constant $C = C_N$ such that

$$\frac{\alpha}{\pi} e^{\frac{\alpha}{N^2}} \left[N^2 + \frac{4N^4|\gamma(v)|}{9} \right] e^{-\frac{\alpha}{2}|v|^2} \le C_N e^{-\frac{\alpha}{3}|v|^2}$$

for all v. This proves the desired result. $\square$

Lemma 8.21. *Suppose f satisfies condition (I_2). If $0 < p \le 2$, then there exists a positive constant $C = C_N$, depending on N and p but not on f, such that*

$$\int_{\mathbb{C}} [MO(f)(z)]^p \, dA(z) \le C_N \sum_{b \in \mathbb{Z}^2/N} [J_N(f \circ t_b)]^{\frac{p}{2}}.$$

Proof. Let us consider the integral

$$I = \int_{\mathbb{C}} [MO(f)(z)]^p \, dA(z).$$

It is clear that

$$\mathbb{C} = \bigcup \left\{ S_N + u : u \in \frac{\mathbb{Z}^2}{N} \right\},$$

and this is a disjoint union. It follows that

$$I = \sum_{u \in \mathbb{Z}^2/N} \int_{S_N + u} [MO(f)(z)]^p \, dA(z)$$

$$\le \frac{1}{N^2} \sum_{u \in \mathbb{Z}^2/N} \sup\{MO(f)^p(z) : z \in S_N + u\}$$

$$= \frac{1}{N^2} \sum_{u \in \mathbb{Z}^2/N} \sup\{MO(f)^p(u + z) : z \in S_N\}$$

$$= \frac{1}{N^2} \sum_{u \in \mathbb{Z}^2/N} \sup\{MO(f \circ t_u)^p(z) : z \in S_N\}.$$

Since $0 < p \le 2$, it follows from Lemma 8.20 and Hölder's inequality that

$$\sup_{z \in S_N} MO(f \circ t_u)^p(z) \le C_N \sum_{v \in \mathbb{Z}^2/N} \sum_{a \in \gamma(v)} e^{-\frac{p\alpha}{6}|v|^2} [J_N(f \circ t_u \circ t_a)]^{\frac{p}{2}}.$$

Since $t_u \circ t_a = t_{u+a}$, we have

$$I \le \frac{C_N}{N^2} \sum_{u \in \mathbb{Z}^2/N} \sum_{v \in \mathbb{Z}^2/N} \sum_{a \in \gamma(v)} e^{-\frac{p\alpha}{6}|v|^2} [J_N(f \circ t_{u+a})]^{\frac{p}{2}}$$

$$= \frac{C_N}{N^2} \sum_{v \in \mathbb{Z}^2/N} e^{-\frac{p\alpha}{6}|v|^2} \sum_{a \in \gamma(v)} \sum_{u \in \mathbb{Z}^2/N} [J_N(f \circ t_{u+a})]^{\frac{p}{2}}$$

$$= \frac{C_N}{N^2} \sum_{v \in \mathbb{Z}^2/N} |\gamma(v)| e^{-\frac{p\alpha}{6}|v|^2} \sum_{b \in \mathbb{Z}^2/N} [J_N(f \circ t_b)]^{\frac{p}{2}},$$

where $|\gamma(v)|$ is the length of the path $\gamma(v)$. Again, since $|\gamma(v)|$ is comparable to $|v|$, the series

$$\sum_{v \in \mathbb{Z}^2/N} |\gamma(v)| e^{-\frac{p\alpha}{6}|v|^2}$$

converges. This proves the desired estimate. $\square$

Lemma 8.22. *There exist a positive integer N and a positive constant C_N such that*

$$I_v(f) \ge C_N J_N(f \circ t_v), \quad v \in \frac{1}{N}\mathbb{Z}^2,$$

for all locally square integrable f, where $I_v(f)$ denotes the integral

$$\int_{Q_N+v} \left| \int_{Q_N+v} (f(z) - f(w)) e^{\alpha z \overline{w} - \frac{\alpha}{2}|w|^2 - \mathrm{i}\mathrm{Im}\,(\alpha v \overline{w})} \, dA(w) \right|^2 e^{-\alpha|z|^2} \, dA(z).$$

Proof. We can write $I_v(f)$ as

$$\int_{Q_N+v} \left| \int_{Q_N+v} (f(z) - f(w)) e^{-\frac{\alpha}{2}|z-w|^2 + \mathrm{i}\alpha \mathrm{Im}\,(z\overline{w} - v\overline{w})} \, dA(w) \right|^2 \, dA(z),$$

which, after a simultaneous change of variables and some simplifications, becomes

$$\int_{Q_N} \left| \int_{Q_N} (f \circ t_v(z) - f \circ t_v(w)) e^{-\frac{\alpha}{2}|z-w|^2 + \mathrm{i}\alpha \mathrm{Im}\,(z\overline{w})} \, dA(w) \right|^2 \, dA(z).$$

Fix any $\delta \in (0, 1/4)$ and choose a positive integer N such that

$$e^{-\frac{\alpha}{2}|z-w|^2 + i\alpha \mathrm{Im}\,(z\bar{w})} = 1 + \gamma_{z,w}, \qquad |\gamma_{z,w}| < \delta,$$

for all $(z, w) \in Q_N \times Q_N$. To compress the expressions below, we write $\gamma = \gamma_{z,w}$. Then, for any $z \in Q_N$, we deduce from the triangle inequality that the quantity

$$\left| \int_{Q_N} (f \circ t_v(z) - f \circ t_v)(1 + \gamma)\,dA \right|^2$$

is greater than or equal to

$$\left[\left| \int_{Q_N} (f \circ t_v(z) - f \circ t_v)\,dA \right| - \left| \int_{Q_N} (f \circ t_v(z) - f \circ t_v)\gamma\,dA \right| \right]^2,$$

which is greater than or equal to

$$\left| \int_{Q_N} (f \circ t_v(z) - f \circ t_v)\,dA \right|^2$$

minus

$$2 \left| \int_{Q_N} (f \circ t_v(z) - f \circ t_v)\,dA \right| \left| \int_{Q_N} (f \circ t_v(z) - f \circ t_v)\gamma\,dA \right|,$$

which is greater than or equal to

$$\left| \int_{Q_N} (f \circ t_v(z) - f \circ t_v)\,dA \right|^2 - 2\delta \left[\int_{Q_N} |f \circ t_v(z) - f \circ t_v|\,dA \right]^2.$$

It follows that

$$I_v(f) \geq \int_{Q_N} \left| \int_{Q_N} (f \circ t_v(z) - f \circ t_v(w))\,dA(w) \right|^2 dA(z)$$

$$- 2\delta \int_{Q_N} \left[\int_{Q_N} |f \circ t_v(z) - f \circ t_v(w)|\,dA(w) \right]^2 dA(z).$$

The first integral above can be written as $[9/(2N^2)]J_N(f \circ t_v)$, and according to the Cauchy–Schwarz inequality, the second integral above is less than or equal to $(9/N^2)J_N(f \circ t_v)$. We conclude that

$$I_v(f) \geq \frac{9}{N^2} \left(\frac{1}{2} - 2\delta \right) J_N(f \circ t_v).$$

This completes the proof of the lemma. $\qquad\qquad\qquad\qquad\qquad\qquad\qquad\qquad$ □

In the remainder of this section, we fix a positive integer N such that Lemma 8.22 holds. We will need to decompose the lattice $\mathbb{Z}^2/N$ into more sparse sublattices. To this end, we fix another positive integer M whose magnitude will be specified later. For any $j = (j_1, j_2)$, where each $j_k \in \{1, 2, \ldots, M\}$, we let

$$\Lambda_j^M = \left\{ v = \left(\frac{v_1}{N}, \frac{v_2}{N} \right) : v_k = j_k \bmod M, k = 1, 2 \right\}.$$

It is clear that

$$\frac{\mathbb{Z}^2}{N} = \bigcup_{j_1, j_2 = 1}^{M} \Lambda_j^M,$$

the sublattices Λ_j^M are disjoint, and the distance between any two points in the same Λ_j^M is at least M/N.

Lemma 8.23. *Suppose $0 < p < \infty$ and f satisfies condition (I_2). Then the Hankel operators H_f and $H_{\bar{f}}$ both belong to the Schatten class S_p if and only if the commutator $[M_f, P] = M_f P - P M_f$ belongs to the Schatten class S_p.*

Proof. It is easy to see that

$$[M_f, P] = [M_f, P]P + [M_f, P](I - P) = H_f - H_{\bar{f}}^*.$$

So the simultaneous membership of H_f and $H_{\bar{f}}$ in S_p implies that $[M_f, P]$ is in S_p. To prove the other direction, note that

$$[M_f, P]P = (M_f P - P M_f)P = M_f P - P M_f P = (I - P)M_f P.$$

So the Hankel operator $H_f : F_\alpha^2 \to L_\alpha^2$ is just the restriction of $[M_f, P]$ on the space F_α^2. It follows that the membership of $[M_f, P]$ in S_p implies the membership of H_f in S_p. But the condition $[M_f, P] \in S_p$ implies $[M_{\bar{f}}, P] \in S_p$, so $[M_f, P] \in S_p$ implies that both H_f and $H_{\bar{f}}$ are in S_p. $\square$

Lemma 8.24. *For any $2 \leq p < \infty$, there exists a positive constant C (depending on N but independent of f) such that*

$$\|[M_f, P]\|_{S_p}^p \leq C \sum_{v \in \mathbb{Z}^2/N} J_N(f \circ t_v)^{p/2}$$

for all $f \in L_{\text{local}}^2(\mathbb{C}, dA)$.

Proof. If $f \in L_{\text{local}}^2(\mathbb{C}, dA)$, then

$$M_{\chi_E}[M_f, P]M_{\chi_E} \in S_2 \subset S_p, \quad 2 \leq p < \infty.$$

Here E is any bounded Borel set in $\mathbb{C}$. Therefore, it suffices to show that there exists a positive constant C, independent of f and E, such that

$$\|M_{\chi_E}[M_f,P]M_{\chi_E}\|_{S_p}^p \le C \sum_{u\in\mathbb{Z}^2/N} J_N(f\circ t_u)^{p/2} \tag{8.7}$$

for all bounded E and $f \in L^2_{\text{local}}(\mathbb{C},dA)$.

Fix a bounded Borel set E and let F be any finite set in $\mathbb{Z}^2$ such that

$$E \subset \bigcup_{u\in F}(S_N+u) =: \widetilde{E}.$$

Since $\|STS\|_{S_p} \le \|S\|\|T\|_{S_p}\|S\|$ for all bounded operators S and all $T \in S_p$, and since $M_{\chi_E}M_{\chi_{\widetilde{E}}} = M_{\chi_E}$, it suffices to estimate the S_p norm of the operator:

$$Y = \sum_{u,u'\in F} M_{\chi_{S_N+u}}[M_f,P]M_{\chi_{S_N+u'}} = \sum_{v\in\mathbb{Z}^2/N} Y_v,$$

where

$$Y_v = \sum_{u\in\mathbb{Z}^2/N} \chi_{F\times F}(u,u+v)M_{\chi_{S_N+u}}[M_f,P]M_{\chi_{S_N+u+v}}.$$

For any given $v \in \mathbb{Z}^2/N$, the family

$$\{\chi_{S_N+u+v}f : f \in L^2_\alpha, u \in \mathbb{Z}^2/N\}$$

of subspaces are pairwise orthogonal in L^2_α. Since $\|T\|_{S_p} \le \|T\|_{S_2}$ when $p \ge 2$, we have

$$\|Y_v\|_{S_p}^p = \sum_{u\in\mathbb{Z}^2/N} \chi_{F\times F}(u,u+v)\|M_{\chi_{S_N+u}}[M_f,P]M_{\chi_{S_N+u+v}}\|_{S_p}^p$$

$$\le \sum_{u\in\mathbb{Z}^2/N} \|M_{\chi_{S_N+u}}[M_f,P]M_{\chi_{S_N+u+v}}\|_{S_2}^p. \tag{8.8}$$

Since $[M_f,P]$ has $(f(z)-f(w))e^{\alpha z\bar{w}}$ as its kernel function, we have

$$\|M_{\chi_{S_N+u}}[M_f,P]M_{\chi_{S_N+u+v}}\|_{S_2}^2$$

$$= \int_{S_N+u}\int_{S_N+u+v} |f(z)-f(w)|^2|e^{\alpha z\bar{w}}|^2\,d\lambda_\alpha(z)\,d\lambda_\alpha(w)$$

$$= \left(\frac{\alpha}{\pi}\right)^2 \int_{S_N+u}\int_{S_N+u+v} |f(z)-f(w)|^2 e^{-\alpha|z-w|^2}\,dA(z)\,dA(w)$$

$$\le \delta(v)\int_{S_N}\int_{S_N+v} |f\circ t_u(z)-f\circ t_u(w)|^2\,dA(z)\,dA(w), \tag{8.9}$$

where t_u is the translation by u and

$$\delta(v) = \exp\left[-\alpha \inf_{w,z \in S_N} |(w-z)+v|^2\right].$$

It follows from the inequalities

$$|(w-z)+v|^2 \geq |v|^2 + |w-z|^2 - 2|w-z||v| \geq \frac{1}{2}|v|^2 - |w-z|^2$$

that there exists a positive constant B such that

$$\delta(v) \leq Be^{-\frac{\alpha}{2}|v|^2}, \qquad v \in \mathbb{Z}^2/N.$$

Because $A(S_N) = 1/N^2$, we have for any $g \in L^2_{\text{local}}(\mathbb{C}, dA)$ that

$$\int_{S_N}\int_{S_N+v} |g(z)-g(w)|^2\, dA(z)\, dA(w)$$

$$\leq 2\int_{S_N}\int_{S_N} \left[|g(z)-g_{S_N}|^2 + |g_{S_N} - g\circ t_v(w)|^2\right] dA(z)\, dA(w)$$

$$= \frac{2}{N^2}\int_{S_N} |g-g_{S_N}|^2\, dA + \frac{2}{N^2}\int_{S_N} |g\circ t_v - g_{S_N}|^2\, dA.$$

It follows from the identity

$$\frac{1}{A(S_N)}\int_{S_N} |g-g_{Q_N}|^2\, dA = \frac{1}{A(S_N)}\int_{S_N} |g-g_{S_N}|^2\, dA + |g_{S_N} - g_{Q_N}|^2$$

that

$$\int_{S_N} |g-g_{S_N}|^2\, dA \leq \int_{S_N} |g-g_{Q_N}|^2\, dA \leq \frac{1}{2}J_N(g).$$

Applying Lemma 8.19 to the integral

$$\int_{S_N} |g\circ t_v - g_{S_N}|^2\, dA,$$

we obtain

$$\int_{S_N}\int_{S_N+v} |g(z)-g(w)|^2\, dA(z)\, dA(w)$$

$$\leq \frac{1}{N^2}J_N(g) + \left(N^2 + \frac{4}{9}N^4|\gamma(v)|\right)\sum_{a\in\gamma(v)} J_N(g\circ t_a)$$

$$\leq \left(N^2 + \frac{1}{N^2} + \frac{4}{9}N^4|\gamma(v)|\right)\sum_{a\in\gamma(v)} J_N(g\circ t_a),$$

where $\gamma(v)$ is the discrete path in $\mathbb{Z}^2/N$ from 0 to v (see Sect. 1.2). Let $g = f \circ t_u$ in the above estimate and use (8.9). We see that

$$\|M_{\chi_{S_N+u}}[M_f, P]M_{\chi_{S_N+u+v}}\|_{S_2}^2$$

is less than or equal to

$$Be^{-\frac{\alpha}{2}|v|^2}\left(N^2 + \frac{1}{N^2} + \frac{4}{9}N^4|\gamma(v)|\right)\sum_{a\in\gamma(v)} J_N(f\circ t_u \circ t_a).$$

Since $p/2 \geq 1$, it follows from Hölder's inequality that

$$\|M_{\chi_{S_N+u}}[M_f, P]M_{\chi_{S_N+u+v}}\|_{S_2}^p \leq h(v)\sum_{a\in\gamma(v)}[J_N(f\circ t_u \circ t_a)]^{\frac{p}{2}},$$

where

$$h(v) = \left[Be^{-\frac{\alpha}{2}|v|^2}\right]^{\frac{p}{2}}\left[N^2 + \frac{1}{N^2} + \frac{4}{9}N^4|\gamma(v)|\right]^{\frac{p}{2}+\frac{p-2}{2}}.$$

Combining this with (8.8), we obtain

$$\|Y_v\|_{S_p}^p \leq h(v)\sum_{u\in\mathbb{Z}^2/N}\sum_{a\in\gamma(v)}[J_N(f\circ t_u\circ t_a)]^{\frac{p}{2}}$$

$$= h(v)\sum_{u\in\mathbb{Z}^2/N}\sum_{b\in\gamma(v)+u}[J_N(f\circ t_b)]^{\frac{p}{2}}. \qquad (8.10)$$

For any $b \in \mathbb{Z}^2/N$, we have $b \in \gamma(v)+u$ if and only if $-u \in \gamma(v)-b$. Thus,

$$\left|\{u\in\mathbb{Z}^2/N : b\in\gamma(v)+u\}\right| = |\gamma(v)-b| = |\gamma(v)| \leq 1+|\gamma(v)|.$$

Therefore,

$$\sum_{u\in\mathbb{Z}^2/N}\sum_{b\in\gamma(v)+u}[J_N(f\circ t_b)]^{\frac{p}{2}}$$

$$= \sum_{b\in\mathbb{Z}^2/N}[J_N(f\circ t_b)]^{\frac{p}{2}}\left|\{u\in\mathbb{Z}^2/N : b\in\gamma(v)+u\}\right|$$

$$= (1+|\gamma(v)|)\sum_{b\in\mathbb{Z}^2/N}[J_N(f\circ t_b)]^{\frac{p}{2}}.$$

A substitution of this in (8.10) gives us

$$\|Y_v\|_{S_p}^p \leq h(v)(1+|\gamma(v)|)\sum_{b\in\mathbb{Z}^2/N}[J_N(f\circ t_b)]^{\frac{p}{2}}.$$

Consequently,

$$\|Y\|_{S_p} \le \sum_{v \in \mathbb{Z}^2/N} \|Y_v\|_{S_p}$$

$$\le \sum_{v \in \mathbb{Z}^2/N} [h(v)(1+|\gamma(v)|)]^{\frac{1}{p}} \left[\sum_{b \in \mathbb{Z}^2/N} [J_N(f \circ t_b)]^{\frac{p}{2}} \right]^{\frac{1}{p}}.$$

From Lemma 1.12, the definition of $h(v)$, and the elementary inequality $|\gamma(v)| \le 2|v|$, we see that the constant

$$C = \sum_{v \in \mathbb{Z}^2/N} [h(v)(1+|\gamma(v)|)]^{\frac{1}{p}}$$

is finite. With this constant C, the inequality in (8.7) holds for any bounded Borel set $E \subset \mathbb{C}$. □

Lemma 8.25. *Suppose $0 < p < 2$ and f satisfies condition (I_2). If both H_f and $H_{\bar{f}}$ are in the Schatten class S_p, then $MO(f) \in L^p(\mathbb{C}, dA)$. Moreover, there exists a positive constant C, independent of f, such that*

$$\int_{\mathbb{C}} [MO(f)(z)]^p \, dA(z) \le C \left[\|H_f\|_{S_p}^p + \|H_{\bar{f}}\|_{S_p}^p \right].$$

Proof. For any

$$j = (j_1, j_2) \in \{1, 2, \dots, M\} \times \{1, 2, \dots, M\},$$

we fix an orthonormal basis $\{e_v : v \in \Lambda_j^M\}$ for L_α^2 and define two sequences $\{h_v\}$ and $\{\zeta_v\}$ in L_α^2 as follows:

$$h_v(w) = e^{\alpha|w|^2/2} e^{-\alpha i \operatorname{Im}(v\bar{w})} \chi_{Q_N+v}(w), \quad v \in \Lambda_j^M,$$

and

$$\zeta_v(z) = \frac{\chi_{Q_N+v}(z)[M_f, P]h_v(z)}{\|\chi_{Q_N+v}[M_f, P]h_v\|}, \quad v \in \Lambda_j^M.$$

We also define two operators A_j and B_j on L_α^2 as follows:

$$A_j e_v = \zeta_v, \qquad B_j e_v = h_v, \quad v \in \Lambda_j^M.$$

It is easy to check that both A_j and B_j extend to bounded linear operators on L^2_α. In fact, since each h_v is supported on $Q_N + v$ and different $Q_N + v$ are disjoint, we have

$$\left\| B_j \left(\sum_{v \in \Lambda^M_j} c_v e_v \right) \right\|^2 = \int_{\mathbb{C}} \left| \sum_{v \in \Lambda^M_j} c_v h_v(w) \right|^2 d\lambda_\alpha(w)$$

$$= \sum_{v \in \Lambda^M_j} |c_v|^2 \int_{Q_N + v} |h_v(w)|^2 d\lambda_\alpha(w)$$

$$= \sum_{v \in \Lambda^M_j} |c_v|^2 \int_{Q_N + v} \frac{\alpha}{\pi} dA(w)$$

$$= \frac{9\alpha}{\pi N^2} \sum_{v \in \Lambda^N_j} |c_v|^2.$$

This shows that $\|B_j\| \le (3\sqrt{\alpha})/(N\sqrt{\pi})$. A similar argument shows that $\|A_j\| \le 1$. Let $W_j = A_j^*[M_f, P]B_j$ for each j. Then,

$$\|W_j\|_{S_p} \le \|A_j\| \|[M_f, P]\|_{S_p} \|B_j\|.$$

Since there are M^2 such j's, we obtain

$$\sum_j \|W_j\|_{S_p}^p \le M^2 \left(\frac{3}{N} \sqrt{\frac{\alpha}{\pi}} \right)^p \|[M_f, P]\|_{S_p}^p$$

$$\le M^2 \left(\frac{6}{N} \sqrt{\frac{\alpha}{\pi}} \right)^p \left(\|H_f\|_{S_p}^p + \|H_{\bar{f}}\|_{S_p}^p \right).$$

Here, we used the first identity in the proof of Lemma 8.23 and the fact that, for any positive p and any Schatten class operators S and T, we always have

$$\|S + T\|_{S_p}^p \le 2^p (\|S\|_{S_p}^p + \|T\|_{S_p}^p). \tag{8.11}$$

Fix a very large natural number R and consider the truncation Z_R of the lattice $\mathbb{Z}^2/N$:

$$Z_R = \left\{ v = (v_1, v_2) \in \mathbb{Z}^2/N : |v_k| \le R, k = 1, 2 \right\}.$$

For any j, we set $Z_j = Z_R \cap \Lambda^M_j$ and denote by P_{Z_j} the orthogonal projection from L^2_α onto the subspace spanned by $\{e_v : v \in Z_j\}$. It is clear that

$$P_{Z_j} W_j P_{Z_j} g = \sum_{v, v' \in Z_j} \langle g, e_v \rangle \langle W_j e_v, e_{v'} \rangle e_{v'}.$$

We are going to decompose $P_{Z_j}W_jP_{Z_j}$ into a "diagonal" part and an "off-diagonal" part. More specifically, we define an operator D_j by

$$D_jg = \sum_{v \in Z_j} \langle g, e_v \rangle \langle W_j e_v, e_v \rangle e_v$$

and set

$$E_j = P_{Z_j}W_jP_{Z_j} - D_j.$$

Both D_j and E_j are finite rank operators, so they both belong to the Schatten class S_p. Also, it follows from (8.11) that

$$2^p\|W_j\|_{S_p}^p \geq 2^p\|P_{Z_j}W_jP_{Z_j}\|_{S_p}^p \geq \|D_j\|_{S_p}^p - 2^p\|E_j\|_{S_p}^p.$$

Since D_j is diagonal, we have

$$
\begin{aligned}
\|D_j\|_{S_p}^p &= \sum_{v \in Z_j} |\langle A_j^*[M_f, P]B_j e_v, e_v \rangle|^p \\
&= \sum_{v \in Z_j} \|\chi_{Q_N+v}[M_f, P]h_v\|^p \\
&= \sum_{v \in Z_j} \left[\int_{Q_N+v} |(M_fP - PM_f)h_v|^2 \, d\lambda_\alpha \right]^{\frac{p}{2}}.
\end{aligned}
$$

Note that

$$
\begin{aligned}
(M_fP - PM_f)h_v(z) &= f(z)Ph_v(z) - P(fh_v)(z) \\
&= \int_{\mathbb{C}} (f(z) - f(w))e^{\alpha z\overline{w}}h_v(w) \, d\lambda_\alpha(w) \\
&= \frac{\alpha}{\pi} \int_{Q_N+v} (f(z) - f(w))e^{\alpha z\overline{w} - \frac{\alpha}{2}|w|^2 - \alpha i\mathrm{Im}(v\overline{w})} \, dA(w).
\end{aligned}
$$

An application of Lemma 8.22 then produces a positive constant C_N such that

$$\|D_j\|_{S_p}^p \geq C_N \sum_{v \in Z_j} [J_N(f \circ t_v)]^{\frac{p}{2}}.$$

Next, we will obtain an upper bound for $\|E_j\|_{S_p}$, which is much more involved than the previous estimates. We begin with the following well-known fact from operator theory: if $0 < p \leq 2$ and T is a compact operator on a separable Hilbert space H, then

$$\|T\|_{S_p}^p \leq \sum_{n,m} |\langle Te_n, e_m \rangle|^p$$

for any orthonormal basis $\{e_n\}$ of H. See Lemma 6.36. Thus,

$$
\begin{aligned}
\|E_j\|_{S_p}^p &\leq \sum_{v,v' \in \Lambda_j^M} |\langle E_j e_v, e_{v'} \rangle|^p \\
&= \sum_{v,v' \in Z_j, v \neq v'} |\langle E_j e_v, e_{v'} \rangle|^p \\
&= \sum_{v,v' \in Z_j, v \neq v'} \left| \frac{\langle [M_f, P]h_v, \chi_{Q_N+v'}[M_f, P]h_{v'} \rangle}{\|\chi_{Q_N+v'}[M_f, P]h_{v'}\|} \right|^p \\
&= \sum_{v,v' \in Z_j, v \neq v'} \left| \frac{\langle \chi_{Q_N+v'}[M_f, P]h_v, \chi_{Q_N+v'}[M_f, P]h_{v'} \rangle}{\|\chi_{Q_N+v'}[M_f, P]h_{v'}\|} \right|^p \\
&\leq \sum_{v,v' \in Z_j, v \neq v'} \|\chi_{Q_N+v'}[M_f, P]h_v\|^p.
\end{aligned}
$$

Write $\|\chi_{Q_N+v'}[M_f, P]h_v\|^p$ as

$$
\left[\int_{Q_N+v'} \left| \int_{Q_N+v} (f(z) - f(w)) e^{\alpha z \overline{w} - \frac{\alpha}{2}|w|^2 - \alpha i \operatorname{Im}(v\overline{w})} \, dA(w) \right|^2 d\lambda_\alpha(z) \right]^{\frac{p}{2}}
$$

and apply the Cauchy–Schwarz inequality in the inner integral. We see that $\|E_j\|_{S_p}^p$ is less than or equal to $[(3\alpha)/(N\pi)]^p$ times

$$
\sum_{v,v' \in Z_j, v \neq v'} \left[\int_{Q_N+v'} \int_{Q_N+v} |f(z) - f(w)|^2 e^{-\alpha|z-w|^2} \, dA(w) \, dA(z) \right]^{\frac{p}{2}}.
$$

It is easy to see that

$$
|z - w| \geq \frac{1}{N}(M - 3)
$$

whenever $z \in Q_N + v'$ and $w \in Q_N + v$ (without loss of generality, we may assume that $M > 3$). Thus $\|E_j\|_{S_p}^p$ is less than or equal to the constant

$$
\left[\frac{3\alpha}{N\pi} \right]^p e^{-\frac{p\alpha}{2}\left(\frac{M-3}{N}\right)^2}
$$

times the infinite sum

$$
\sum_{v,v' \in Z_j, v \neq v'} \left[\int_{Q_N+v'} \int_{Q_N+v} |f(z) - f(w)|^2 e^{-\frac{\alpha}{2}|z-w|^2} \, dA(w) \, dA(z) \right]^{\frac{p}{2}}.
$$

Making the simultaneous change of variables

$$z \mapsto z + v', \qquad w \mapsto w + v,$$

and estimating the resulting exponential function with the help of the triangle inequality, we obtain a positive constant C_N such that

$$\|E_j\|_{S_p}^p \leq C_N e^{-\frac{p\alpha}{2}\left(\frac{M-3}{N}\right)^2} \sum_{v,v' \in Z_j, v \neq v'} e^{-\frac{p\alpha|v-v'|^2}{5}} \left[I(v,v')\right]^{\frac{p}{2}},$$

where

$$I(v,v') = \int_{Q_N} \int_{Q_N} |f \circ t_{v'}(z) - f \circ t_v(w)|^2 \, dA(w) \, dA(z).$$

We enumerate the points in the path $\gamma(v',v) \subset Z_R$ as $\{a_0, \ldots, a_l\}$ in such a way that $a_0 = v'$, $a_l = v$, and

$$(S_N + a_{k-1}) \cup (S_N + a_k) \subset Q_N + a_{k-1}, \qquad 1 \leq k \leq l,$$

where $l+1$ is the length of the path $\gamma(v',v)$. By the triangle inequality,

$$|f \circ t_{v'}(z) - f \circ t_v(w)| \leq |f \circ t_{v'}(z) - (f \circ t_{v'})_{Q_N}|$$

$$+ |(f \circ t_v)_{Q_N} - f \circ t_v(w)|$$

$$+ \sum_{k=1}^{l} |(f \circ t_{a_{k-1}})_{Q_N} - (f \circ t_{a_k})_{Q_N}|.$$

By Cauchy–Schwarz, the integrand $|f \circ t_{v'}(z) - f \circ t_v(w)|^2$ in $I(v,v')$ is less than or equal to $(l+2)$ times

$$|f \circ t_{v'}(z) - (f \circ t_{v'})_{Q_N}|^2 + |(f \circ t_v)_{Q_N} - f \circ t_v(w)|^2$$

$$+ \sum_{k=1}^{l} |(f \circ t_{a_{k-1}})_{Q_N} - (f \circ t_{a_k})_{Q_N}|^2.$$

Therefore, if we also assume $N \geq 3$, the double integral $I(v,v')$ is less than or equal to $9(l+2)/N^2$ times

$$\int_{Q_N} |f \circ t_{v'}(z) - (f \circ t_{v'})_{Q_N}|^2 \, dA(z)$$

$$+ \int_{Q_N} |f \circ t_v(w) - (f \circ t_v)_{Q_N}|^2 \, dA(w)$$

$$+ \sum_{k=1}^{l} |(f \circ t_{a_{k-1}})_{Q_N} - (f \circ t_{a_k})_{Q_N}|^2.$$

Since $0 < p/2 < 1$, $I(v, v')^{p/2}$ is less than or equal to $[9(l+2)/N^2]^{p/2}$ times

$$\left[\int_{Q_N} |f \circ t_{v'} - (f \circ t_{v'})_{Q_N}|^2 \, dA\right]^{\frac{p}{2}} \tag{8.12}$$

$$+\left[\int_{Q_N} |f \circ t_v - (f \circ t_v)_{Q_N}|^2 \, dA\right]^{\frac{p}{2}} \tag{8.13}$$

$$+\left[\sum_{k=1}^{l} |(f \circ t_{a_{k-1}})_{Q_N} - (f \circ t_{a_k})_{Q_N}|^2\right]^{\frac{2}{p}}. \tag{8.14}$$

It follows that $\|E_j\|_{S_p}^p$ is less than or equal to

$$C_N e^{-\frac{p\alpha}{4}\left(\frac{M-3}{N}\right)^2} \left(\frac{9(l+2)}{N^2}\right)^{\frac{p}{2}}$$

times

$$\sum_{v,v'\in Z_j, v\neq v'} e^{-\frac{\alpha}{5}|v'-v|^2} \left[\int_{Q_N} |f \circ t_{v'} - (f \circ t_{v'})_{Q_N}|^2 \, dA\right]^{\frac{p}{2}} \tag{8.15}$$

$$+ \sum_{v,v'\in Z_j, v\neq v'} e^{-\frac{\alpha}{5}|v'-v|^2} \left[\int_{Q_N} |f \circ t_v - (f \circ t_v)_{Q_N}|^2 \, dA\right]^{\frac{p}{2}} \tag{8.16}$$

$$+ \sum_{v,v'\in Z_j, v\neq v'} e^{-\frac{\alpha}{5}|v'-v|^2} \left[\sum_{k=1}^{l} |(f \circ t_{a_{k-1}})_{Q_N} - (f \circ t_{a_k})_{Q_N}|^2\right]^{\frac{p}{2}}. \tag{8.17}$$

Since l is comparable to $|v' - v|$, we can find another constant C_N such that

$$[9(l+1)/N^2]^{\frac{p}{2}} e^{-\frac{p\alpha}{5}|v'-v|^2} \leq C_N e^{-\frac{p\alpha}{6}|v'-v|^2}.$$

So the quantity in (8.15) is dominated by (up to a multiplicative constant that only depends on N)

$$e^{-\frac{p\alpha}{4}\left(\frac{M-3}{N}\right)^2} \sum_{v,v'\in Z_j, v\neq v'} e^{-\frac{p\alpha}{6}|v-v'|^2} \left[\int_{Q_N} |f \circ t_{v'} - (f \circ t_{v'})_{Q_N}|^2 \, dA\right]^{\frac{p}{2}},$$

which is equal to

$$Ce^{-\frac{p\alpha}{4}\left(\frac{M-3}{N}\right)^2} \sum_{v'\in Z_j} [J_N(f \circ t_{v'})]^{\frac{p}{2}},$$

where

$$C = \left(\frac{N^2}{18}\right)^{\frac{p}{2}} \sum_{v \in Z_j} e^{-\frac{p\alpha}{6}|v-v'|^2}$$

$$\leq \left(\frac{N^2}{18}\right)^{\frac{p}{2}} \sum_{v \in \mathbb{Z}^2/N} e^{-\frac{p\alpha}{6}|v-v'|^2}$$

$$= \left(\frac{N^2}{18}\right)^{\frac{p}{2}} \sum_{v \in \mathbb{Z}^2/N} e^{-\frac{p\alpha}{6}|v|^2}.$$

By symmetry, we get exactly the same estimate for the quantity in (8.16).

Since $0 < p/2 < 1$, we can apply Hölder's inequality in (8.17) and reduce our estimate to the following quantity:

$$S_j = e^{-\frac{p\alpha}{4}\left(\frac{M-3}{N}\right)^2} \sum_{v,v' \in Z_j, v \neq v'} \sum_{k=1}^{l} e^{-\frac{p\alpha}{6}|v-v'|^2} |(f \circ t_{a_{k-1}})_{QN} - (f \circ t_{a_k})_{QN}|^p.$$

Just like the computation we performed in the proof of Lemma 8.19, we have

$$|(f \circ t_{a_{k-1}})_{QN} - (f \circ t_{a_k})_{QN}|^p = |f_{QN+a_{k-1}} - f_{QN+a_k}|^p$$

$$\leq 2^p \left[|f_{QN+a_{k-1}} - f_{SN+a_k}|^p + |f_{SN+a_k} - f_{QN+a_k}|^p\right]$$

$$\leq C_N \left[\left(J_N(f \circ t_{a_{k-1}})\right)^{\frac{p}{2}} + \left(J_N(f \circ t_{a_k})\right)^{\frac{p}{2}}\right].$$

Thus,

$$S_j \leq C_N \sum_{v,v' \in Z_j, v \neq v'} e^{-\frac{p\alpha}{6}|v-v'|^2} \sum_{u \in \gamma(v,v')} [J_N(f \circ t_u)]^{\frac{p}{2}}$$

$$= C_N \sum_{v,v' \in Z_j, v \neq v'} e^{-\frac{p\alpha}{6}|v-v'|^2} \sum_{u \in Z_j} [J_N(f \circ t_u)]^{\frac{p}{2}} \chi_{\gamma(v,v')}(u)$$

$$= C_N \sum_{u \in Z_j} [J_N(f \circ t_u)]^{\frac{p}{2}} \sum_{v,v' \in Z_j, v \neq v'} e^{-\frac{p\alpha}{6}|v-v'|^2} \chi_{\gamma(v,v')}(u).$$

By Lemma 1.15, there exists a constant $C > 0$, independent of u and R, such that

$$\sum_{v,v' \in Z_j, v \neq v'} e^{-\frac{p\alpha}{6}|v-v'|^2} \chi_{\gamma(v,v')}(u) \leq C$$

for all $u \in Z$. Therefore,

$$\|E_j\|_{S_p}^p \leq C_N e^{-\frac{p\alpha}{4}\left(\frac{M-3}{N}\right)^2} \sum_{u \in Z_j} [J_N(f \circ t_u)]^{\frac{p}{2}}$$

for all j, where C_N is yet another constant that depends on N only. Combining this with the earlier lower estimate for $\|D_j\|_{S_p}$, we see that there exist two constants C_N^1 and C_N^2, which are both independent of M and R, such that

$$M^2\left[\|H_f\|_{S_p}^p + \|H_{\bar{f}}\|_{S_p}^p\right] \geq \left[C_N^1 - C_N^2 M^2 e^{-\frac{p\alpha}{4}\left(\frac{M-3}{N}\right)^2}\right] \sum_{u \in Z_j} [J_N(f \circ t_u)]^{\frac{p}{2}}.$$

If we pick M such that

$$C_N^1 - C_N^2 M^2 e^{-\frac{p\alpha}{4}\left(\frac{M-3}{N}\right)^2} > 0,$$

then we obtain a constant $C > 0$, independent of f and R, such that

$$\|H_f\|_{S_p}^p + \|H_{\bar{f}}\|_{S_p}^p \geq C \sum_{u \in Z_j} [J_N(f \circ t_u)]^{\frac{p}{2}}$$

for all $j \in \{1,2,\ldots,M\} \times \{1,2,\ldots,M\}$. Summing over all such j, we obtain a constant $C > 0$, independent of the truncating constant R, such that

$$\|H_f\|_{S_p}^p + \|H_{\bar{f}}\|_{S_p}^p \geq C \sum_{u \in Z_R} [J_N(f \circ t_u)]^{\frac{p}{2}}.$$

Let $R \to \infty$. We obtain

$$\|H_f\|_{S_p}^p + \|H_{\bar{f}}\|_{S_p}^p \geq C \sum_{u \in \mathbb{Z}^2/N} [J_N(f \circ t_u)]^{\frac{p}{2}}.$$

This, along with Lemma 8.21, completes the proof of Lemma 8.26.   □

Theorem 8.26. *Suppose $0 < p < \infty$, $r > 0$, N is any positive integer, and f satisfies condition (I_2). Then the following conditions are equivalent:*

(a) The operators H_f and $H_{\bar{f}}$ both belong to the Schatten class S_p.
(b) The function

$$MO(f)(z) = [\widetilde{|f|^2}(z) - |\tilde{f}(z)|^2]^{1/2}$$

is in $L^p(\mathbb{C}, dA)$.
(c) The function

$$MO_r(f)(z) = [\widehat{|f|^2}_r(z) - |\hat{f}_r(z)|^2]^{1/2}$$

is in $L^p(\mathbb{C}, dA)$.

(d) The sequence

$$\left\{[J_N(f \circ t_v)]^{\frac{1}{2}} : v \in \mathbb{Z}^2/N\right\}$$

belongs to l^p.

Proof. That (a) implies (b) follows from Lemmas 8.14 and 8.25. Lemmas 8.15 and 8.18 show that condition (b) implies (a). So (a) and (b) are equivalent.

By the double integral representations for $MO(f)$ and $MO_r(f)$, it is easy for us to find a positive constant $C = C(\alpha, r)$ such that

$$MO_r(f)(z) \leq CMO(f)(z), \qquad z \in \mathbb{C},$$

which shows that (b) implies (c).

To show that (c) implies (d), we fix any positive r and choose a sufficiently large positive integer N such that

$$Q_N + v \subset B(\zeta, r), \qquad v \in \mathbb{Z}^2/N, \zeta \in S_N + v. \tag{8.18}$$

This is possible because of the triangle inequality for the Euclidean metric.

Consider the function:

$$F_r(z) = \left[\int_{B(z,r)} \int_{B(z,r)} |f(u) - f(v)|^2 \, dA(u) \, dA(v)\right]^{\frac{1}{2}}.$$

Since $MO_r(f)$ and F_r differ only by a multiplicative constant, condition (c) implies that $F_r \in L^p(\mathbb{C}, dA)$.

Let

$$I = \int_{\mathbb{C}} F_r(z)^p \, dA(z).$$

Since the complex plane is the disjoint union of $S_N + v$, $v \in \mathbb{Z}^2/N$, it follows from the mean value theorem and (8.18) that

$$I = \sum_{v \in \mathbb{Z}^2/N} \int_{S_N + v} F_r(z)^p \, dA(z) = \frac{1}{N^2} \sum_{v \in \mathbb{Z}^2/N} F_r(\zeta_v)^p$$

$$= \frac{1}{N^2} \sum_{v \in \mathbb{Z}^2/N} \left[\int_{B(\zeta_v,r)} \int_{B(\zeta_v,r)} |f(u) - f(v)|^2 \, dA(u) \, dA(v)\right]^{\frac{p}{2}}$$

$$\geq \frac{1}{N^2} \sum_{v \in \mathbb{Z}^2/N} \left[\int_{Q_N + v} \int_{Q_N + v} |f(u) - f(v)|^2 \, dA(u) \, dA(v)\right]^{\frac{p}{2}}$$

$$= \frac{1}{N^2} \sum_{v \in \mathbb{Z}^2/N} [J_N(f \circ t_v)]^{\frac{p}{2}}.$$

Thus, condition (c) implies (d).

When $0 < p \leq 2$, Lemma 8.21 shows that condition (d) implies (b). When $2 \leq p < \infty$, Lemmas 8.23 and 8.24 show that condition (d) implies (a). Since (a) and (b) are already equivalent, we see that condition (d) implies (a) for all $0 < p < \infty$. This completes the proof of the theorem. $\square$

8.4 Notes

The study of Hankel operators on the Fock space goes back to [28] at least, where the compactness was studied for Hankel operators induced by bounded symbols. This compactness problem is equivalent to the symbol calculus for Toeplitz operators with bounded symbols modulo compact operators.

The introduction of BMO (and VMO) defined with a fixed radius into the study of Hankel and Toeplitz operators was first made in [257] in the context of Bergman spaces in the unit disk. The extension to Fock spaces was first carried out in [32].

One of the unique features of the Fock space theory is the following: when f is bounded, the Hankel operator H_f is compact on F_α^2 if and only if $H_{\bar{f}}$ is compact. This result is due to Berger and Coburn [28, 29], and it is not true for Hankel operators on the Bergman space or the Hardy space. A partial explanation for this difference is probably the lack of bounded analytic or harmonic functions on the entire complex plane.

The material in Sect. 8.3 concerning membership of the Hankel operators H_f in Schatten classes is mostly from [131, 242]. Again, there is a key difference between the Fock and Bergman theories. In the Bergman space setting, there is a cutoff point when the invariant mean oscillation $MO(f)$ is used to describe the membership of H_f and $H_{\bar{f}}$ in S_p, while in the Fock space setting, this cutoff point disappears because of the exponential decay of the Fock kernel $e^{-\alpha|z|^2}$.

8.5 Exercises

1. Show that on the space F_α^2, we have $W_a = e^{iT_\psi}$ for any $a \in \mathbb{C}$, where $\psi(z) = 2\text{Im}(\bar{a}z)$.

2. Show that H_f and $H_{\bar{f}}$ both belong to the Schatten S_p if and only if the sequence $\{MO_r(f)(v) : v \in \mathbb{Z}^2/N\}$ belongs to l^p, where $r > 0$ and N is any positive integer.

3. For $f \in L^\infty(\mathbb{C})$, show that H_f is Hilbert–Schmidt if and only if $H_{\bar{f}}$ is Hilbert–Schmidt. See [12].

4. Show that Theorems 8.4 and 8.5 remain valid with the weaker assumption that $\varphi \in L_\alpha^2$.

5. Show that $H_\varphi^* H_\varphi = T_{|\varphi|^2} - T_{\bar{\varphi}} T_\varphi$.

6. If $\|fk_z\|^2 \le C$ as for all $z \in \mathbb{C}$, show that H_f and $H_{\bar{f}}$ are both bounded. Similarly, if $\|fk_z\| \to 0$ as $z \to \infty$, then H_f and $H_{\bar{f}}$ are both compact.

7. Show that $|f(z) - Pf(z)| \le 2\|f\|_\infty e^{\frac{\alpha}{4}|z|^2}$ for almost all $z \in \mathbb{C}$ and $f \in L^\infty(\mathbb{C})$.

8. Define and study Hankel operators on the Fock space F_α^p when $1 \le p \le \infty$.

References

1. P. Ahern, M. Flores, W. Rudin, An invariant volume-mean-value property. J. Funct. Anal. **111**, 380–397 (1993)
2. A. Alexandrov, G. Rozenblum, Finite rank Toeplitz operators: some extensions of D. Luecking's theorem. J. Funct. Anal. **256**, 2291–2303 (2009)
3. N. Aronszajn, Theory of reproducing kernels. Trans. Amer. Math. Soc. **68**, 337–404 (1950)
4. G. Ascensi, Y. Lyubarskii, K. Seip, Phase space distribution of Gabor expansions. Appl. Comput. Harmon. Anal. **26**, 277–282 (2009)
5. S. Axler, The Bergman space, the Bloch space, and commutators of multiplication operators. Duke Math. J. **53**, 315–332 (1986)
6. S. Axler, D. Zheng, Compact operators via the Berezin transform. Indiana Univ. Math. J. **47**, 387–400 (1998)
7. H. Bacry, A. Grossmann, J. Zak, Proof of the completeness of lattice states in the kq-representation. Phy. Rev. **B12**, 1118–1120 (1975)
8. V. Bargmann, On a Hilbert space of analytic functions and an associated integral transform I. Comm. Pure Appl. Math. **14**, 187–214 (1961)
9. V. Bargmann, On a Hilbert space of analytic functions and an associated integral transform II. Comm. Pure. Appl. Math. **20**, 1–101 (1967)
10. V. Bargmann, Remarks on a Hilbert space of analytic functions, Proc. N.A.S. **48**, 199–204 (1962)
11. V. Bargmann, P. Butera, L. Girardello, J. Klauder, On the completeness of coherent states. Rep. Math. Phys. **2**, 221–228 (1971)
12. W. Bauer, Hilbert–Schmidt Hankel operators on the Segal–Bargmann space. Proc. Amer. Math. Soc. **132**, 2989–2998 (2004)
13. W. Bauer, Mean oscillation and Hankel operators on the Segal–Bargmann space. Integr. Equat. Operat. Theor. **52**, 1–15 (2005)
14. W. Bauer, Berezin–Toeplitz quantization and composition formulas. J. Funct. Anal. **256**, 3107–3142 (2009)
15. W. Bauer, L. Coburn, J. Isralowitz, Heat flow, BMO, and the compactness of Toeplitz operators. J. Funct. Anal. **259**, 57–78 (2010)
16. W. Bauer, K. Furutani, Compact operators and the pluriharmonic Berezin transform. Int. J. Math. **19**, 645–669 (2008)
17. W. Bauer, K. Furutani, Hilbert–Schmidt Hankel operators and berezin iteration. Tokyo J. Math. **31**, 293–319 (2008)
18. W. Bauer, H. Issa, Commuting toeplitz operators with quasi-homogeneous symbols on the Segal–Bargmann space, J. Math. Anal. Appl. **386**, 213–235 (2012)
19. W. Bauer, T. Le, Algebraic properties and the finite rank problem for toeplitz operators on the Segal–Bargmann space. J. Funct. Anal. **261**, 2617–2640 (2011)

K. Zhu, *Analysis on Fock Spaces*, Graduate Texts in Mathematics 263,
DOI 10.1007/978-1-4419-8801-0, © Springer Science+Business Media New York 2012

20. W. Bauer, Y.J. Lee, Commuting toeplitz operators on the Segal–Bargmann space. J. Funct. Anal. **260**, 460–489 (2011)

21. D. Bekolle, C. Berger, L. Coburn, K. Zhu, BMO in the Bergman metric on bounded symmetric domains. J. Funct. Anal. **93**, 310–350 (1990)

22. C. Bénéteau, B. Carswell, S. Kouchedian, Extremal problems in the Fock space. Comput. Methods Funct. Theory **10**, 189–206 (2010)

23. F.A. Berezin, Covariant and contravariant symbols of operators. Math. USSR-Izv. **6**, 1117–1151 (1972)

24. F.A. Berezin, Quantization. Math. USSR-Izv. **8**, 1109–1163 (1974)

25. F.A. Berezin, Quantization in complex symmetric spaces. Math. USSR-Izv. **9**, 341–379 (1975)

26. F.A. Berezin, General concept of quantization. Comm. Math. Phys. **40**, 153–174 (1975)

27. F.A. Berezin, The relation between co- and contra-variant symbols of operators on classical complex symmetric spaces. Soviet Math. Dokl. **19**, 786–789 (1978)

28. C. Berger, L. Coburn, Toeplitz operators and quantum mechanics. J. Funct. Anal. **68**, 273–299 (1986)

29. C. Berger, L. Coburn, Toeplitz operators on the Segal–Bargmann space. Trans. Amer. Math. Soc. **301**, 813–829 (1987)

30. C. Berger, L. Coburn, Heat flow and Berezin–Toeplitz estimates. Amer. J. Math. **116**, 563–590 (1994)

31. C. Berger, L. Coburn, A symbol calculus for Toeplitz operators. Proc. Nat. Acad. Sci. USA **83**, 3072–3073 (1986)

32. C. Berger, L. Coburn, K. Zhu, Toeplitz operators and function theory in n-dimensions. Springer Lect. Notes Math. **1256**, 28–35 (1987)

33. J. Bergh, J. Löfstrom, Interpolation Spaces. Grundlehren Math. Wiss. **223**, Springer, (1976)

34. S. Bergman, *The Kernel Function and Conformal Mapping*, Math. Surveys V, (American Mathematical Society, Providence, RI, 1950)

35. B. Berndtsson, J. Ortega-Cerdá, On interpolation and sampling in Hilbert spaces of analytic functions. J. Reine Angew. Math. **464**, 109–128 (1995)

36. A. Beurling, *The Collected Works of Arne Beurling*, vol. 2, (Harmonic Analysis, Boston, 1989)

37. O. Blasco, A. Galbis, On taylor coefficients of entire functions integrable against exponential weights. Math. Nachr. **223**, 5–21 (2001)

38. R. Boas, *Entire Functions*, (Academic Press, New York, 1954)

39. H. Bommier-Hato, H. Youssfi, Hankel operators on weighted Fock space. Integr. Equ. Oper. Theory **59**, 1–17 (2007)

40. H. Bommier-Hato, H. Youssfi, Hankel operators and the Stieltjes moment problem. J. Funct. Anal. **258**, 978–998 (2010)

41. A. Borichev, R. Dhuez, K. Kellay, Sampling and interpolation in large Bergman and Fock spaces. J. Funct. Anal. **242**, 563–606 (2007)

42. A. Borichev, Y. Lyubarskii, Riesz bases of reproducing kernels in Fock-type spaces. (English summary) J. Inst. Math. Jussieu **9**, 449–461 (2010)

43. O. Bratteli, D. Robinson, *Operator Algebras and Quantum Statistical Mechanics*, I, (Springer, New York, 1979)

44. O. Bratteli, D. Robinson, *Operator Algebras and Quantum Statistical Mechanics*, II, (Springer, New York, 1981)

45. S. Brekke, K. Seip, Density theorems for sampling and interpolation in the Bargmann–Fock spaces III. Math. Scand. **73**, 112–126 (1993)

46. B. Carswell, B. MacCluer, A. Schuster, Composition operators on the Fock space. Acta. Sci. Math. (Szeged) **69**, 871–887 (2003)

47. J. Cheeger, M. Gromov, M. Taylor, Finite propagation speed, kernel estimates for functions of the Laplace operator. J. Differ. Geom. **17**, 15–53 (1982)

48. X. Chen, K. Guo, Analytic hilbert spaces over the complex plane. J. Math. Anal. Appl. **268**, 684–700 (2002)

49. X. Chen, S. Hou, A Beurling type theorem for the Fock space. Proc. Amer. Math. Soc. **131**, 2791–2795 (2003)

50. H. Cho, B. Choe, H. Koo, Linear combinations of composition operators on the Fock–Sobolev spaces, preprint, (2011)

51. H. Cho, K. Zhu, Fock–Sobolev spaces and their Carleson measures, to appear in *J. Funct. Anal.*

52. B. Choe, On higher dimensional Luecking's theorem. J. Math. Soc. Japan **61**, 213–224 (2009)

53. B. Choe, K. Izuchi, H. Koo, Linear sums of two composition operators on the Fock space. J. Math. Anal. Appl. **369**, 112–119 (2010)

54. L. Coburn, A sharp berezin lipschitz estimate. Proc. Amer. Math. Soc. **135**, 1163–1168 (2007)

55. L. Coburn, A Lipschitz estimate for Berezin's operator calculus. Proc. Amer. Math. Soc. **133**, 127–131 (2005)

56. L. Coburn, *The Bargmann isometry and Gabor–Daubechies wavelet localization operators*, in *Systems, Approximation, Singular Integral Operators, and Related Topics*, (Bordeaux, 2000), 169–178; Oper. Theory Adv. Appl. **129**, (Birkhauser, Basel, 2001)

57. L. Coburn, On the Berezin–Toeplitz calculus. Proc. Amer. Math. Soc. **129**, 3331–3338 (2001)

58. L. Coburn, The measure algebra of the Heisenberg group. J. Funct. Anal. **161**, 509–525 (1999)

59. L. Coburn, *Berezin–Toeplitz quantization*, in *Algebraic Methods in Operator Theory*, (Birkhauser, Boston, 1994) pp. 101–108

60. L. Coburn, Deformation estimates for the Berezin–Toeplitz quantization. Comm. Math. Phys. **149**, 415–424 (1992)

61. L. Coburn, J. Isralowitz, B. Li, Toeplitz operators with BMO symbols on the Segal–Bargmann space. Trans. Amer. Math. Soc. **363**, 3015–3030 (2011)

62. L. Coburn, B. Li, Directional derivative estimates for Berezin's operator calculus. Proc. Amer. Math. Soc. **136**, 641–649 (2008)

63. L. Coburn, J. Xia, Toeplitz algebras and Rieffel deformation. Comm. Math. Phys. **168**, 23–38 (1995)

64. M. Christ, On the $\overline{\partial}$ equation in weighted L^2 norms in $\mathbb{C}^1$. J. Geom. Anal. **1**, 193–230 (1991)

65. R. Coifman, R. Rochberg, Representation theorems for holomorphic and harmonic functions in L^p. Astérisque **77**, 11–66 (1980)

66. R. Coifman, R. Rochberg, G. Weiss, Factorization theorems for Hardy spaces in several complex variables. Ann. Math. **103**, 611–635 (1976)

67. J. Conway, *Functions of One Complex Variable*, (Springer, New York, 1973)

68. I. Daubechies, Time-frequency localization operators–a geometric phase space approach. IEEE Trans. Inform. Th. **34**, 605–612 (1988)

69. I. Daubechies, The wavelet transform, time-frequency localization, and signal analysis. IEEE Trans. Inform. Th. **36**, 961–1005 (1990)

70. I. Daubechies, *Ten Lectures on Wavelets*, (Society for Industrial and Applied Mathematics, Philladephia, 1992)

71. I. Daubechies, A. Grossmann, Frames in the Bargmann space of entire functions. Comm. Pure. Appl. Math. **41**, 151–164 (1988)

72. I. Daubechies, A. Grossmann, Y. Meyer, Painless nonorthogonal expansions. J. Math. Phys. **27**, 1271–1283 (1986)

73. I. Daubechies, T. Paul, Time-frequency localization operators– a geometric phase space approach II. Inverse Probl. **4**, 661–680 (1988)

74. M. Dostanić, K. Zhu, Integral operators induced by the Fock kernel. Integr. Equat. Operat. Theor. **60**, 217–236 (2008)

75. R.J. Duffin, A.C. Schaeffer, A class of nonharmonic Fourier series. Trans. Amer. Math. Soc. **72**, 341–366 (1952)

76. P. Duren, *Theory of H^p Spaces*, 2nd edn, (Dover Publications, New York, 2000)

77. P. Duren, B. Romberg, A. Shields, Linear functionals on H^p spaces with $0 < p < 1$. J. Reine Angew. Math. **238**, 32–60 (1969)

78. P. Duren, A. Schuster, *Bergman Spaces*, (American Mathematical Society, Providence, RI, 2004)

79. P. Duren, A. Schuster, D. Vukotić, On Uniformly Discrete Sets in the Unit Disk, in Quadrature Domains and Applications, Oper. Theory Adv. Appl. **156**, (Birkhaüser, Basel, 2005) pp. 131–150

80. P. Duren, G. Taylor, Mean growth and coefficients of H^p functions. Illinois J. Math. **14**, 419–423 (1970)

81. A.E. Dzhrbashyan, Integral representation and continuous projections in certain spaces of harmonic functions, Mat. Sb. **121**, 259–271 (1983); (Russian). English translation: Math. USSR-Sb **49**, 255–267 (1984)

82. M. Englis, *Toeplitz Operators on Bergman-Type Spaces*, Ph.D. thesis, (MU CSAV, Prague, 1991)

83. M. Englis, Some density theorems for Toeplitz operators on Bergman spaces. Czechoslovak Math. J. **40**, 491–502 (1990)

84. M. Englis, Functions invariant under the Berezin transform. J. Funct. Anal. **121**, 233–254 (1994)

85. M. Englis, Compact Toeplitz operators via the Berezin transform on bounded symmetric domains. Integr. Equat. Operat. Theor. **33**, 326–355 (1999)

86. M. Englis, Berezin transform on the harmonic Fock space. J. Math. Anal. Appl. **367**, 75–97 (2010)

87. M. Englis, Toeplitz operators and localization operators. Trans. Amer. Math. Soc. **361**, 1039–1052 (2009)

88. H.G. Feichtinger, On a new segal algebra. Monatsh. f. Math. **92**, 269–289 (1981)

89. E. Fischer, Über algebraische Modulsysteme und lineare homogene partielle Differentialgleichungen mit konstanten Koeffizienten. J. Reine Angew. Math. **141**, 48–81 (1911)

90. V. Fock, Verallgemeinerung und Lösung der Diracschen statistischen Gleichung. Z. *Physik* **49**, 339–357 (1928)

91. V. Fock, Konfigurationsraum und zweite Quantelung. Z. *Phys.* **75**, 622–647 (1932)

92. G. Folland, *Harmonic Analysis in Phase Space*, Ann. Math. Studies **122**, (Princeton University Press, Princeton, NJ, 1989)

93. G. Folland, *Fourier Analysis and its Applications*, Brooks/Cole Publishing Company, (Pacific Grove, California, 1992)

94. F. Forelli, W. Rudin, Projections on spaces of holomorphic functions in balls. Indiana Univ. Math. J. **24**, 593–602 (1974)

95. O. Furdui, Norm calculations of composition operators on Fock spaces. Acta Sci. Math. (Szeged) **74**, 281–288 (2008)

96. O. Furdui, On a class of integral operators. Integr. Equat. Operat. Theor. **60**, 469–483 (2008)

97. D. Gabor, Theory of communication. J. Inst. Elect. Eng. **93**, 429–457 (1946)

98. D. Garling, P. Wojtaszczyk, Some Bargmann spaces of analytic functions. Lect. Notes Pure Appl. Math. **172**, 123–138 (1995)

99. H.F. Gautrin, Toeplitz operators in Bargmann spaces. Integr. Equat. Operat. Theor. **11**, 173–185 (1988)

100. I.C. Gohberg, M.G. Krein, *Introduction to the Theory of Linear Nonselfadjoint Operators*, (Nauka, Moscow, 1965); (Russian). English translation: Trans. Math. Monographs **18**, (American Mathematical Society, Providence, RI, 1969)

101. A. Grishin, A. Russakovskii, Free interpolation by entire functions. J. Sov. Math. **48**, 267–275 (1990)

102. K. Grochenig, H. Razafinjatovo, On Landau's necessary conditions for sampling and interpolation of band-limited functions. J. London Math. Soc. **54**, 557–565 (1996)

103. K. Grochenig, D. Walnut, A Riesz basis for the Bargmann–Fock space related to sampling and interpolation. Ark. Math. **30**, 283–295 (1992)

104. A. Grossmann, J. Morlet, Decomposition of hardy functions into square integrable wavelets of constant shape. SIAM J. Math. Anal. **15**, 723–736 (1984)

105. S.M. Grudsky, N.L. Vasilevski, Toeplitz operators on the Fock space: radial component effects. Integr. Equat. Operat. Theor. **44**, 10–37 (2002)

106. W. Gryc, T. Kemp, Duality in Segal-Bargmann spaces. J. Funct. Anal. **261**, 1591–1623 (2011)

107. V. Guillemin, Toeplitz operators in n-dimensions. Integr. Equat. Operat. Theor. **7**, 145–205 (1984)

108. K. Guo, Quasi-invariant subspaces generated by polynomials with nonzero leading terms. Studia Math. **164**, 231–241 (2004)

109. K. Guo, Homogeneous quasi-invariant subspaces of the Fock space. J. Aust. Math. Soc. **75**, 399–407 (2003)

110. K. Guo, K. Izuchi, Composition operators on Fock type spaces. Acta Sci. Math. (Szeged) **74**, 807–828 (2008)

111. K. Guo, D. Zheng, Invariant subspaces, quasi-invariant subspaces, and Hankel operators. J. Funct. Anal. **187**, 308–342 (2001)

112. B. Hall, W. Lewkeeratiyutkul, Holomorphic sobolev spaces and the generalized Segal–Bargmann transform. J. Funct. Anal. **217**, 192–220 (2004)

113. P. Halmos, V. Sunder, *Bounded Integral Operator on L^2 Spaces*, (Springer, Berlin, 1978)

114. G. Hardy, J. Littlewood, Some new properties of fourier constants. Math. Ann. **97**, 159–209 (1926)

115. G. Hardy, J. Littlewood, Some properties of conjugate functions. J. Reine Angew. Math. **167**, 405–423 (1931)

116. W.W. Hastings, A carleson measure theorem for Bergman spaces. Proc. Amer. Math. Soc. **52**, 237–241 (1975)

117. W. Hayman, On a conjecture of korenblum. Anal. (Munich) **19**, 195–205 (1999)

118. H. Hedenmalm, An invariant subspace of the Bergman space having the co-dimension 2 property. J. Reine Angew. Math. **443**, 1–9 (1993)

119. H. Hedenmalm, B. Korenblum, K. Zhu, *Theory of Bergman Spaces*, (Springer, New York, 2000)

120. H. Hedenmalm, S. Richter, K. Seip, Interpolating sequences and invariant subspaces of given index in the Bergman space. J. Reine Angew. Math. **477**, 13–30 (1996)

121. H. Hedenmalm, K. Zhu, On the failure of optimal factorization for certain weighted Bergman spaces. Complex Variables Theor. Appl. **19**, 165–176 (1992)

122. A. Hinkkanen, On a maximum principle in Bergman space. J. Anal. Math. **79**, 335–344 (1999)

123. F. Holland, R. Rochberg, Bergman kernels and Hankel forms on generalized Fock spaces. Contemp. Math. **232**, 189–200 (1999)

124. F. Holland, R. Rochberg, Bergman kernel asymptotics for generalized Fock spaces. J. Anal. Math. **83**, 207–242 (2001)

125. L. Hörmander, *The Analysis of Linear Partial Differential Operators*, IV, (Springer, Berlin, 1985)

126. L. Hörmander, *An Introduction to Complex Analysis in Several Complex Variables*, 3rd edn. (Van Nostrand, Princeton, NJ, 1990)

127. C. Horowitz, Zeros of functions in the Bergman spaces, Ph. D. thesis, (University of Michigan, Ann Arbor, 1974)

128. C. Horowitz, Zeros of functions in the Bergman spaces. Duke Math. J. **41**, 693–710 (1974)

129. C. Horowitz, Factorization theorems for functions in the Bergman spaces. *Duke Math. J.* **44**, 201–213 (1977)

130. R. Howe, Quantum mechanics and partial differential equations. *J. Funct. Anal.* **38**, 188–254 (1980)

131. J. Isralowitz, Schatten p class Hankel operators on the Segal–Bargmann space $H^2(\mathbb{C}^n, d\mu)$ for $0 < p < 1$. J. Operat. Theor. **66**, 145–160 (2011)

132. J. Isralowitz, K. Zhu, Toeplitz operators on the Fock space. Integr. Equat. Operat. Theor. **66**, 593–611 (2010)

133. K. Izuchi, Cyclic vectors in the Fock space over the complex plane. Proc. Amer. Math. Soc. **133**, 3627–3630 (2005)

134. K. Izuchi, K. Izuchi, Polynomials having leading terms over $\mathbb{C}^2$ in the Fock space. J. Funct. Anal. **225**, 439–479 (2005)

135. J. Janas, Unbounded Toeplitz operators in the Segal–Bargmann space. Studia Math. **99**, 8799 (1991)

136. J. Janas, J. Stochel, Unbounded Toeplitz operators in the Segal–Bargmann space II. J. Funct. Anal. **126**, 418-447 (1994)

137. S. Janson, P. Jones, Interpolation between H^p spaces: the complex method. J. Funct. Anal. **48**, 58–80 (1982)

138. S. Janson, J. Peetre, R. Rochberg, Hankel forms and the Fock space. Revista Mat. Ibero-Amer. **3**, 61–138 (1987)

139. J.R. Klauder, B.S. Skagerstam, *Coherent States–Applications in Physics and Mathematical Physics*, (World Scientific Press, Singapore, 1985)

140. W. Knirsch, G. Schneider, Continuity and Schatten–von Neumann p-class membership of Hankel operators with anti-holomorphic symbols on generalized Fock spaces. J. Math. Anal. Appl. **320**, 403–414 (2006)

141. B. Korenblum, A maximum principle for the Bergman space. Publ. Mat. **35**, 479–486 (1991)

142. S. Krantz, *Function Theory of Several Complex Variables*, 2nd edn. (American Mathematical Society, Providence, RI, 2001)

143. O. Kures, K. Zhu, A class of integral operators on the unit ball of C^n. Integr. Equ. Oper. Theory **56**, 71–82 (2006)

144. H. Landau, Necessary density conditions for sampling and interpolation of certain entire functions. Acta Math. **117**, 37–52 (1967)

145. H. Landau, Sampling, data transmission, and the Nyquist rate. Proc. IEEE **55**, 1701–1706 (1967)

146. S. Lang, *Algebra, Graduate Texts in Mathematics*, **211**, (Springer, New York, 2002)

147. B. Levin, *Lectures on Entire Functions, Transl. Math. Monographs* **150**, (American Mathematical Society, Providence, RI, 1996)

148. B. Levin, Y. Lyubarskii, Interpolation by special classes of entire functions and related expansions in exponential series, *Izv. Akad. Nauk SSSR Ser. Mat.* **39**, 657–702 (1975); (Russian). English translation: *Math. USSR-Izv* **9**, 621–662 (1975)

149. H. Li, BMO, VMO, and Hankel operators on the Bergman space of strongly pseudo-convex domains. J. Funct. Anal. **106**, 375–408 (1992)

150. N. Lindholm, Sampling in weighted L^p spaces of entire functions in $\mathbb{C}^n$ and estimates of the Bergman kernel. J. Funct. Anal. **182**, 390–426 (2001)

151. D. Luecking, Forward and reverse Carleson inequalities for functions in Bergman spaces and their derivatives. Amer. J. Math. **107**, 85–111 (1985)

152. D. Luecking, Trace ideal criteria for Toeplitz operators. J. Funct. Anal. **73**, 345–368 (1987)

153. D. Luecking, Finite rank Toeplitz operators on the Bergman space. Proc. Amer. Math. Soc. **136**, 1717–1723 (2008)

154. W. Lusky, On the Fourier series of unbounded harmonic functions. J. London Math. Soc. **61**, 568–580 (2000)

155. Y. Lyubarskii, Frames in the Bargmann space of entire functions. Adv. Soviet Math. **429**, 107–113 (1992)

156. T. MacGregor, K. Zhu, Coefficient multipliers between the Bergman and Hardy spaces. Mathematika **42**, 413–426 (1995)

157. M. Morris, *Understanding Quantum Physics*, (Prentice Hall, New Jersey, 1990)

158. Y. Lyubarskii, K. Seip, Sampling and interpolation of entire functions and exponential systems in convex domains. Ark. Mat. **32**, 157–193 (1994)

159. Y. Lyubarskii, K. Seip, Complete interpolating sequences for Paley–Wiener spaces and Muckenhoupt's A_p-condition. Rev. Mat. Iberoamer. **13**, 361–376 (1997)

160. N. Marco, X. Massaneda, J. Ortega-Cerdá, Interpolating and sampling sequences for entire functions. Geom. Funct. Anal. **13**, 862–914 (2003)

161. X. Massaneda, P. Thomas, Interpolating sequences for Bargmann–Fock spaces in $\mathbb{C}^n$. Indag. Math. **11**, 115–127 (2000)

162. A. Nakamura, F. Ohya, H. Watanabe, On some properties of functions in weighted Bergman spaces. Proc. Fac. Sci. Tokai Univ. **15**, 33–44 (1979)

163. D. Newman, H. Shapiro, Certain Hilbert spaces of entire functions. Bull. Amer. Math. Sco. **72**, 971–977 (1966)

164. D. Newman, H. Shapiro, *Fischer Spaces of Entire Functions*, in *1968 Entire Functions and Related Parts of Analysis*, (American Mathematical Society, Providence, RI, 1966)

165. D. Newman, H. Shaprio, *A Hilbert Space of Entire Functions Related to the Operational Calculus*, unpublished manuscript, (1964, 1971)

166. J. von Neumann, *Foundations of Quantum Mechanics*, (Princeton University Press, Princeton, 1955)

167. V.L. Oleinik, Carleson measures and the heat equation. J. Math. Sci. **101**, 3133–3188 (2000)

168. J. Ortega-Cerdá, K. Seip, Beurling type density theorems for sampling and interpolation in weighted L^p spaces of entire functions. J. Anal. Math. **75**, 247–266 (1998)

169. J. Ortega-Cerdá, K. Seip, Multipliers for entire functions and an interpolation problem of Beurling. J. Funct. Anal. **161**, 400–415 (1999)

170. J. Ortega-Cerdá, K. Seip, Fourier frames. Ann. Math. **155**, 789–806 (2002)

171. J. Peetre, Paracommutators and Minimal Spaces, in *Operators and Function Theory* ed. S.C. Power (Reidel, Dordrecht, 1985) pp. 163–224

172. J. Peetre, Invariant function spaces and Hankel operators–a rapid survey. Exposition. Math. **5**, 5–16 (1987)

173. V.V. Peller, Hankel operators of the class S_p and their applications (rational approximation, Gaussian processes, the problem of majorizing operators). Mat. Sb. **113**, 538–581 (1980); (Russian). English translation: Math. USSR-Sb. **41**, 443–479 (1980)

174. V.V. Peller, *Hankel Operators and Their Applications*, (Springer, New York, 2003)

175. L. Peng, Paracommutators of Schatten–Von Neumann class S_p, $0 < p < 1$. Math Scand. **61**, 68–92 (1987)

176. A.M. Perelomov, On the completeness of a system of coherent states. Theor. Math. Phys. **6**, 156–164 (1971)

177. A.M. Perelomov, *Generalized Coherent States and Their Applications*, (Springer, Berlin, 1986)

178. S. Pichorides, On the best values of the constants in the theorems of M. Riesz, Zygmund, and Kolmogorov. Studia Math. **44**, 165–179 (1972)

179. J. Pool, Mathematical aspects of the Weyl correspondence. J. Math. Phys. **7**, 66–76 (1966)

180. S.C. Power, Hankel operators on Hilbert space. Bull. London Math. Soc. **12**, 422–442 (1980)

181. S.C. Power, Hankel operators on Hilbert space. Res. Notes Math. **64**, 87 (1982)

182. S.C. Power, Finite rank multivariable Hankel forms. Lin. Algebra Appl. **48**, 237–244 (1982)

183. D. Quillen, On the representation of Hermitian forms as sums of squares. Invent. Math. **5**, 237–242 (1968)

184. E. Ramirez de Arellano, N. Vasilevski, Toeplitz operators on the Fock space with presymbols discontinuous on a thick set. Math. Nachr. **180**, 299–315 (1996)

185. E. Ramirez de Arellano, N. Vasilevski, Bargmann projection, three-valued functions, and corresponding Toeplitz operators. Contemp. Math. **212**, 185–196 (1998)

186. R. Rochberg, Trace ideal criteria for Hankel operators and commutators. Indiana Univ. Math. J. **31**, 913–925 (1982)

187. R. Rochberg, Decomposition Theorems for Bergman Spaces and Their Applications, in *Operators and Function Theory* ed. by S.C. Power (Reidel, Dorddrecht, 1985) pp. 225–277

188. C. Rondeaux, Classes de Schatten d'opérateurs pseudo-différentiels. Ann. Sci. École Norm. Sup. **4**, 67–81 (1984)

189. G. Rozenblum, N. Shirokov, Finite rank Bergman–Toeplitz and Bargmann–Toeplitz operators in many dimensions. Complex Anal. Oper. Theor. **4**, 767-775 (2010)

190. W. Rudin, *Functional Analysis*, 2nd edn. (McGraw–Hill, New York, 1991)

191. W. Rudin, *Function Theory in the Unit Ball of C^n*, (Springer, New York, 1980)

192. Sangadji, K. Stroethoff, Compact Toeplitz operators on generalized Fock spaces. Acta Sci. Math. (Szeged) **64**, 657–669 (1998)

193. A. Schuster, On Seip's description of sampling sequences for Bergman spaces. Complex Variables **42**, 347–367 (2000)

194. A. Schuster, The maximum principle for the Bergman space and the Möbius pseudodistance for the annulus. Proc. Amer. Math. Soc. **134**, 3525–3530 (2006)

195. A. Schuster, K. Seip, A Carleson type condition for interpolation in Bergman spaces. J. Reine Angew. Math. **497**, 223–233 (1998)
196. A. Schuster, K. Seip, Weak conditions for interpolation in holomorphic spaces. Publ. Mat. **44**, 277–293 (2000)
197. A. Schuster, D. Varolin, Sampling sequences for Bergman spaces, $0 < p < 1$. Complex Variables **47**, 243–253 (2002)
198. I.E. Segal, *Lectures at the Summer Seminar on Applied Mathematics*, (Boulder, Colorado, 1960)
199. I.E. Segal, *Mathematical Problems of Relativistic Physics*, (American Mathematical Society, Providence, RI, 1963)
200. I.E. Segal, The Complex Wave Representation of the Free Boson Field, in *Topics in Functional Analysis: Essays dedicated to M.G. Krein on the occasion of his 70th birthday*, ed. by I. Gohberg, M. Kac (Academic Press, New York, 1978) pp. 321–343
201. K. Seip, Reproducing formulas and double orthogonality in Bargmann and Bergman spaces. SIAM J. Math. Anal. **22**, 856–876 (1991)
202. K. Seip, Regular sets of sampling and interpolation for weighted Bergman spaces. Proc. Amer. Math. Soc. **117**, 213–220 (1993)
203. K. Seip, *Interpolation and Sampling in Spaces of Analytic Functions*, (American Mathematical Society, Providence, RI, 2004)
204. K. Seip, Density theorems for sampling and interpolation in the Bargmann–Fock space I. J. Reine Angew. Math. **429**, 91–106 (1992)
205. K. Seip, Beurling type density theorems in the unit disk. Invent. Math. **113**, 21–39 (1993)
206. K. Seip, Density theorems for sampling and interpolation in the Bargmann–Fock space. Bull. Amer. Math. Soc. **26**, 322–328 (1992)
207. K. Seip, On Korenblum's density condition for the zero sequences of $A^{-\alpha}$. J. Analyse Math. **67**, 307–322 (1995)
208. K. Seip, Interpolating and sampling in small Bergman spaces, to appear in Collect. Math.
209. K. Seip, R. Wallstén, Density theorems for sampling and interpolation in the Bargmann–Fock space II. J. Reine Angew. Math. **429**, 107–113 (1992)
210. K. Seip, H. Youssfi, Hankel operators on Fock spaces and related Bergman kernel estimates. preprint, (2010)
211. S. Semmes, Trace ideal criteria for Hankel operators and applications to Besov spaces. Integr. Equat. Operat. Theor. **7**, 241–281 (1984)
212. M. Shubin, *Pseudodifferential Operators and Spectral Theory*, (Springer, Berlin, 1987)
213. B. Simon, *Trace Ideals and Their Applications*, London Math. Soc. Lecture Notes Series **35**, (Cambridge University Press, Cambridge, 1979)
214. P. Sjögren, Un contre-exemple pour le noyau reproduisant de la mesure gaussienne dans le plan complexe, Seminaire Paul Krée (Equations aux dérivées partienlles en dimension infinite) 1975/76, Paris
215. E. Stein, Interpolation of linear operators. Trans. Amer. Math. Soc. **83**, 482–492 (1956)
216. E. Stein, G. Weiss, Interpolation of operators with change of measures. Trans. Amer. Math. Soc. **87**, 159–172 (1958)
217. R. Strichartz, L^p contractive projections and the heat semigroup for differential forms. J. Funct. Anal. **65**, 348–357 (1986)
218. K. Stroethoff, Hankel and Toeplitz operators on the Fock space. Mich. Math. J. **39**, 3–16 (1992)
219. R. Supper, Zeros of functions of finite order. J. Inequal. Appl. **7**, 49–60 (2002)
220. M. Tatari, S. Vaezpour, A. Pishinian, On some properties of Fock space F_α^2 by frame theory. Int. J. Contemp. Math. Sci. **5**, 1107–1114 (2010)
221. D. Timotin, C_p estimates for certain kernels: the case $0 < p < 1$, J. Funct. Anal. **72**, 368–380 (1987)
222. H. Triebel, *Interpolation Theory, Function Spaces, and Differential Operators*, (VEB, Berlin, 1977)
223. J.Y. Tung, Fock spaces, Ph.D. thesis, (University of Michigan, Ann Arbor, 2005)

224. J.Y. Tung, Taylor coefficients of functions in Fock spaces. J. Math. Anal. Appl. **318**, 397–409 (2006)
225. J.Y. Tung, Zero sets and interpolating sets in Fock spaces. Proc. Amer. Math. Soc. **134**, 259–263 (2005)
226. J.Y. Tung, On Taylor coefficients and multipliers in Fock spaces. Contemp. Math. **454**, 135–147 (2008)
227. G. Valiron, Sur la formule d'interpolation de Lagrange. Bull. Sci. Math. **49**, 181–192 (1925)
228. N. Vasilevski, V. Kisil, E. Ramirez, R. Trujilo, Toeplitz operators with discontinuous presymbols in the Fock space. Dokl. Math. **52**, 345–347 (1995)
229. D. Vukotić, A sharp estimate for A^p functions in C^n. Proc. Amer. Math. Soc. **117**, 753–756 (1993)
230. D. Vukotić, On the coefficient multipliers of Bergman spaces. J. London Math. Soc. **50**, 341–348 (1994)
231. R. Wallsten, The S^p-criterion for Hankel forms on the Fock space, $0 < p < 1$. Math. Scand. **64**, 123–132 (1989)
232. C. Wang, Some results on Korenblum's maximum principle J. Math. Anal. Appl. **373**, 393–398 (2011)
233. C. Wang, Domination in the Bergman space and Korenblum's constant. Integr. Equat. Operat. Theor. **61**, 423–432 (2008)
234. C. Wang, Behavior of the constant in Korenblum's maximum principle. Math. Nachr. **281**, 447–454 (2008)
235. C. Wang, On a maximum principle for Bergman spaces with small exponents. Integr. Equat. Operat. Theor. **59**, 597–601 (2007)
236. C. Wang, On Korenblum's maximum principle. Proc. Amer. Math. Soc. **134**, 2061–2066 (2006)
237. C. Wang, An upper bound on Korenblum's maximum principle. Integr. Equat. Operat. Theor. **49**, 561–563 (2004)
238. C. Wang, On Korenblum's constant. J. Math. Anal. Appl. **296**, 262–264 (2004)
239. C. Wang, Refining the constant in a maximum principle for the Bergman space. Proc. Amer. Math. Soc. **132**, 853–855 (2004)
240. E.T. Whittaker, On the functions which are represented by the expansions of interpolation theory. Proc. R. Soc. Edinburgh **35**, 181–194 (1915)
241. E.T. Whittaker, G.N. Watson, *A Course of Modern Analysis*, 4th edn. (Cambridge University Press, Cambridge, 1996)
242. J. Xia, D. Zheng, Standard deviation and Schatten class Hankel operators on the Segal–Bargmann space. Indiana Univ. Math. J. **53**, 1381–1399 (2004)
243. J. Xia, D. Zheng, Two-variable Berezin transform and Toeplitz operators on the Fock space, preprint
244. R.M. Young, *An Introduction to Nonharmonic Fourier Series*, (Academic Press, New York, 1980)
245. K. Zhu, Positive Toeplitz operators on weighted Bergman spaces of bounded symmetric domains. J. Operat. Theor. **20**, 329–357 (1988)
246. K. Zhu, A Forelli–Rudin type theorem. Complex Variables **16**, 107–113 (1991)
247. K. Zhu, Schatten class Hankel operators on the Bergman space of the unit ball. Amer. J. Math. **113**, 147–167 (1991)
248. K. Zhu, BMO and Hankel operators on Bergman spaces. Pacific J. Math. **155**, 377–397 (1992)
249. K. Zhu, Zeros of functions in Fock spaces. Complex Variables **21**, 87–98 (1993)
250. K. Zhu, *Operator Theory in Function Spaces*, 2nd edn. (American Mathematical Society, Providence, RI, 2007)
251. K. Zhu, *Spaces of Holomorphic Functions in the Unit Ball*, (Springer, New York, 2005)
252. K. Zhu, Interpolating and recapturing in reproducing Hilbert spaces. Bull. Hong Kong Math. Soc. **1**, 21–33 (1997)
253. K. Zhu, Evaluation operators on the Bergman space. Math Proc. Cambridge Philos. Soc. **117**, 513–523 (1995)

254. K. Zhu, Interpolating sequences for the Bergman space. Michigan Math. J. **41**, 73–86 (1994)
255. K. Zhu, Invariance of Fock spaces under the action of the Heisenberg group. Bull. Sci. Math. **135**, 467–474 (2011)
256. K. Zhu, Duality of Bloch spaces and norm convergence of Taylor series. Michigan Math. J. **38**, 89–101 (1991)
257. K. Zhu, VMO, ESV, and Toeplitz operators on the Bergman space. Trans. Amer. Math. Soc. **302**, 617–646 (1987)
258. K. Zhu, Maximal zero sequences for Fock spaces, preprint, (2011)
259. N. Zorboska, Toeplitz operators with BMO symbols and the Berezin transform. Int. J. Math. Math. Sci. **46**, 2929–2945 (2003)

Index

Symbols

$B(a,r)$, Euclidean disk, 63

BA^p, functions of bounded averages, 125

BA^p_r, functions of bounded averages, 125

BO, functions of bounded oscillation, 124

BO_r, functions of bounded oscillation, 124

$B_\alpha f$, Berezin transform of f, 101

$C_0(\mathbb{C})$, space of continuous functions vanishing at ∞, 23

$C_c(\mathbb{C})$, space of continuous functions with compact support, 23

D, diffferential operator, 19

$D^+(Z)$, upper (uniform) density, 139

$D^-(Z)$, lower (uniform) density, 139

$E_n(z)$, elementary factor, 4

F_α^p, Fock space, 36

H_f, Hankel operator on F_α^2, 287

$H_n(x)$, Hermite polynomials, 221

H_t, heat transform, 101

I, identity operator, 19

$K(z,w)$, reproducing kernel in F_α^2, 34

$K_H(z,w)$, reproducing kernel for H, 78

$K_S(z,w)$, kernel function induced by S, 99

$K_\alpha(z,w)$, reproducing kernel in F_α^2, 34

K_w, reproducing kernel in F_α^2, 34

L_α^p, the space $L^p(\mathbb{C}, d\lambda_{p\alpha/2})$, 36

$MO(f)(z)$, invariant mean oscillation of f at z, 127, 290

$MO_{p,r}(f)(z)$, mean oscillation of f on $B(z,r)$, 123

$M_p(Z)$, stable sampling constant, 145

$M_p(Z,\alpha)$, stable sampling constant, 145

$N_p(Z)$, stable interpolation constant, 144

$N_p(Z,\alpha)$, stable interpolation constant, 144

P_α, orthogonal projection from $L^2(\mathbb{C}, d\lambda_\alpha)$ onto F_α^2, 34

Q_α, integral operator, 43

$S(w,r)$, square centered at w with side length r, 139

S_1, trace class, 24

S_2, Hilbert–Schmidt class, 24

S_p, Schatten class, 24

T_μ, Toeplitz operator on F_α^2, 216

T_φ, Toeplitz operator on F_α^2, 215

U_a, weighted translation operator, 76

VA^p, functions of vanishing averages, 130

VA^p_r, functions of vanishing averages, 130

VO, functions of vanishing oscillation, 130

VO_r, functions of vanishing oscillation, 130

$W(Z)$, weak limits of translates of Z, 165

W_a, Weyl operator, 76

X, multiplication operator, 19

Z, the operator $X + iD$, 19

Z^*, the operator $X - iD$, 19

$[A,B]$, Hausdorff distance between two sets, 151

$[D,X]$, commutator, 20

$[X,Y]_\theta$, complex interpolation space, 59

$\Gamma(a,z)$, incomplete gamma function, 167

$\mathbb{H}$, Heisenberg group, 25

$\mathbb{H}_n$, Heisenberg group, 25

Λ, lattice, 9

$\Lambda(\omega,\omega_1,\omega_2)$, lattice, 9

Λ_α, square lattice, 16

$\|f|Z\|_{p,\alpha}$, sequence norm, 144

$\mathbb{Z}$, integer group, 9

$\mathbb{Z}^2$, integer lattice, 9

BMO, bounded mean oscillation, 123

BMO^p, bounded mean oscillation, 123

BMO^p_r, bounded mean oscillation, 123

χ_S, characteristic function, 11

$\delta(Z)$, separation constant, 143

$\delta(x)$, δ function, 22

δ_z, point mass at z, 148

K. Zhu, *Analysis on Fock Spaces*, Graduate Texts in Mathematics 263, DOI 10.1007/978-1-4419-8801-0, © Springer Science+Business Media New York 2012

341

$\gamma(z)$, discrete path between 0 and z, 11
$\gamma(z,w)$, discrete path between z and w, 11
$\hat{\sigma}(p,q)$, Fourier transform, 22
λ_α, Gaussian measure, 33
$\langle f,g \rangle_\alpha$, inner product in $L^2(\mathbb{C}, d\lambda_\alpha)$, 33
$\omega_r(f)(z)$, oscillation of f over $B(z,r)$, 124
ω_{mn}, lattice points, 9
$\rho_p(z,Z)$, certain "distance" from z to Z, 181
$\sigma(D,X)$, pseudodifferential operator, 19
τ_a, translation by $-a$, 75
$\text{tr}(T)$, trace of T, 96
φ_a, $\varphi_a(z) = a - z$, 75
VMO, vanishing mean oscillation, 130
VMO^p, vanishing mean oscillation, 130
VMO$_r^p$, vanishing mean oscillation, 130
$\hat{f}_r(z)$, mean of f over $B(z,r)$, 123
$\hat{\mu}_r$, averaging function of μ, 246
$\tilde{T}$, Berezin transform of T on F_α^2, 95
$\tilde{f}$, Berezin transform of f, 101
$\tilde{\mu}(z)$, Berezin transform of μ or T_μ, 216
$d(z,S)$, Euclidean distance from z to S, 18
f_α^∞, Fock space, 39
f_r, dilation of f by r, 23
h_f, small Hankel operator on F_α^2, 269
$h_n(x)$, Hermite functions, 222
k_z, normalized reproducing kernel, 35
$n(Z,S)$, number of points in $Z \cap S$, 139
$r\mathbb{Z}^2$, square lattice, 11
t_a, translation by a, 75
$\mathcal{B}_\alpha$, (parametrized) Bargmann transform, 222
$\mathcal{B}_\alpha^{-1}$, inverse Bargmann transform, 223

A

anti-Wick correspondence, 20
anti-Wick pseudodifferential operator, 226
antisymmetric function, 255
antisymmetric polynomial, 255
antisymmetrization, 255
arithmetic mean, 60
atomic decomposition, 63, 277

B

Bargmann isometry, 221
Bargmann transform, 221
Berezin symbol, 93
Berezin transform, 93
Berezin transform of functions, 101
Berezin transform of operators, 93
Bergman space, 4, 57, 293
big Hankel operator, 287
bounded mean oscillation, 123
bounded oscillation, 125

C

Calderón–Vaillancourt theorem, 23
canonical decomposition, 129
Carleson measure, 117, 148
closed-graph theorem, 144
commutator, 312
complex interpolation, 59
Condition (I_1), 101
Condition (I_2), 101
Condition (I_p), 101
Condition (M), 216
congruent parallelogram, 10

D

decomposition, 10, 64
density, 139
diagonal operator, 251, 278, 318
diagonalization argument, 152
diagonalization process, 152
dilation, 23
dilation operator, 36
discriminant, 48
dominated convergence theorem, 39
double pole, 13
doubly periodic, 13
dual space, 53
duality, 53

E

eigenvalue, 98
eigenvector, 98
embedding, 56
equivalence relation, 258
even function, 14
extremal function, 38

F

Fatou's lemma, 39
finite genus, 6
finite order, 6
finite rank, 5
finite rank Hankel operator, 281
finite rank operator, 255
finite rank Toeplitz operator, 255
fixed points of the Berezin transform, 113
Fock projection, 61
Fock spaces, 33
Fock–Carleson measure, 117
Fourier inversion formula, 22
Fourier transform, 20
fundamental region, 9, 64, 142, 202

G
Gaussian measure, 33
Gaussian weights, 87
genus, 5
geometric mean, 60

H
Hadamard factorization, 6
Hankel operator, 287
Hardy space, 4, 57, 293
Hausdorff distance, 151
heat equation, 102
heat transform, 102, 229
Heisenberg group, 25, 76
Hermite polynomials, 221
Hilbert–Schmidt class, 96
Hilbert–Schmidt integral operator, 302

I
ideal, 281
identity operator, 20
identity theorem, 3
infinite order, 6
infinite rank, 5
infinite type, 6
initial condition, 102
integer group, 9
integer lattice, 9
integral operator, 43
integral pairing, 53
integral representation, 35
intermediate value theorem, 166
interpolating sequence for F_α^p, 143
inverse Bargmann transform, 223
inverse Fourier transform, 20
isometry, 76
iterates of the Berezin transform, 110

J
Jensen's formula, 4
John–Nirenberg correspondence, 20

K
Korenblum's maximum principle, 87

L
Lagrange-type interpolation formula, 164
Laplacian, 104
lattice, 9, 142, 277, 312

Lindelöf's theorem, 7, 200
Liouville's theorem, 3
Lipschitz, 99
Lipschitz estimate, 99
Lipschitz functions, 289
local oscillation, 124
lower density, 139

M
maximal invariant Fock space, 77
maximum modulus principle, 7
maximum order, 41
maximum principle for Fock spaces, 81
maximum type, 6, 41
mean oscillation, 123
mean value theorem, 3
minimal invariant Fock space, 77
modified Weierstrass σ-function, 159

N
Nevanlinna–Fock class, 211
normalized reproducing kernel, 35

O
odd function, 14
optimal rate of growth, 36
order, 6
orthogonal projection, 34
orthonormal basis, 33

P
parallelogram, 9
parametrized Bargmann transform, 222
parametrized Berezin transforms, 105
pathological properties, 199
period, 13
periodicity, 13
permutation, 255
permutation invariance, 255
permutation invariant, 257
perturbation, 154
Planck's constant, 19, 87
pseudodifferential operator, 19

Q
quantum physics, 19
quasi-periodic, 15
quasi-periodicity, 13, 201, 202

R
rank, 5
rank of an operator, 258
rank-one operator, 98
rank-two operator, 98
rate of growth, 36
relatively closed set, 151
reproducing formula, 36
reproducing kernel, 34
Riemann ζ-function, 14
Riesz representation, 34

S
sampling sequence for F_α^p, 144
sampling set, 145
Schatten class, 96, 97, 275
Schatten class Hankel operator, 301
Schatten class operator, 96
Schatten class Toeplitz operator, 96
Schatten classes, 24
Schrödinger representation, 26
Schur's test, 43
Schwarz lemma, 84
semi-group property, 101
separated sequence, 143
separation constant, 143
set of uniqueness, 165
small Hankel operator, 267
square lattice, 11, 16
stability, 151
stable interpolation, 144
standar factorization, 5
Stein–Weiss interpolation theorem, 59
Stirling's formula, 169
Stone–Weierstrass theorem, 258
strong convergence, 151
strong limit, 165
strong operator topology, 265
sub-lattice, 10
symbol, 19
symbol calculus, 19
symbol function, 19
symmetric function, 255
symmetric polynomial, 255
symmetrization, 255

T
telescoping decomposition, 306
Toeplitz operator, 213

trace, 98
trace class, 96
trace class operator, 96
trace formula, 213
translation, 75
translation invariance, 10, 75, 145
translation operator, 76

U
uniform density, 139
uniformly close, 160
uniqueness sequence, 165
uniqueness set, 165
unit mass, 22
unitary operator, 26, 76
unitary representation, 25, 76
upper density, 139

V
Vandermonde determinant, 258
vanishing average, 130
vanishing Carleson measure, 118
vanishing Fock–Carleson measure, 118
vanishing mean oscillation, 130
vanishing oscillation, 130
vertices, 9

W
weak convergence, 151
weak limit, 165
Weierstrass σ-function, 13, 201
Weierstrass factorization, 4
Weierstrass factorization theorem, 202
Weierstrass functions, 15
Weierstrass product, 16, 197
weighted translation operator, 76
Weyl pseudodifferential operator, 20
Wick correspondence, 20
Wyle operator, 76

Z
zero sequence, 193
zero set, 193